Springer Series in Optical Sciences Volume 12

Edited by David L. MacAdam

# Springer Series in Optical Sciences

Edited by David L. MacAdam



1 **Solid-State Laser Engineering**
By W. Koechner

2 **Table of Laser Lines in Gases and Vapors** 2nd Edition
By R. Beck, W. Englisch, and K. Gürs

3 **Tunable Lasers and Applications**
Editors: A. Mooradian, T. Jaeger, and P. Stokseth

4 **Nonlinear Laser Spectroscopy**
By V. S. Letokhov and V. P. Chebotayev

5 **Optics and Lasers** An Engineering Physics Approach
By M. Young

6 **Photoelectron Statistics** With Applications to Spectroscopy and Optical Communication
By B. Saleh

7 **Laser Spectroscopy III**
Editors: J. L. Hall and J. L. Carlsten

8 **Frontiers in Visual Science**
Editors: S. J. Cool and E. J. Smith III

9 **High-Power Lasers and Applications** 2nd Printing
Editors: K.-L. Kompa and H. Walther

10 **Detection of Optical and Infrared Radiation**
By R. H. Kingston

11 **Matrix Theory of Photoelasticity**
By P. S. Theocaris and E. E. Gdoutos

12 **The Monte Carlo Methods in Atmospheric Optics**
By G. I. Marchuk, G. A. Mikhailov, M. A. Nazaraliev, R. A. Darbinjan, B. A. Kargin, and B. S. Elepov

13 **Physiological Optics**
By Y. Le Grand and S. G. El Hage

14 **Laser Crystals** Physics and Properties
By A. A. Kaminskii

15 **X-Ray Spectroscopy**
By B. K. Agarwal

16 **Holographic Interferometry** From the Scope of Deformation Analysis of Opaque Bodies
By W. Schumann and M. Dubas

17 **Nonlinear Optics of Free Atoms and Molecules**
By D. C. Hanna, M. A. Yuratich, D. Cotter

18 **Holography in Medicine and Biology**
Editor: G. von Bally

19 **Color Theory and Its Application in Art and Design**
By G. A. Agoston

20 **Interferometry by Holography**
By J. Ostrowskij, M. Butussov, G. Ostrowskaja

21 **Laser Spectroscopy IV**
Editors: H. Walther, K. W. Rothe

# The Monte Carlo Methods in Atmospheric Optics

By G.I. Marchuk G.A. Mikhailov
M.A. Nazaraliev R.A. Darbinjan B.A. Kargin
B.S. Elepov

With 44 Figures

Springer-Verlag Berlin Heidelberg GmbH 1980

Professor GURI I. MARCHUK, Ph. D., Professor GENNADI A. MIKHAILOV, Ph. D., MAGAMEDSHAFI A. NAZARALIEV, RADZMIK A. DARBINJAN, BORIS A. KARGIN, BORIS S. ELEPOV

Computing Center, Siberian Branch of the Academy of Sciences of the USSR
SU-630090 Novosibirsk, USSR



Title of the original Russian edition:
*Metod Monte-Karlo v Atmosfernoy Optike*


ISBN 978-3-662-13503-7 ISBN 978-3-540-35237-2 (eBook)
DOI 10.1007/978-3-540-35237-2

Library of Congress Cataloging in Publication Data. Main entry under title: Monte Carlo methods in atmospheric optics. (Springer series in optical sciences ; v. 12) Includes bibliographical references and index. 1. Meteorological optics. 2. Monte Carlo method. I. Marchuk, Guriĭ Ivanovich. QC975.2.M66 551.5'27 79-12122


Originally published by Springer-Verlag Berlin Heidelberg New York in 1980
Softcover reprint of the hardcover 1st edition 1980



2153/3130-543210

# Preface

This monograph is devoted to urgent questions of the theory and applications of the Monte Carlo method for solving problems of atmospheric optics and hydrooptics. The importance of these problems has grown because of the increasing need to interpret optical observations, and to estimate radiative balance precisely for weather forecasting. Inhomogeneity and sphericity of the atmosphere, absorption in atmospheric layers, multiple scattering and polarization of light, all create difficulties in solving these problems by traditional methods of computational mathematics. Particular difficulty arises when one must solve nonstationary problems of the theory of transfer of narrow beams that are connected with the estimation of spatial location and time characteristics of the radiation field. The most universal method for solving those problems is the Monte Carlo method, which is a numerical simulation of the radiative-transfer process. This process can be regarded as a Markov chain of photon collisions in a medium, which result in scattering or absorption. The Monte Carlo technique consists in computational simulation of that chain and in constructing statistical estimates of the desired functionals.

The authors of this book have contributed to the development of mathematical methods of simulation and to the interpretation of optical observations. A series of general method using Monte Carlo techniques has been developed. The present book includes theories and algorithms of simulation. Numerical results corroborate the possibilities and give an impressive prospect of the applications of Monte Carlo methods.

As a rule, complicated problems of transfer theory cannot be solved sufficiently accurately by direct simulation. Therefore, variance-reduction methods and algorithms that take into account the specific character of a problem are developed. The authors have carried out similar investigations in atmospheric optics.

The book deals with general applications of the Monte Carlo method to radiative-transfer problems. A series of effective algorithms is given for estimating the linear functionals that depend on the solution of the transfer equation. In order to reduce statistical errors, modifications based on asymptotic solutions of the Milne problem are elaborated. General algorithms are proposed for solving systems of integral equations of the second kind and also algorithms for estimating the plarization characteristics of the light. Use of symmetry and other peculiarities of problems enable the authors to construct effective local estimates for calculating the multiple-scattering radiation field at desired points of the phase space. The corresponding algorithms of the dependent-sampling method are proposed. The general formulation of inverse problems is given and numerical algorithms are proposed for solving those problems by the use of linearization, for which the required derivatives are calculated by use of the Monte Carlo method. Algorithms are also given for estimating the correlation function of the strong random fluctuation of light in a turbulent medium. How the radiation field characteristics depend on the various parameters of the optical model, as well as on observation and illumination conditions, is investigated. The book is directed to specialists in applied mathematics and physics, and to students and post-graduates studying Monte Carlo methods.

The authors are greatly indebted to G. V. Rosenberg, K. J. Kondratjev, W. E. Zuev, K. S. Shiphrin, M. S. Malkevič, W. I. Tatarsky, I. N. Minin, E. M. Feigelson, A. P. Ivanov, L. M. Romanova, and others for consultations and collaboration.

Novosibirsk
October 1979

G. I. Marchuk G. A. Mikhailov
M. A. Nazaraliev R. A. Darbinjan
B. A. Kargin B. S. Elepov

# Contents

# 1. Introduction

## 1.1 Atmospheric Optics Problems and the Monte Carlo Method

There is a class of physical problems that require precise calculation of radiative transfer in the atmosphere and ocean, taking into account multiple scattering and a detailed radiation model of the medium. One of these is the problem of the interpretation of optical observations, for which the sphericity of the atmosphere, the transmission function and the polarization of light must all be taken into consideration. Another important class of problems is connected with the theory of narrow-beam propagation. In this case it is necessary to estimate fine characteristics of the radiation field, for instance the time distribution of the intensity of a pencil of light for a local collimated detector, or the perturbation of the observed intensity when a body is inserted into the medium.

In the geometrical-optics approximation, the above-mentioned problems are described by the integro-differential equation of transfer with corresponding boundary conditions. This equation is very difficult to solve by classical methods of computation (for example by finite-difference methods or by the method of spherical harmonics) when it is necessary to estimate the space-time characteristics of the radiation field and if the real phase functions, inhomogeneity of the medium and polarization are used.

Use of the transmission function for calculating the field of radiation scattered in an inhomogeneous medium makes it impossible to use the integro-differential equation; therefore, the problem must be solved by successive calculation of the intensity of multiply scattered light. In certain cases, it can be carried out by the Monte Carlo method.

Light propagation can be regarded as a Markov chain of photon collisions in a medium in which it is scattered or absorbed. The Monte Carlo technique consists in computational simulation of that chain and in calculating a statistical estimate for the desired functionals. Construction of the random trajectories for the physical model of a process is called direct simulation. Here the mathematical problem is to find an optimal way of computational sampling. Direct simulation of the paths of the photons is the same as the simulation of neutron and gamma-ray trajectories used in nuclear physics. As a rule, the complicated problems of transfer theory cannot be solved sufficiently accurately by direct simulation, so variance-reduction methods and computational algorithms that take into account the specific character of each problem are developed. The efficiency of the variance-reduction method depends essentially on the specific character of the problem.

Consideration of a new series of problems usually requires special investigations of various combinations of well-known algorithms and construction of new modifications of the Monte Carlo method. Such investigations have been carried out by the authors for atmospheric-optics problems. This work has stimulated development of general areas of the Monte Carlo technique: the use of adjoint equations, a method of dependent sampling, calculation of partial derivatives with respect to parameters of the model of the medium, the local-calculation method, and a method for solving systems of integral equations.

## 1.2 Equation of Transfer

Optically, the atmosphere is a turbid medium, i.e., a medium in which scattering affects essentially the conditions of light propagation. The most important characteristic of a light beam is its intensity $I(\boldsymbol{r}, \omega)$ which is the radiance at the point defined by the vector $\boldsymbol{r}$, for propagation in the direction $\omega$. The radiance of the beam decreases in the medium on account of absorption and scattering of the photons. This decrease is characterized by the extinction coefficient $\sigma$, which depends on the nature of the medium, the wavelength $\lambda$, and, in the case of an inhomogeneous medium, on the position coordinates $x$, $y$, $z$. Further, $\sigma = \sigma_s + \sigma_c$, where $\sigma_s$ is the scattering coefficient and $\sigma_c$, the absorption coefficient. In turn, $\sigma_s = \sum_i \sigma_s^{(i)}$, and $\sigma_c = \sum_i \sigma_c^{(i)}$, where $\sigma_s^{(i)}$, $\sigma_c^{(i)}$ are the scattering and absorption coefficients of the $i$th component of the medium. By definition, the extinction of the beam along the path is given by $dI = -\sigma \cdot I \cdot dl$. The value $\tau = \int_0^L \sigma[\lambda, \boldsymbol{r}(l)]dl$, where $\boldsymbol{r} = (x, y, z)$, is called the optical depth of the layer corresponding to the path $L$. The reciprocal of $\sigma$ is called the free-path length.

Scattering results not only in light extinction but also causes propagation of light in all other directions $\omega' \in \Omega$. Consequently, the radiance in a direction $\omega$ can be increased by a beam with direction $\omega'$, due to multiple scattering. This increase is proportional to the radiance $I(\omega')$ and is determined by

$$\sigma_s I(\omega') g(\omega', \omega).$$

$g(\omega', \omega)$ is called a phase function (indicatrix). The quantity $g(\omega', \omega)d\omega$ is the probability that the photon that before being scattered had the direction $\omega'$ will travel in the direction $\omega \in d\omega$ after being scattered.

Assume that

$$\int_\Omega g(\omega', \omega)\, d\omega = 1;$$

then we can write the transfer equation

$$\begin{aligned}\omega \cdot \operatorname{grad} I(\boldsymbol{r}, \omega) = & -\sigma(\lambda, \boldsymbol{r}) I(\boldsymbol{r}, \omega) \\ & + \sigma_s(\lambda, \boldsymbol{r}) \int_\Omega I(\boldsymbol{r}, \omega') g(\boldsymbol{r}, \omega', \omega)\, d\omega' + \Phi_0(\boldsymbol{r}, \omega).\end{aligned} \tag{1.1}$$

Here $x = (r, \omega)$ is a point in the phase space $X = R \times \Omega$ of coordinates $r \in R$ and directions $\omega \in \Omega$, and $\Phi_0(r, \omega)$ = source distribution density.

Equation (1.1) indicates that the change of radiance of a light beam is caused by scattering and absorption (the first expression of the right-hand side of the equation); the photons which before scattering have direction $\omega'$, and travel in the direction $\omega$ after scattering (the second expression); and, finally, photon emission (the last expression). The steady-state radiative transfer is described by (1.1). In order to obtain the nonstationary equation, we must write (in units in which the speed of light is unity)

$$\frac{\partial I(r, \omega, t)}{\partial t} + \omega \cdot \operatorname{grad} I(r, \omega, t) = -\sigma(\lambda, r) I(r, \omega, t) + \sigma_s(\lambda, r) \int_{\Omega} I(r, \omega', t) g(r, \omega', \omega)\, d\omega' + \omega_0(r, \omega, t).$$

Real light sources emit photons in a certain spectral wavelength range $\Delta\lambda$. When this range is small we may assume that the scattering coefficients do not depend on $\lambda$. But we cannot assume that the coefficient $\sigma_c(\lambda)$ can be represented by a single mean value, because it may fluctuate strongly. Such a situation arises for atmospheric steam, carbonic gas, ozone, and other atmospheric gases. In these cases, it is possible to use a transmission function $Q(L)$ which is the probability of photon survival at the part of the path $L$. But then the transfer process in an inhomogeneous medium cannot be described by (1.1); the corresponding problems must be solved by successive calculation of the radiance of the multiply scattered light. The most convenient and often the only possible method to solve this kind of problem is the Monte Carlo method.

Light propagation in a scattering medium, for example, in the terrestrial atmosphere, leads to light scattering in all directions. Continuous mixing of the light beams occurs as a result of incoherent scattering.

In each collision, radiance is redistributed not only in angle but also in polarization. In addition, the polarization depends on the local characteristics of the medium. As a result, scattered light is an incoherent statistical mixture of beams that have various radiances with highly different polarization characteristics. When polarization is taken into account, the parameters of the medium (absorption and scattering coefficients, among others) depend, in general, on the polarization characteristics of the incident beam. Therefore, the result of scattering depends to a certain degree on the polarization. Thus, (1.1), which includes only the radiance, cannot adequately describe a radiation field of general form. Other parameters are needed to describe not only the radiance but also the state of polarization. There are many methods of describing the state of polarization, but the most commonly used and most convenient is the method due to Stokes. He introduced four parameters $I$, $Q$, $U$, $V$, each having the dimensions of intensity, which determine the degree of polarization, the plane of polarization and the degree of ellipticity of the radiation; they will be regarded as components of *Stokes's vector* $\boldsymbol{I} = (I_1, I_2, I_3, I_4)$ in a four dimensional space. The radiative-transfer equation that takes polarization into account can be written in the form [1, 2]

$$\omega \cdot \operatorname{grad} I_i(\mathbf{r}, \omega) = \sum_j \left\{ -\varkappa_{ij} I_j(\mathbf{r}, \omega) + \frac{\sigma_s(\lambda, \mathbf{r})}{4\pi} \left[ \int_\Omega f_{ij}(\mathbf{r}, \omega', \omega) \cdot I_j(\mathbf{r}, \omega') d\omega' \right] \right\} + \Phi_i^{(0)}(\mathbf{r}, \omega), \qquad i, j = 1, 2, 3, 4 \tag{1.2}$$

Here $\varkappa_{ij}$ is the extinction matrix (in the isotropic case, $\varkappa_{ij} = \sigma \cdot \delta_{ij}$, where $\delta_{ij}$ is the Kronecker symbol), and $F = \{f_{ij}\}$ is the scattering matrix [1]. $F$ depends essentially on the nature of the scattering medium; in particular, the structure of the medium, the size, form and orientation of the suspended particles, and the wavelength of the light all affect the form of the matrix $F$. In general, all components of the scattering are different; its form is known explicitly only for a few cases. The most precise form of the scattering matrix is given by *van de Hulst* [3] for molecular collisions as well as for collisions with spherical-shell particles of arbitrary size. For an isotropic medium the system takes the form

$$\begin{aligned}
\frac{dI_1(\mathbf{r}, \omega)}{d\tau} &= -I_1(\mathbf{r}, \omega) + \frac{\lambda_0}{4\pi} \int_\Omega (f_{11} I_1 + f_{12} I_2)\, d\omega' + \frac{\phi_1^{(0)}}{\sigma} \\
\frac{dI_2(\mathbf{r}, \omega)}{d\tau} &= -I_2(\mathbf{r}, \omega) + \frac{\lambda_0}{4\pi} \int_\Omega (f_{21} I_1 + f_{22} I_2)\, d\omega' + \frac{\phi_2^{(0)}}{\sigma} \\
\frac{dI_3(\mathbf{r}, \omega)}{d\tau} &= -I_3(\mathbf{r}, \omega) + \frac{\lambda_0}{4\pi} \int_\Omega (f_{33} I_3 + f_{34} I_4)\, d\omega' + \frac{\phi_3^{(0)}}{\sigma} \\
\frac{dI_4(\mathbf{r}, \omega)}{d\tau} &= -I_4(\mathbf{r}, \omega) + \frac{\lambda_0}{4\pi} \int_\Omega (f_{43} I_3 + f_{44} I_4)\, d\omega' + \frac{\phi_4^{(0)}}{\sigma},
\end{aligned} \tag{1.3}$$

where the optical-depth element $d\tau = \sigma dl$ is taken along the direction $\omega$, and $\lambda_0$ is the probability of survival of a photon.

# 2. Elements of Radiative-Transfer Theory Used in the Monte Carlo Methods

## 2.1 The Process of Radiative Transfer; Collision-Density Function; Photon Flux

The transfer process is regarded here as the stationary Markov chain whose states are photon collisions with atoms of the medium. When a collision takes place, absorption or scattering may occur; i.e., the photon may travel no further, or it may leave the point of collision in a new direction.

The straight path between two successive collisions is called a free path. The angle between the previous and the new direction is called the scattering angle. If the photon undergoes absorption or escapes from the medium, its trajectory is terminated. For the sake of simplicity, stationary problems will be considered without taking into account polarization or transmission. More-complicated models are studied in Chaps. 4 and 6, where realistic atmospheric-optics problems are solved. We introduce the notation:

$\boldsymbol{r} = (x, y, z)$: position vector ($r = |\boldsymbol{r}|$);
$\boldsymbol{\omega} = (a, b, c)$: direction of propagation of the radiation ($a^2 + b^2 + c^2 = 1$);
$\mu = \boldsymbol{\omega}\cdot\boldsymbol{\omega}'$: cosine of the scattering angle;
$g(\mu, \boldsymbol{r})$: phase function (indicatrix), normalized by

$$\int_{-1}^{+1} g(\mu, \boldsymbol{r})\, d\mu = 1;$$

$\sigma_s(\boldsymbol{r})$: scattering coefficient (cross section);
$\sigma_c(\boldsymbol{r})$: absorption coefficient;
$\sigma = \sigma_s + \sigma_c$: extinction coefficient,
$f(\boldsymbol{r}, \boldsymbol{\omega})$: collision density;
$\Phi(\boldsymbol{r}, \boldsymbol{\omega})$: photon flux (irradiance).

By definition, $\Phi(\boldsymbol{r}, \boldsymbol{\omega})dS\cdot d\boldsymbol{\omega}$ is the average number of photons that cross a plane element $dS$ oriented perpendicular to the direction $\boldsymbol{\omega}$, in a direction $\boldsymbol{\omega}' \in [\boldsymbol{\omega}, \boldsymbol{\omega} + d\boldsymbol{\omega}]$. It is well known that

$$f(\boldsymbol{r}, \boldsymbol{\omega}) = \sigma(\boldsymbol{r})\cdot\Phi(\boldsymbol{r}, \boldsymbol{\omega}). \tag{2.1}$$

The photon flux across an arbitrary plane element $dS$ is defined by

$$\Phi_s(\boldsymbol{r}, \boldsymbol{\omega}) = |\boldsymbol{n}_s\cdot\boldsymbol{\omega}|\cdot\Phi(\boldsymbol{r}, \boldsymbol{\omega}), \tag{2.2}$$

where $\boldsymbol{n}_s$ is the normal to $dS$. Hence, the average number of photons that cross a surface S in the direction $\boldsymbol{\omega} \in \Omega_k$ is given by

$$\int_S dS \int_{\Omega_k} \Phi[\boldsymbol{r}(S), \boldsymbol{\omega}] |\boldsymbol{\omega} \cdot \boldsymbol{n}_s| \, d\boldsymbol{\omega}, \tag{2.3}$$

where $\Omega_k$ is an arbitrary subset of the sphere of directions.

We now define probability distributions of elements of the transfer process. The free-path length $l$ in the direction $\boldsymbol{\omega}$ has the density function

$$f_l(t) = \sigma[\boldsymbol{r}(t)] \exp\left\{-\int_0^t \sigma[\boldsymbol{r}(t_1)] \, dt_1\right\}, \tag{2.4}$$

where $\boldsymbol{r}(t) = \boldsymbol{r}_0 + t\boldsymbol{\omega}$, $\boldsymbol{r}_0$ is the initial position. The quantity $\tau(t) = \int_0^t \sigma[\boldsymbol{r}(t_1)] dt_1$ is called the optical depth of $[\boldsymbol{r}_0, \boldsymbol{r}(t)]$. To normalize (2.4), we suppose that the medium is bounded by a convex surface outside which $\sigma = \sigma_c \neq 0$. The cosine of the scattering angle at the point $\boldsymbol{r}$ has the distribution density $g(\mu, \boldsymbol{r})$. For several types of scattering with cross sections $\sigma_s^{(i)}(\boldsymbol{r})$ and indicatrixes $g_i(\mu, \boldsymbol{r})$, we put

$$\sigma_s(\boldsymbol{r}) = \sum_i \sigma_s^{(i)}(\boldsymbol{r}), \qquad g(\mu, \boldsymbol{r}) = \frac{\sum_i \sigma_s^{(i)}(\boldsymbol{r}) g_i(\mu, \boldsymbol{r})}{\sigma_s(\boldsymbol{r})}, \tag{2.5}$$

The probabilities of absorption and scattering at a point $\boldsymbol{r}$ are defined by

$$p(\boldsymbol{r}) = \sigma_c(\boldsymbol{r})/\sigma(\boldsymbol{r}), \qquad q(\boldsymbol{r}) = \sigma_s(\boldsymbol{r})/\sigma(\boldsymbol{r}). \tag{2.6}$$

It follows from (2.4) that the probability of photon escape from the medium is given by

$$P[l > t^*(\boldsymbol{r}_0, \boldsymbol{\omega})] = \exp\{-\tau[t^*(\boldsymbol{r}_0, \boldsymbol{\omega})]\},$$

where $t^*(\boldsymbol{r}_0, \boldsymbol{\omega})$ is the distance from $\boldsymbol{r}_0$ to the surface in the direction $\boldsymbol{\omega}$.

## 2.2 Outline of the Simulation of the Transfer Process

The Monte Carlo method for solving problems of transfer theory consists of computational simulation of photon trajectories according to the following scheme:

1) The original position is simulated by a source distribution $\Psi(\boldsymbol{r}, \boldsymbol{\omega})$.
2) The free-path length $l$ is simulated.

3) The escape from the medium is examined.
4) The coordinates of the new collision are calculated:

$$x = x' + al, \qquad y = y' + bl, \qquad z = z' + cl$$

5) The type of collision (absorption or scattering) is simulated.
6) The cosine of the scattering angle $\mu$ is simulated.
7) The coordinates of a new direction are calculated:

$$a = a'\mu - (b' \sin\varphi + a'c' \cos\varphi)[(1 - \mu^2)/(1 - c'^2)]^{1/2}$$

$$b = b'\mu + (a' \sin\varphi - b'c' \cos\varphi)[(1 - \mu^2)/(1 - c'^2)]^{1/2}$$

$$c = c'\mu + (1 - c'^2) \cos\varphi[(1 - \mu^2)/(1 - c'^2)]^{1/2},$$

in which $\varphi$ is the azimuth of the scattering see Fig. 2.1. $\varphi$ is isotropic, so we have $\varphi = 2\pi\alpha$. Here and subsequently, $\alpha$ denotes a random number uniformly distributed between 0 and 1. However, more effective is a rejection method in which $\cos\varphi$ and $\sin\varphi$ are simulated as the coordinates of the isotropic unit vector,

i) $W_1 = 1 - 2\alpha_1$, $W_2 = 1 - 2\alpha_2$
ii) $d = W_1^2 + W_2^2$; if $d > 1$, go to i);
iii) otherwise $\cos\varphi = W_1 \cdot d^{-1/2}$, $\sin\varphi = W_2 \cdot d^{-1/2}$

8) Go to 2)

If escape occurs in 3) or absorption in 5), then 1 is stored and a new trajectory is simulated. The most complicated step is the simulation of the free-path length if $\sigma$ is not constant for the medium. By integration of (2.4) we obtain the distribution function of $l$,

$$F_l(t) = 1 - \exp[-\tau(t)], \qquad t > 0.$$

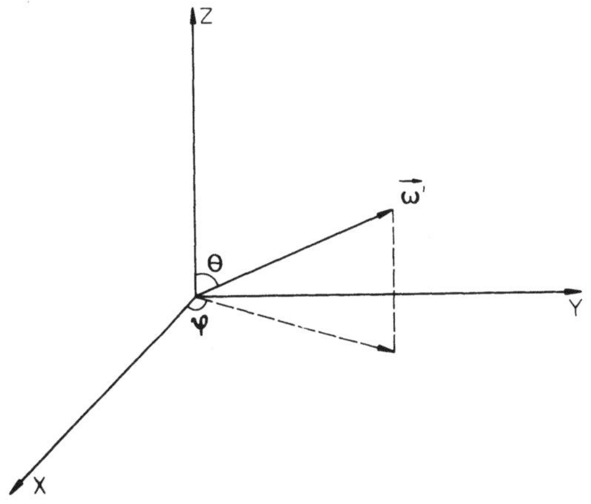


**Fig. 2.1.** Coordinate system

Thus $l$ could be obtained from

$$\tau(l) = \int_0^l \sigma[r(t)]\,dt = -\ln \alpha, \tag{2.7}$$

which is easy to solve if $\sigma[\boldsymbol{r}(t)]$ is a step function, by use of an algorithm similar to that used for solving $F(\xi) = \alpha$ where the distribution function $F$ is a step function. Therefore the medium is usually divided into sufficiently small regions each having constant $\sigma$.

Universal algorithms for calculating distances to the boundary of these regions and simulations of $l$ are given in [4, 5]. This type of algorithm for atmospheric-optics problems is treated also in Chaps. 4–6. Another method for an arbitrary $\sigma(\boldsymbol{r})$ is considered in Sect. 2. 3, but it is effective only for the case of slowly fluctuating $\sigma(\boldsymbol{r})$.

Let us consider the question of how numerical characteristics of the transfer process could be estimated by computational simulation. The integral

$$\int_{\mathscr{D}_i} f(\boldsymbol{x})\,d\boldsymbol{x}, \qquad \boldsymbol{x} = (\boldsymbol{r}, \boldsymbol{\omega})$$

is estimated by the average number of photon collisions in the domain $\mathscr{D}_i$.

If we associate with a collision a weight $[\sigma(\boldsymbol{r})]^{-1}$, then from (2.1) we obtain an estimate of the integral

$$\int_{\mathscr{D}_i} \Phi(\boldsymbol{x})\,d\boldsymbol{x}.$$

Another well-known estimate of the integral is

$$\int_{R_i} d\boldsymbol{r} \int_{\Omega} \Phi(\boldsymbol{r}, \boldsymbol{\omega})\,d\boldsymbol{\omega},$$

where $R_i$ is the coordinate space, and $\Omega$ is the space of directions $\boldsymbol{\omega}$. This integral can be shown to be equal to the average length $L_i$ of the photon trajectory inside the region $R_i$. Indeed, intuitively, the free-path length is equal to $\sigma^{-1}$; hence, the average number of collisions inside the region $R_i$ is equal to $\sigma \cdot L_i$. Consequently, the average number of collisions with weight $\sigma^{-1}$ is $L_i$. This algorithm is particularly convenient for regions of small volume but of great length, i.e., when the region is crossed frequently but collisions seldom occur.

The integral (2.3) is estimated by the average number of photons that cross a surface $S$ in the direction $\omega_i \in \Omega_i$. If we assign to these crossings the weight $|\boldsymbol{n}_s \cdot \boldsymbol{\omega}|^{-1}$, we obtain by (2.2) an estimate of the integral

$$\int_S dS \int_{\Omega_i} \Phi[\boldsymbol{r}(S), \boldsymbol{\omega}]\,d\boldsymbol{\omega}.$$

Thus the mean values of $f(\boldsymbol{x})$ and $\Phi(\boldsymbol{x})$ can be estimated. For large regions, this estimate is not sufficiently accurate; but for small regions, the statistical error is great. Therefore the local radiance estimates that will be introduced in Chap. 3 on the basis of the integral equation of transfer are of great importance.

It should be noted that to solve some physical problems, it is sufficient to obtain the integral characteristics of the transfer process.

## 2.3 Maximal Cross-Section Method for Simulating the Free-Path Length

*Coleman* [6] has described the following method for simulating $l$ under the condition $\sigma(\boldsymbol{r}) \leqslant \sigma_{\mathrm{m}}$, where $\sigma_{\mathrm{m}}$ is some maximal cross section. Two sequences of independent samplings are constructed:

$\{\xi_k\}_{k=1}^n$ from the density $\sigma_{\mathrm{m}} \cdot \exp\{-\sigma_{\mathrm{m}} \cdot t\}$,
$\{\alpha_k\}_{k=1}^n$ from the uniform density on (0, 1).

Let $\zeta_n = \sum_{k=1}^n \xi_k$, and set $N = \min\{n: \alpha_n \leqslant \sigma(\boldsymbol{r} + \zeta_n \boldsymbol{\omega})/\sigma_{\mathrm{m}}\}$. Then $l = \zeta_N$. This method greatly simplifies Monte Carlo calculations for many complicated systems.

Because Coleman's justification is very unwieldy, we present another quite simple proof of this fact.

Let $W(\boldsymbol{\omega}, \boldsymbol{\omega}', \boldsymbol{r})$ be the distribution density of directions $\boldsymbol{\omega}$ after scattering at the point $\boldsymbol{r}$ when $\boldsymbol{\omega}'$ is the direction before the scattering.

The photon flux satisfies

$$\boldsymbol{\omega} \cdot \operatorname{grad} \Phi + \sigma(\boldsymbol{r})\Phi(\boldsymbol{r}, \boldsymbol{\omega}) = \int \Phi(\boldsymbol{r}, \boldsymbol{\omega}')\sigma_{\mathrm{s}}(\boldsymbol{r}) W(\boldsymbol{\omega}, \boldsymbol{\omega}' \ \boldsymbol{r})\, d\boldsymbol{\omega}' + \omega_0(\boldsymbol{r} \ \boldsymbol{\omega}). \quad (2.8)$$

After adding (2.8) and the equality

$$[\sigma_m - \sigma(\boldsymbol{r})]\Phi(\boldsymbol{r}, \boldsymbol{\omega}) = \int \Phi(\boldsymbol{r}, \boldsymbol{\omega}')[\sigma_m - \sigma(\boldsymbol{r})]\delta(\boldsymbol{\omega}' - \boldsymbol{\omega})\, d\boldsymbol{\omega}',$$

and combining the right-hand integrals, we may consider the equation obtained as the equation of transfer for a fictitious medium in which $\sigma_{\mathrm{m}}$ is the extinction coefficient, $\sigma_{\mathrm{s}}(\boldsymbol{r})$ the cross section of scattering with indicatrix $W(\boldsymbol{\omega}, \boldsymbol{\omega}', \boldsymbol{r})$, and $[\sigma_{\mathrm{m}} - \sigma(\boldsymbol{r})]$ the cross section of the scattering with constant $\boldsymbol{\omega}'$. Thus the direct simulation of this process obviously yields the desired simulation of the free-path length $l$.

This justification shows how this method could be applied to certain parts of the system. It also becomes clear how such a simulation can be combined with the weight method.

It is well known that the average number of collisions is given by $(\sigma, \Phi) = \iint \sigma(\boldsymbol{r})\Phi(\boldsymbol{r}, \boldsymbol{\omega}) d\boldsymbol{r} d\boldsymbol{\omega}$. Consequently, the average number of collisions for the transformed equation is equal to $(\sigma_{\mathrm{m}}, \Phi)$.

These speculations could be applied when the method of simulation of the free-path length is chosen.

## 2.4 Exponential Transformation

In this section, a Monte Carlo modification based on exponential transformation of the photon flux is considered.

In some cases, for instance, when propagation through a thick layer $0 \leqslant x \leqslant H$ is considered, the photon flux decreases approxomately as $\exp\{-cx\}$. Hence an attempt can be made to substitute $\Phi_1(\boldsymbol{r}, \omega) = \exp(cx)\Phi(\boldsymbol{r}, \omega)$ in (2.8). We obtain then

$$\int \Phi_1 \sigma_s W(\omega, \omega', \boldsymbol{r})\, d\omega' + \exp(cx)\Phi_0(\boldsymbol{r}, \omega) = \omega \cdot \operatorname{grad} \Phi_1 + (\sigma - c \cos v)\Phi_1, \tag{2.9}$$

where $v$ is the angle between $\omega$ and the $x$ axis.

Equation (2.9) describes the transfer process with $\sigma_1 = \sigma - c\cos v$, and such that each collision results in $\mathscr{W} = \sigma_s(\sigma - c\cos v)^{-1}$ new photons.

The quantity $\mathscr{W}$ is called the survival coefficient; if $\mathscr{W} \leqslant 1$, it could be regarded as the scattering probability, and $1 - \mathscr{W}$ as the absorption probability. If $\mathscr{W} > 1$, then either $\mathscr{W}$ new photons appear after the collision or the weight $\mathscr{W}$ is associated with the single photon. This method simplifies the calculations but the statistical error is increased and may become infinite (see Sects. 3, 4). It is possible to correct the weight fluctuations by special modification of the scattering, i.e., by use of a new importance sampling (see Sect. 3, 5) in which the importance function is approximated by an asymptotic solution of the Milne problem. Algorithms of this kind and their efficiency will be studied in Chap. 5.

Sometimes it is neccessary to calculate photon propagation in an optically thick medium in the neigborhood of a point, for example, at $\boldsymbol{r} = 0$. In this case, it is expedient to use the transformation

$$\Phi_1(\boldsymbol{r}, \omega) = \exp(-cr)\Phi(\boldsymbol{r}, \omega).$$

Substitution of this expression in (2.8) leads to an analogous modification of the process, but with $\sigma_1 = \sigma + c\cos v(t)$, where $v(t)$ is the angle between $\boldsymbol{r}(t) = \boldsymbol{r}_0 + t\,\omega$ and $\omega$, i.e., $\sigma_1$ depends on the photon position. Consequently,

$$\int_0^l [\sigma + c\cos v(t)]\, dt = -\ln \alpha.$$

This equation is easy to solve if $\sigma$ is constant, because

$$\cos v(t) = \frac{\partial r(t)}{\partial t}, \qquad \int_0^l \cos v(t)\, dt = r(l) - r_0,$$

$$r(l) = [r_0^2 + l^2 + 2lr_0 \cos v(0)]^{1/2}.$$

Hence $l = [B - (B^2 - AC)^{1/2}]/A$, where $A = \sigma^2 - c^2$,

$$B = \sigma(-\ln \alpha + cr_0) + c^2 r_0 \cos v(0),$$

$$C = (-\ln \alpha + cr_0)^2 - c^2 r_0^2.$$

## 2.5 The Integral Equation of Transfer (with Generalized Kernel Function)

In this section, we obtain the integral equation of the second kind for the photon collision density.

It was mentioned above that the transfer process is regarded as a Markov chain of photon collisions in a medium. We obtain now the transition density $k(\boldsymbol{x}', \boldsymbol{x})$ of this chain, proceeding from the fact that the quantity $k(\boldsymbol{x}', \boldsymbol{x})d\boldsymbol{x}$ is equal to the number of photons leaving a collision at $\boldsymbol{x}'$ and having their next collisions in $(\boldsymbol{x}, \boldsymbol{x} + d\boldsymbol{x})$.

Let us consider a collision at the phase point $\boldsymbol{x}' = (\boldsymbol{r}', \boldsymbol{\omega}')$ and the right circular cylinder of height $dh$ and base perpendicular to $\boldsymbol{\omega}$, with area $dS$ about the point $\boldsymbol{r}'$. The probability that the photon will cross the area element $dS$ after scattering is equal to

$$\frac{\sigma_s(\boldsymbol{r}')}{\sigma(\boldsymbol{r}')} \frac{g(\mu)}{2\pi} \frac{dS}{|\boldsymbol{r} - \boldsymbol{r}'|^2}, \quad \text{where} \quad \mu = \frac{\boldsymbol{\omega}' \cdot (\boldsymbol{r} - \boldsymbol{r}')}{|\boldsymbol{r} - \boldsymbol{r}'|}. \tag{2.10}$$

Indeed, the element of solid angle is given by

$$d\omega = d\mu\, d\varphi.$$

We have

$$P(\mu_0 \leqslant \mu < \mu_0 + d\mu; \varphi_0 \leqslant \varphi \leqslant \varphi_0 + d\varphi) = g(\mu_0)\, d\mu\, d\varphi / 2\pi,$$

because $\mu$ and $\varphi$ are independent.

Hence $g(\mu)/2\pi$ is the density of scattering distribution for the solid angle. Then, keeping in mind that $dS/|\boldsymbol{r} - \boldsymbol{r}'|^2$ is the value of the solid angle, corresponding to $dS$, we obtain (2.10).

Further, we have by (2.4) that the conditional probability (under the condition $\boldsymbol{\omega}' \in d\boldsymbol{\omega}$, where $\boldsymbol{\omega}'$ is a new direction) of the collision in the cylinder is equal to $\sigma(\boldsymbol{r}) \exp\{-\tau(\boldsymbol{r}, \boldsymbol{r}')\}dh$, where $\tau(\boldsymbol{r}, \boldsymbol{r}')$ is the optical length between $\boldsymbol{r}'$ and $\boldsymbol{r}$.

Hence the absolute probability of collision in this cylinder is given by

$$\frac{\sigma_s(\boldsymbol{r}')g(\mu)\exp[-\tau(\boldsymbol{r}',\boldsymbol{r})]\sigma(\boldsymbol{r})}{\sigma(\boldsymbol{r}')2\pi|\boldsymbol{r}-\boldsymbol{r}'|^2}\,dh\,dS. \tag{2.11}$$

We obtain the absolute density of distribution of $\boldsymbol{r}$ by dividing (2.11) by $dhdS$ (= volume of the cylinder). To obtain $k(\boldsymbol{x}', \boldsymbol{x})$, we note that if $\boldsymbol{r}$ is fixed then $\boldsymbol{\omega}$ is determined uniquely.

Therefore,

$$k(\boldsymbol{x}', \boldsymbol{x}) = \frac{\sigma_s g(\mu)\exp[-\tau(\boldsymbol{r}',\boldsymbol{r})]\sigma(\boldsymbol{r})}{\sigma(\boldsymbol{r})\cdot 2\pi|\boldsymbol{r}-\boldsymbol{r}'|^2}\,\delta\left(\boldsymbol{\omega}-\frac{\boldsymbol{r}-\boldsymbol{r}'}{|\boldsymbol{r}-\boldsymbol{r}'|}\right) \tag{2.12}$$

where $\mu = \boldsymbol{\omega}'\cdot(\boldsymbol{r}-\boldsymbol{r}')/|\boldsymbol{r}-\boldsymbol{r}'|$. The expressions just obtained can be verified in the following way. The expression (2.11) in polar coordinates with the center at $\boldsymbol{r}'$ yields the densities for $\mu$, $\varphi$ and $l$. Let $f_n(\boldsymbol{x})$ be a collision density of the $n$th order. Clearly, $f_n = Kf_{n-1}$, where $K$ is the integral operator with the kernel function $k(\boldsymbol{x}', \boldsymbol{x})$, i.e.,

$$f_n(\boldsymbol{x}) = \int_X k(\boldsymbol{x}', \boldsymbol{x}) f_{n-1}(\boldsymbol{x}')\,d\boldsymbol{x}'.$$

Since $f = f_0 + f_1 + f_2 + \cdots$, we have

$$f = \sum_{n=0}^{\infty} K^n\psi, \quad \text{where} \quad \psi = f_0. \tag{2.13}$$

Consequently,

$$f(\boldsymbol{x}) = \int_X k(\boldsymbol{x}', \boldsymbol{x}) f(\boldsymbol{x}')\,d\boldsymbol{x}' + \psi(\boldsymbol{x}) \quad \text{or} \quad f = Kf + \psi, \tag{2.14}$$

which is called an integral equation of transfer.

It is well known that series (2.13) converges if, for some $n_0$, $\|K^{n_0}\| < 1$. It is easy to show that $K \in (L_1 \to L_1)$.

Therefore, by (2.12)

$$\int_X k(\boldsymbol{x}', \boldsymbol{x})\,d\boldsymbol{x} \leqslant \sigma_s(\boldsymbol{r}')/\sigma(\boldsymbol{r}') = q(\boldsymbol{r}').$$

Hence $\|K\|_{L_1} \leqslant \sup\limits_{\boldsymbol{r}'} q(\boldsymbol{r}')$.

Thus, if $q(\boldsymbol{r}') \leqslant q_0 < 1$, then $\|K\|_{L_1} < 1$.

In practice, it may appear that $\|K\|_{L_i} = 1$, but $\|K^{n_0}\|_{L_1} < 1$ for sufficiently large $n_0$.

It is easy to see that for a bounded medium $\|K^2\|_{L_1} < 1$ even if $q(\boldsymbol{r}') = 1$. The second power appears here because of the assumption that the medium is bounded by a convex surface outside which $\sigma(\boldsymbol{r}) = \sigma_c(\boldsymbol{r}) > 0$ (see Sect. 2.1).

It is reasonable to consider the integral equation of transfer in the space $N$ of generalized measure densities of bounded variation, containing the space $L_1$. This is expedient when the distribution of the initial collisions has a generalized density $\Psi(\boldsymbol{x})$. For instance, for the monodirectional source, the density has the factor $\delta(\boldsymbol{\omega} - \boldsymbol{\omega}_0)$. It is easy to see that $K \in [N_1 \to N_1]$ and $\|K\|_{N_1} < 1$ if $q_0 < 1$.

The Monte Carlo method is usually used to estimate linear functionals of the form

$$I_\varphi = (f, \varphi) = \int_X f(\boldsymbol{x})\varphi(\boldsymbol{x})\, d\boldsymbol{x}.$$

The integral exists if $f \in L_1$, $\varphi \in L_\infty$ or if $f \in N$, $\varphi \in C(X)$. For example, to estimate the integral $\int_{\mathscr{D}} f(\boldsymbol{x}) d\boldsymbol{x}$, where $\mathscr{D}$ is a bounded domain, we must put

$$\varphi(\boldsymbol{x}) = \begin{cases} 1 & \boldsymbol{x} \in \mathscr{D} \\ 0 & \boldsymbol{x} \notin \mathscr{D}. \end{cases}$$

Because the elements of the transfer equation are nonnegative, the Monte Carlo method is applicable for an arbitrary $\varphi \geqslant 0$ provided that $(f, \varphi) < +\infty$. If $\{\boldsymbol{x}_n\}$ is the chain of collisions, then

$$I_\varphi = M\xi, \qquad \xi = \sum_{n=0}^{N} \varphi(\boldsymbol{x}_n).$$

To solve the problem under study it is possible to simulate another Markov chain if only $r(\boldsymbol{x}', \boldsymbol{x})$ contains a $\delta$ function as in (2.12). When the weight $Q_n$ is constructed (see Sect. 3.3), these $\delta$ functions can be, formally, canceled. It is easy to show, as in Sect. 3.3, that the estimate $\xi = \sum_{n=0}^{N} Q_n \varphi(\boldsymbol{x}_n)$ is unbiased. Algorithms with $Q_n \not\equiv 1$ are called weight methods. The weight method enables us to solve problems of the theory of transfer in various media by simulation of a particular Markov chain. By use of results of Chap. 3, in some cases a weight method can be constructed that has less variance than the direct simulation method.

For example, we may multiply the weight of a photon by the scattering probability $\sigma_s(\boldsymbol{r})/\sigma(\boldsymbol{r})$ instead of simulating absorption or we may multiply the weight by $1 - \exp[-\tau(t^*)]$ (see Sect. 2.1) instead of simulating escape. It is easy to see from (3.12) that the variance is reduced but the time $t$ required for computational simulaiotn of the trajectory is increased. Consefluently, if the quantity $t \cdot \mathscr{D}\xi$ decreases, the modification is expedient (see Chap. 3). The general principles and various modifications of the weight method will be considered in the chapters that follow.

## 2.6 Derivation of the Integral Transfer Equation from the Linear Integro-Differential Boltzmann Equation

We have used in the above the Boltzmann equation for the photon flux,

$$\omega \cdot \operatorname{grad} \Phi + \sigma(\boldsymbol{r})\Phi(\boldsymbol{r}, \omega) = \int \Phi(\boldsymbol{r}, \omega')\sigma_s(\boldsymbol{r}) W(\omega, \omega', \boldsymbol{r})\, d\omega' + \Phi_0(\boldsymbol{r}, \omega), \quad (2.15)$$

where $W(\omega, \omega', \boldsymbol{r}) = g(\mu, \boldsymbol{r})/2\pi$. In this section, we obtain the integral transfer equation from (2.15), using some results due to *Vladimirov* [7]. For the sake of simplicity, let us suppose that the domain in which the transfer process is considered, $G$, is bounded by a convex surface $\Gamma$ and that $\sigma(\boldsymbol{r}) \geqslant \sigma_0 > 0$ for $\boldsymbol{r} \in G$. Dividing (2.15) by $\sigma(\boldsymbol{r})$, we get

$$(\omega \cdot \operatorname{grad} \Phi)/\sigma(\boldsymbol{r}) + \Phi(\boldsymbol{r}, \omega) = \int \frac{\sigma_s(\boldsymbol{r})}{\sigma(\boldsymbol{r})} \Phi(\boldsymbol{r}, \omega') W(\omega, \omega', \boldsymbol{r}) d\omega' + \Phi_0(\boldsymbol{r}, \omega)/\sigma(\boldsymbol{r}). \quad (2.16)$$

The boundary conditions for (2.16) are

$$\Phi(\boldsymbol{r}, \omega) = 0 \quad \text{if} \quad \boldsymbol{r} \in \Gamma, \quad \text{and} \quad \omega \cdot \boldsymbol{n}_r > 0,$$

where $\boldsymbol{n}_r$ is the inner normal to the surface $\Gamma$ at the point $\boldsymbol{r}$.

Rewrite (2.16) in the form

$$L\Phi = S\Phi + \frac{\Phi_0}{\sigma},$$

where $L$ is the differential transfer opeator, and $S$ the integral scattering operator. Then

$$\Phi = (L^{-1}S)\Phi + L^{-1}\left(\frac{\Phi_0}{\sigma}\right), \qquad f = \sigma(L^{-1}S)\Phi + \sigma L^{-1}\left(\frac{\Phi_0}{\sigma}\right), \quad (2.17)$$

because $f = \sigma\Phi$.

Substituting the expression for the operator $L^{-1}S$ given in [7], p. 28, in (2.17), and replacing $\sigma\Phi$ by $f$, we obtain

$$f(\boldsymbol{r}, \omega) = \int_0^{t^*} \int_\Omega \sigma(\boldsymbol{r}) \exp\left\{-\int_0^\xi \sigma(\boldsymbol{r}' + \omega t)\, dt\right\} \cdot W(\omega, \omega', \boldsymbol{r}) \times \frac{\sigma_s(\boldsymbol{r}')}{\sigma(\boldsymbol{r}')} f(\boldsymbol{r}', \omega')\, d\omega' d\xi + \psi(\boldsymbol{r}, \omega), \quad (2.18)$$

where $r' = r - \omega\xi$, $t^*$ is the distance from $r$ to the surface $\Gamma$ measured along the direction $\omega$, $\Omega$ is the direction space, and $\Psi(r, \omega)$ is the density of initial collisions. The outer integral is taken here over the interval $r' = r - \omega\xi$, $0 \leqslant \xi \leqslant t^*$. Consequently, it could be rewritten as the double integral in polar coordinates with the center at $r$ by introducing in the integrand the factor $\delta(s - \omega)$. Using cartesian coordinates, we obtain

$$f(r, \omega) = \int_G \int_\Omega \frac{\sigma(r) W(\omega, \omega', r') \exp[-\tau(r, r')] \sigma_s(r')}{|r - r'|^2 \sigma(r')} \times f(r', \omega') \delta\left(\omega - \frac{r - r'}{|r - r'|}\right) dr' d\omega' + \psi(r),$$

which coincides with the integral transfer equation. From the assumptions, the boundary condition for this equation obviously holds.

It should be noted that this is the case when the integrals of the function $\Phi$ over $r' = r - \omega\xi$, $\xi > 0$ are continuous in $\omega$. Therefore, the derivation of the integral transfer equation obtained in Sect. 2.5 from physical premises is more general.

## 2.7 Adjoint Transfer Equation; Theorem of Optical Mutuality

Let $I_p = (\Phi, p)$ be the functional to be estimated, where $p$ is some nonnegative function. It is well known that

$$I_p = (\Phi, p) = (\Phi^*, \Phi_0), \tag{2.19}$$

where $\Phi^*$ is the solution of the adjoint transfer equation

$$-\omega \cdot \operatorname{grad} \Phi^* + \sigma \Phi^* = \int W(\omega', \omega, r) \sigma_s(r) \cdot \Phi^*(r, \omega') d\omega' + p,$$

with the boundary condition

$$\Phi^*(r, \omega) = 0 \quad \text{if} \quad r \in \varphi, \qquad \omega \cdot n_r < 0.$$

Hence for $\Phi^*(r, \omega) = \Phi_1^*(r, -\omega)$ we get

$$\omega \cdot \operatorname{grad} \Phi_1^* + \sigma \Phi_1^* = \int W(\omega', \omega, r) \sigma_s(r) \Phi_1^*(r, \omega') \, d\omega' + p(r, -\omega),$$

with the boundary condition

$$\Phi_1^*(r, \omega) = 0 \quad \text{if} \quad r \in \Gamma, \omega \cdot n_r > 0.$$

This is the integro-differential equation with source density $p_1(\boldsymbol{r}, \boldsymbol{\omega}) = p(\boldsymbol{r}, -\boldsymbol{\omega})$. Now

$$I_p = (\Phi^*, \Phi_0) = \int_R \int_\Omega \Phi^* \Phi_0 \, d\boldsymbol{r} \, d\boldsymbol{\omega} = \int_R \int_\Omega \Phi_1^*(\boldsymbol{r}, \boldsymbol{\omega}) \Phi_0(\boldsymbol{r}, -\boldsymbol{\omega}) \, d\boldsymbol{r} \, d\boldsymbol{\omega}. \tag{2.20}$$

The relations just obtained show that to estimate $I_p$, the transfer from a source with density $p(\boldsymbol{r}, -\boldsymbol{\omega})$ may be simulated and (2.20), which defines the detector readings with the weight function $\Phi_0(\boldsymbol{r}, -\boldsymbol{\omega})$, may be calculated. This is the statement of the theorem of optical mutuality.

For practical use of this theorem, (2.20) must be rewritten as a functional of the collision density,

$$I_p = \int_R \int_\Omega f_1^*(\boldsymbol{r}, \boldsymbol{\omega}) \frac{\Phi_0(\boldsymbol{r}, -\boldsymbol{\omega})}{\sigma(\boldsymbol{r})} \, d\boldsymbol{r} \, d\boldsymbol{\omega}.$$

Thus, at each collision, the quantity $\Phi_0(\boldsymbol{r}, -\boldsymbol{\omega})/\sigma(\boldsymbol{r})$ must be calculated.

As an example, we consider a nonhomogeneous, spherically symmetric, pure-scattering medium illuminated with parallel radiation flux. The problem is to calculate the integral $I_0$ of the radiance over all directions, at the center of the sphere. The medium is bounded by a sphere $S$ of radius $R$. Let the radiance be $(\pi R^2)^{-1}$, i.e., there is a unit power source on the surface of the sphere. The density of the surface source is equal to (see [8]) the radiance of the incident flux multiplied by the cosine of the angle between the photon direction and the inner normal to the surface at the point of photon incidence. Consequently, in our case

$$\Phi_0(\boldsymbol{r}, \boldsymbol{\omega}) = \frac{1}{\pi R^2} \Delta_S(\boldsymbol{r}) \delta(\boldsymbol{\omega} - \boldsymbol{\omega}_0) \frac{|\boldsymbol{r} \cdot \boldsymbol{\omega}|}{R},$$

where $\Delta_S(\boldsymbol{r})$ is a generalized function that corresponds to integration over the sphere $S$. The functional to be found is

$$I_0 = \int_\Omega \Phi(0, \boldsymbol{\omega}) \, d\boldsymbol{\omega} = (\Phi, p),$$

where $p(\boldsymbol{r}, \boldsymbol{\omega}) = \delta(\boldsymbol{r})$. Hence, by the theorem of optical mutuality,

$$I_0 = (\Phi^*, \Phi_0) = \frac{1}{\pi R^2} \int_S \Phi^*(\boldsymbol{r}(S), \boldsymbol{\omega}_0) \frac{|\boldsymbol{r}(S) \cdot \boldsymbol{\omega}_0|}{R} \, dS.$$

Here $\Phi^*$ is the solution of the transfer equation for an isotropic source of density $p(\boldsymbol{r}, \boldsymbol{\omega}) = \delta(\boldsymbol{r})$. For this source, the function

$$Q(\boldsymbol{\omega}_0) = \int_S \Phi^*(\boldsymbol{r}(S), \boldsymbol{\omega}_0) \frac{|\boldsymbol{r}(S) \cdot \boldsymbol{\omega}_0|}{R} \, dS,$$

is the integral angular density of photons that escape from the medium. In this case, $Q(\omega_0) \equiv Q = \text{const}$ and

$$\int_{\Omega} Q\, d\omega = 4\pi Q = \int_{R}\int_{\Omega} p(\mathbf{r}, \omega)\, d\mathbf{r}\, d\omega = 4\pi; \text{ hence } Q \equiv 1.$$

Finally, we have $I_0 = (\pi R^2)^{-1}$. This result is obvious in the case when the scattering coefficient is zero. It remains true for an arbitrary, pure-scattering medium. For the net flux $I_s(0)$ of photons having at least one collision,

$$I_s(0) = \frac{1 - e^{-\tau}}{\pi R^2}, \tag{2.21}$$

where $\tau$ is the optical length of the radius of the sphere $S$. The theorem of optical mutuality is useful when the detector is fixed and the source is extended in phase space.

# 3. General Questions About the Monte Carlo Technique for Solving Integral Equations of Transfer

## 3.1 Preliminary Remarks on Integral Equations of the Second Kind and Markov Chains

Let us consider the integral equation of the second kind,

$$f(x) = \int_X k(x', x) f(x')\, dx' + \psi(x) \quad \text{or} \quad f = Kf + \psi, \tag{3.1}$$

where $X$ is the $n$-dimensional euclidean space, $f,\ \psi \in L$, the Banach space of integrable functions. For many applications (for example in transfer theory), we put $L = L_1$, where

$$\|f\| = \int_X |f(x)|\, dx, \qquad \|K\| \leqslant \sup_x \int_X |k(x, x')|\, dx'.$$

The Monte Carlo algorithms constructed here and subsequently are based on representation of the solution of (3.1) by the Neumann series,

$$f = \sum_{n=0}^{\infty} K^n \psi, \tag{3.2}$$

where

$$[K^n \psi](x) = \overbrace{\int \cdots \int}^{n} \psi(x_0) k(x_0, x_1) \cdots k(x_{n-1}, x)\, dx_0 \cdots dx_{n-1}.$$

It is easy to show that (3.2) converges (in the norm) and the the solution of (3.1) exists if $\|K\| < 1$. But, for convergence of the Neumann series and existence of a solution, it is sufficient to require $\|K^{n_0}\| < 1$ for some integer $n_0 \geqslant 1$. This follows from $f = K^{n_0} f + K^{n_0 - 1}\psi + \cdots + K\psi + \psi$, which is equivalent to (3.1).

$$\text{Let} \quad f^* = K^* f^* + \varphi, \tag{3.3}$$

be the equation adjoint to (3.1), where $f^*$, $\varphi \in L^*$, $K^* \in [L^* \to L^*]$, $L^*$ is the space adjoint to $L$, $K^*$ is operator adjoint to $K$.

We recall that $|(\varphi, \psi)| \leqslant \|\psi\|_L \cdot \|\varphi\|_{L^*}$,

where $(\psi, \varphi) = \int_X \psi(x)\varphi(x)\,dx$,

$$\|K\| = \|K^*\|, \qquad (K\psi, \varphi) = (\psi, K^*\varphi),$$

$$[K^*\varphi](x) = \int_X k(x, x')\varphi(x')\,dx'.$$

By definition, $L_\infty$ is a space adjoint to $L$, i.e., the space of bounded (almost everywhere) functions provided with a norm

$$\|\varphi\|_{L_\infty} = \text{vrai sup}\,|\varphi(x)|, \qquad x \in X.$$

In this chapter we consider Monte Carlo algorithms for estimating the functionals of the form

$$I_\varphi = (f, \varphi) = \sum_{n=0}^{\infty} (K^n\psi, \varphi). \tag{3.4}$$

The convergence of (3.4) is assured by the convergence (in norm) of the Neumann series.

In order to show that $(f, \varphi) = (\psi, f^*)$, where $f^* = K^*f^* + \psi$, we multiply (3.1) by $f^*$ and (3.3) by $f$ and then compare these expressions, keeping in mind that $(Kf, f^*) = (f, K^*f^*)$.

The stationary Markov chain is defined as the sequence of random points (states) $x_0, x_1, \ldots, x_n$ such that the distribution of $x_n$ is independent of all previous states exept its immediate predecessor $x_{n-1}$ or, more formally (for distribution densities)

$$P(x_n = x|x_{n-1} = x', \ldots, x_2 = s_2, x_1 = s_1) = P(x_n = x|x_{n-1} = x')$$
$$= r(x', x).$$

The function $r(x', x)$ is called the transition density and is sometimes denoted by $r(x' \to x)$. The distribution of the initial state is defined by the initial density $r_0(x)$. We may extend this definition by introducing a termination probability $p(x')$ at the point $x'$. The random number of the immediate predecessor of the terminal state is denoted by $N$. The Monte Carlo method deals, clearly, only with Markov chains, having a finite number of steps with probability one. Moreover, it is usually supposed that the mean value $E(N)$ is finite. We give a sufficient condition for this in Sect. 3.2. We emphasize again that the Markov chain is completely defined by the initial density $r_0(x)$, the transition density $r(x', x)$ and the probability of termination $p(x')$.

## 3.2 Sufficient Conditions for $E(N)$ to be Finite.

We use the notation $k_p(x', x) = r(x', x)[1 - p(x')]$, $K_p \in [L_1 \to L_1]$ with kernel function $k_p(x', x)$.

First, we obtain the probability of the event $\{N = n\}$ using Bayes's theorem on conditional probabilities.

$$
\begin{aligned}
P(N = n) &= E_{(x_0,\ldots,x_n)}[P(N = n | x_0, \ldots, x_n)] \\
&= E_{(x_0,\ldots,x_n)}\left[ p(x_n) \times \prod_{k=0}^{n-1} (1 - p(x_k)) \right] \\
&= \overbrace{\int_X \cdots \int_X}^{n+1} r_0(x_0) p(x_n) \left[ \prod_{k=0}^{n-1} (1 - p(x_k)) \times r(x_k, x_{k+1}) \right] dx_0, \ldots, dx_n \\
&= \overbrace{\int_X \cdots \int_X}^{n+1} r_0(x_0) p(x_n) \left[ \prod_{k=0}^{n-1} k_p(x_k, x_{k+1}) \right] dx_0, \ldots, dx_n = (K_p^n r_0, p).
\end{aligned} \tag{3.5}
$$

To obtain $P(N \geqslant n)$, we must replace $p(x)$ with $\delta(x) \equiv 1$; hence $P(N \geqslant n) = (K_p r_0, \delta)$.

Now $\{N = \infty\} \supset \{N \geqslant n\} \Rightarrow P(N = \infty) \leqslant P(N \geqslant n)$.

The relations just obtained show that $P(N > n) \to 0$ as $n \to \infty$; hence $P(N = \infty) = 0$, provided that the Neumann series for $f = K_p f + r_0$ converges. But, as was mentioned previously, for this, it is sufficient to require

$$
\|K_p^{n_0}\| < 1 \quad \text{for some} \quad n_0. \tag{3.6}
$$

Thus if (3.6) holds, the Markov chain terminates after a finite number of steps with probability one. It is easy to show that (3.6) is a sufficient condition for $E(N)$ to be finite.

Indeed,

$$
\begin{aligned}
E(N) &= \sum_{n=0}^{\infty} n(K_p^n r_0, p) = \left( \sum_{n=0}^{\infty} n K_p^n r_0, p \right) \\
&= \left( \sum_{n=1}^{\infty} \sum_{k=n}^{\infty} K_p^k r_0, p \right) = \left( \sum_{n=1}^{\infty} K_p^n f, p \right) = (f_p, p),
\end{aligned}
$$

where $f = K_p f + r_0, f_p = K_p f_p + K_p f$.

By (3.6), we have $f, f_p \in L_1, (f_p, p) < +\infty$. In particular, $E(N) < +\infty$ if $p(x) \geqslant \varepsilon > 0$, because

$$\|K_p\| \leqslant \sup_x \int_X r(x, x')[1 - p(x)]\, dx' \leqslant (1 - \varepsilon) \sup_x \int_X r(x, x')\, dx' = 1 - \varepsilon < 1.$$

## 3.3 Basic Estimate of $(f, \varphi)$

Let $I_\varphi = (f, \varphi)$ be the quantity that is to be estimated, where $f = Kf + \varphi$ and $\|K^{n_0}\| < 1$. Consider now the Markov chain with initial density $r(x_0)$, transition density $r(x', x)$ and probability of termination $p(x)$; $N$ is the random number of the last state. Introduce auxiliary random weights by use of the recursion formula

$$Q_0 = \frac{\psi(x_0)}{r_0(x_0)}, \qquad Q_n = Q_{n-1} \frac{k(x_{n-1}, x_n)}{r(x_{n-1}, x_n)} \cdot \frac{1}{1 - p(x_n)},$$

and consider the random variable

$$\xi = \sum_{n=0}^{N} Q_n \cdot \varphi(x_n).$$

We shall see, under certain assumptions about the functions $r_0(x)$, $k(x', x)$, $p(x)$, that

$$E[\xi] = I_\varphi = (f, \varphi).$$

The nature of these asumptions is clear: the paths of the chain must have a nonzero probability of originating at points $x$ with $\psi(x) \neq 0$, and for the transition $x' \to x$, $k(x', x) \neq 0$. That is, we must require

$$\begin{aligned} &r_0(x) \neq 0 && \text{if} \quad \psi(x) \neq 0, \\ &r(x', x) \neq 0 && \text{if} \quad k(x', x) \neq 0, \\ &p(x') \neq 1 && \text{if} \quad k(x', x) \neq 0. \end{aligned} \tag{3.7}$$

The random variable $\xi$ is a commonly used and most convenient estimator of the functional $I_\varphi = (f, \varphi)$; we shall therefore call it a basic estimator of $(f, \varphi)$. Let $K_1$ be an operator with the kernel function

$$k_1(x', x) = |k(x', x)|.$$

*Theorem.* Under the conditions (3.7) and $\|K_1^{n_0}\| < 1$ for some integer $n_0 \geqslant 1$,

$$E\xi = E \sum_{n=0}^{N} Q_n \varphi(x_n) = I_\varphi = (f, \varphi).$$

*Proof.* To average the series termwise let us reconstruct the Markov chain by introducing a new state coordinate,

$$\delta_n = \begin{cases} 0 & \text{if the transition } x_{n-1} \to x_n \text{ leads to a break of the trajectory,} \\ 1 & \text{otherwise,} \end{cases}$$

so that this chain is formally of infinite length.

Let $\Delta_n = \prod_{k=0}^{n} \delta_k = \begin{cases} 1 & \text{until the first break,} \\ 0 & \text{after the first break.} \end{cases}$

Then we can write

$$\xi = \sum_{n=0}^{\infty} \Delta_n Q_n \varphi(x_n). \tag{3.8}$$

Suppose for the moment that the functions $k(x', x)$, $\psi(x)$, $\varphi(x)$ are nonnegative. Then

$$E\xi = \sum_{n=0}^{\infty} E[\Delta_n \cdot Q_n \cdot \varphi(x_n)]. \tag{3.9}$$

Now, by Bayes's theorem on conditional probabilities,

$$\begin{aligned} E[\Delta_n \cdot Q_n \cdot \varphi(x_n)] &= E_{(x_0, \ldots, x_n)} E[\Delta_n \cdot Q_n \varphi(x_n) | x_0, \ldots, x_n] \\ &= E_{(x_0, \ldots, x_n)} [Q_n \varphi(x_n) E(\Delta_n | x_0, \ldots, x_n)] \\ &= E_{(x_0, \ldots, x_n)} \left[ Q_n \varphi(x_n) \prod_{k=0}^{n-1} (1 - p(x_k) \right] \\ &= \overbrace{\int_X \cdots \int_X}^{n+1} \varphi(x_n) r_0(x_0) \left[ \prod_{k=0}^{n-1} r(x_k, x_{k+1}) \right] \\ &\quad \times \left[ \prod_{k=0}^{n-1} \frac{k(x_k, x_{k+1})}{r(x_k, x_{k+1})[1 - p(x_k)]} \right] \frac{\psi(x_0)}{r(x_0)} \left[ \prod_{k=0}^{n-1} [1 - p(x_k)] \, dx_0, \ldots, dx_n \right] \\ &= \overbrace{\int_X \cdots \int_X}^{n+1} \psi(x_0) \varphi(x_n) \left[ \prod_{k=0}^{n-1} k(x_k, x_{k+1}) \right] dx_0, \ldots, dx_n = (K^n \psi, \varphi), \end{aligned}$$

because

$$\begin{aligned} E[\Delta_n | x_0, \ldots, x_n] &= P(\Delta_n = 1 | x_0, \ldots, x_n) \\ &= P(\delta_0 = \delta_1 = \cdots = \delta_n = 1 | x_0, \ldots, x_n) = \prod_{k=0}^{n-1} [1 - p(x_k)]. \end{aligned}$$

Thus, for the nonnegative functions $k(x', x)$, $\varphi(x)$, $\psi(x)$, and by use of (3.9), the theorem is proved.

Let us turn to the general case of alternating functions $k(x', x)$, $\psi(x)$, $\varphi(x)$. Let $Q_n^{(1)}$ be the weights that correspond to (3.1) with $k_1(x', x) = |k(x', x)|$, $\psi_1(x) = |\psi(x)|$, $\varphi_1(x) = |\varphi(x)|$.
Now

$$|\eta_m| = \left|\sum_{n=0}^{m} \Delta_n Q_n \varphi(x_n)\right| \leqslant \sum_{n=0}^{m} |\Delta_n Q_n \varphi(x_n)|$$

$$= \sum_{n=0}^{m} \Delta_n Q_n^{(1)} \varphi_1(x_n) = \eta_m^{(1)} \to \sum_{n=0}^{\infty} \Delta_n Q_n^{(1)} \varphi_1(x_n) = \xi_1.$$

By the previous assumptions, $E\xi_1 = (f_1, \varphi_1)$ is finite. Here $f_1 = K_1 f_1 + \psi_1$. Thus, by the Lebesgue-dominated convergence theorem,

$$\lim_{m\to\infty} E\eta_m = E(\lim_{m\to\infty} \eta_m) = E\xi.$$

Next $E(\eta_m) = E \sum_{k=0}^{m} \Delta_k Q_k \varphi(x_k) = \sum_{k=0}^{m} E(\Delta_k Q_k \varphi(x_k)) = \sum_{k=0}^{m} (K^k \psi, \varphi)$,

because the relation $E(\Delta_n Q_n \varphi(x_n)) = (K^n \psi, \varphi)$ holds in the general case. Hence, since $\|K^{n_0}\| \leqslant \|K_1^{n_0}\| < 1$,

$$E\xi = \lim_{m\to\infty} E\eta_m = \sum_{n=0}^{\infty} (K^n \psi\ \ \varphi) = I_\varphi < \infty, \qquad \text{Q.E.D.}$$

Notice that formally substituting $\psi(x') = \delta(x' - x)$, $r_0(x') = \delta(x' - x)$ into

$$E\xi = E\left[\sum_{n=0}^{N} Q_n \varphi(x_n)\right] = (f, \varphi) = (\psi, f^*), \tag{3.10}$$

and putting $Q_0 = 1$ yields

$$f(x) = \varphi(x) + E \sum_{n=1}^{N} Q_n \cdot \varphi(x_n). \tag{3.11}$$

Thus we may apply (3.11) to estimate the solution of the adjoint equation at a given point. We can prove the equality (3.11) (as the theorem above was proved) by use of the expansion $f^* = \sum_{n=0}^{\infty} K^{*n} \varphi$.

The expression (3.11) is widely used in Monte Carlo technique to calculate the variances of estimators and to construct estimators for the bilinear functionals of the form $(f, f^* \chi)$ and, in particular, for the functionals of the perturbation theory.

## 3.4 Additional Remarks

For the case of nonnegative functions $\psi(x)$, $\varphi(x)$, $k(x', x)$, the variance of the basic estimate is given by

$$\mathscr{D}\xi = (\chi, \varphi[2f^* - \varphi]) - I_\varphi^2, \tag{3.12}$$

where $\chi$ is the Neumann series for the equation

$$f_1(x) = \int_X \frac{k^2(x', x) f_1(x')}{r(x', x)[1 - p(x')]} dx' + \frac{\psi^2(x)}{r_0(x)}.$$

In the case of alternating $f$, $\varphi$, this result is also valid if $\mathscr{D}\xi_1 < \infty$. If $\chi$ is not convergent, the variance may become infinite.

The variance of the $\xi$ estimate (3.12) was first obtained by *Ermakov* and *Zolotukhin* [9].

$$\text{If} \quad \psi(x) \geqslant 0 \quad \text{and} \quad \int_X \psi(x)\,dx = 1,$$

$$k(x', x) \geqslant 0 \quad \text{and} \quad q(x') = \int_X k(x', x)\,dx \leqslant 1,$$

then, taking

$$r_0(x) = \psi(x), \qquad r(x', x) = \frac{k(x', x)}{q(x')}, \qquad p(x) = 1 - q(x),$$

we obtain $Q_n = 1, n = 0, 1, \ldots$ and $\xi = \sum_{n=0}^{N} \varphi(x_n)$.

This type of kernel is called a "substochastic kernel", which occurs when the physical process may be regarded as a Markov chain, for example a chain of photon collisions in a medium.

The algorithm is a direct simulation of this chain. For the direct-simulation method, the variances of the corresponding estimators are always finite, because, for this case, $\chi = f$. However, these variances are frequently too large to permit sufficiently accurate calculation of the result, even on high-speed digital computers.

Therefore, modifications of the direct-simulation method based on the appropriate integral equation are of great importance.

A great many functionals of the Markov chain with expectation equal to $I_\varphi$ (see [10]) can be constructed. There is, for instance, a well-known absorption estimate

$$\eta = \frac{Q_N \cdot \varphi(x_N)}{p(x_N)},$$

which is unbiased (i.e., $E\eta = I_\varphi$), if $p(x) \neq 0$ for $\{x: \varphi(x) \neq 0\}$. The absorption estimator has been proposed specially for use in constructing an ideal Monte Carlo algorithm, i.e., an algorithm with zero variance, for the case of nonnegative functions $k(x', x)$, $\psi(x)$, $\varphi(x)$. The ideal Markov chain is then defined by

$$r_0(x) = \frac{\psi(x)\cdot f^*(x)}{(\psi, f^*)}, \quad r(x', x) = \frac{k(x', x)\cdot f^*(x)}{[K^*f^*](x')}, \quad p(x) = \frac{\varphi(x)}{f^*(x)},$$

Substituting these quantities in the expression for $\eta$ immediately yields $\eta \equiv I_\varphi$.

Less obviously, an ideal $\xi$ estimate can be constructed. *Hisamutdinov* [11] has proposed a method for describing a certain class of unbiased estimates and has constructed some examples.

The absorption estimates and all estimates other than the $\xi$ estimate have a general disadvantage: they include quantities that are either difficult to calculate or fluctuate excessively. Practical calculations show that for atmospheric-optics problems, it is more effective to develop modifications of the basic estimate.

## 3.5 $\xi$ Estimate with Zero Variance; Importance Sampling

In this section, we shall take the functions $k(x', x)$, $\psi(x)$, $\varphi(x)$ to be nonnegative. For the sake of simplicity, we suppose also that $[K^*f^*](x) = f^*(x) - \varphi(x) > 0$ for all $x \in X$.

*Theorem.* If

$$r_0(x) = \frac{\psi(x)\cdot f^*(x)}{(\psi, f^*)}, \quad p(x) \equiv 0, \quad \text{and} \quad r(x', x) = \frac{k(x', x)\cdot f^*(x)}{[K^*f^*](x')}, \tag{3.13}$$

then $\mathscr{D}\xi = 0, \quad E\xi = I_\varphi$.

*Proof.* It is easy to show that

$$\eta_m \equiv \sum_{n=0}^{m-1} Q_n\varphi(x_n) + Q_m\cdot f^*(x_m) = I_\varphi,$$

with probability one, and consequently $\mathscr{D}\xi_m = 0$ for $m \geqslant 1$. Putting

$$\xi_m = \sum_{n=0}^{m} Q_n\cdot\varphi(x_n),$$

we get

$$\eta_m - \xi_m = Q_m[f^*(x_m) - \varphi(x_m)] = Q_m[K^*f^*](x_m)$$
$$= \frac{(\psi, f^*)}{f^*(x_0)} \cdot \frac{[K^*f^*](x_0)}{f^*(x_1)} \cdots \frac{[K^*f^*](x_{m-1})}{f^*(x_m)} [K^*f^*](x_m).$$

Hence $0 \leqslant \eta_m - \xi_m \leqslant (\psi, f^*) = I_\varphi$,

because $K^*f^* = f^* - \varphi \leqslant f^*$.

By (3.1) and from

$$E\eta_m = \sum_{n=0}^{m-1} (K^n\psi, \varphi) + E[Q_m \cdot f^*(x_m)] = I_\varphi,$$

it follows that

$$E[Q_m \cdot f^*(x_m)] \xrightarrow[m\to\infty]{} 0.$$

Hence $\rho_m^2 = E(\eta_m - \xi_m)^2 \leqslant I_\varphi \cdot E(\eta_m - \xi_m) \xrightarrow[m\to\infty]{} 0$.

Writing $\xi_m = \eta_m + (\xi_m - \eta_m)$, we have $\sigma[\xi_m] \leqslant \sigma[\eta_m] + \sigma[\xi_m - \eta_m]$.

But $\sigma[\xi_m - \eta_m] \leqslant \rho_m$ and $\sigma[\eta_m] = 0$.

Consequently $\lim_{m\to\infty} \sigma[\xi_m] = 0$.

Thus, we have a nonnegative sequence of random variables $\xi_m$ which converges monotonically to the random variable $\xi$ with probability one; therefore

$$\mathscr{D}\xi = \lim_{m\to\infty} \mathscr{D}\xi_m, \quad E\xi = \lim_{m\to\infty} E\xi_m, \qquad \text{Q.E.D.}$$

The peculiarity of the Markov chain (3.11) is that

$$r(x', x) = 0 \quad \text{if} \quad f^*(x) = 0.$$

The last expression does not bias the estimate because, in view of (3.11), this estimator scores zero for paths that originate at points $\{x: f^*(x) = 0\}$. *Holton* [12] obtained an analogous result for linear systems with a symmetrical matrix using the restriction $(f^* - \varphi)/f^* \leqslant c < 1$.

This restriction simplifies the proof. Indeed, it is clear that

$$\xi_m = I_\varphi\left[1 - \prod_{k=0}^{m} \frac{f^*(x_k) - \varphi(x_k)}{f^*(x_k)}\right],$$

Hence, by Holton's condition, $\xi = \lim_{m\to\infty} \xi_m = I_\varphi$ with probability one.

This method of proof is not applicable to the general case when $\varphi$ may be zero for $f^* \neq 0$ (or when $\varphi/f^*$ is arbitrarily close to zero). Most Monte Carlo applications in the theory of photon transfer are concerned with this general case.

We call $f^*(x)$ an importance function (of the point $x$ with respect to the functional $I_\varphi$), and, Monte Carlo techniques similar to the ideal are called importance sampling. Further, we shall consider some practical uses of the importance sampling principle. First it should be noted that zero-variance importance sampling cannot be used to solve practical problems for, first, the function $f^*$ is unknown and, second, because $p \equiv 0$, it is necessary to simulate chains of infinite length. Therefore, in practice, an approximate importance sampling based on the chain $f^*$ with $g \approx f^*$ is used.

The remarkable peculiarity of this algorithm is its independence of the constant factor of the function $g$. Also, absorption is introduced in the $m$th state, which almost certainly results in a break in the trajectory.
If

$$g = \text{const}\,(1 + \varepsilon)f^*, \qquad |\varepsilon(x)| \leqslant \delta < 1,$$

$$p(x) \leqslant \delta_p, \qquad \{p(x) = 0 \text{ if } n \leqslant m\},$$

$$q' = \|K\| \cdot \frac{1 + \delta}{(1 - \delta)(1 - \delta_p)} < 1,$$

this algorithm gives

$$\mathscr{D}\xi \leqslant \frac{2I_\varphi \cdot \|\psi\| \cdot \|f^*\| \cdot (1 + \delta)^2}{1 - q'} \cdot \left\{\left(\frac{2\delta}{1 - \delta} + \frac{\delta_p}{1 - \delta_p} \cdot \frac{1 + \delta}{1 - \delta}\right)^2 \cdot \frac{1 + q}{1 - q}\right.$$
$$\left. + \delta_p\left(\frac{1 + \delta}{1 - \delta}\right)^2 \cdot \left(\frac{1 + \delta}{1 - \delta} q\right)^m\right\}, \quad q = \|K\|.$$

The efficiency of the Monte Carlo methods is measured by $s = 1/t \cdot \mathscr{D}\xi$, where $t$ is the average computer time per sample value $\xi$. Let $p(x) \equiv \delta_p$ for $n > m$. It follows from the estimate above that for appropriate $\sigma_p$ and $m$, $s \to 0$ and $\delta \to 0$ (for example, when $\delta_p \sim \delta$, $m \sim |\ln \delta|$). This conclusion could be used when $p$ and $m$ are to be chosen.

Because information is usually available about a function proportional to $f^*$ only, the absence of dependence on the constant factor in $g = \text{const}\,(1 + \varepsilon)f^*$ is very useful. The cited relations show that the Monte Carlo techniques under study may be improved by making use of a priori information (even though crude) about $f^*$. The large $m$ is used for the good approximation of $f^*$. The most important applications of this method are those used in the problem of photon transfer theory. In these problems, $X$ is the phase space of coordinates and velocities and the quantity $K^*g$ is in fact not computable for realistic models.

However, the kernel function of the integral equation may be represented as the product of the conditional densities of the scattering angle, the free-path length, and other quantities. This makes it possible to use the approximation of the importance function at each step of simulation.

It is not difficult to formulate and prove the corresponding zero-variance theorem. This technique often results in simpler formulas than the method of direct utilization of the quantity $K^*g$. In a similar manner, applications of the asymptotic solution of the Milne problem to the calculation of photon transfer through thick layers of a medium have been developed.

It should be noted that there are also other ways of using importance functions, for instance, a direct realization of the importance sampling for estimating the path-integral representation $I_\varphi = E\xi$ [13]. However, the dependence of the importance function on the constant factor, the necessity of calculating the quantity $K^*g$, and the severe restriction $K^*g/g \leqslant 1$ all create difficulties in the practical realization of these techniques.

For example, the unbiased estimate

$$\zeta = (\psi, g) + \sum_{n=0}^{N} Q_n\{\varphi(x_n) + [K^*g](x_n) - g(x_n)\},$$

which is, clearly, analogous to the control-variance method for calculating an integral, seems to be of particular interest, because we may simultaneously use various values of $g$ when a number of $I_\varphi$ functionals are calculated. But, as we have mentioned, it is difficult to use this estimate in complicated calculations.

## 3.6 Use of Importance Sampling in Estimation of a Number of Functionals

Let $\varphi(x)$ be a function of $x$ and a parameter $t = 1, 2, \ldots, s$, and let $I(t) = (f, \varphi_t)$, where $\varphi_t = \varphi(x, t)$.

Let $\mu(t) \geqslant 0$ be a weight function,

$$\sum_{t=1}^{s} \mu(t) = 1, \quad \text{and let} \quad \xi(t) = \sum_{n=0}^{N} Q_n \cdot \varphi(x_n, t).$$

The function $g$ will be called best (see Sect. 3.5) if it minimizes the average variance

$$\mathscr{D}_1 = \sum_{t=1}^{s} \mathscr{D}\xi(t)\mu(t).$$

Define the function $\varphi_0$ and $f^*$ by

$$\varphi_0(x) = \left[\sum_{t=1}^{s} \varphi^2(x, t)\mu(t)\right]^{1/2}, \qquad f_0^* = K^* f_0^* + \varphi_0$$

and put $I_0 = (f, \varphi_0)$.

By the Cauchy–Schwarz-Buniakowski inequality and Fubini's theorem we have

$$\begin{aligned}\sum_t E\xi^2(t)\cdot\mu(t) &= \sum_t E\left[\sum_{n=0}^{N}\sum_{m=0}^{N} Q_n Q_m \varphi(x_n, t)\varphi(x_m, t)\right]\mu(t)\\ &= E\sum_{n=0}^{N}\sum_{m=0}^{N} Q_n Q_m \sum_t \varphi(x_n, t)\varphi(x_m, t)\mu(t)\\ &\leqslant E\sum_{n=0}^{N}\sum_{m=0}^{N} Q_n Q_m \varphi_0(x_n)\varphi_0(x_m) = E\xi_0^2.\end{aligned}$$

By the theorem of Sect. 3.5, the function $g = f^*$, $(p = 0)$ makes the quantity $E\xi_0^2$ a minimum $I_0^2$; consequently,

$$\mathscr{D}_1 \leqslant I_0^2 - \sum_t I^2(t)\mu(t).$$

The ability to make use of prior information about $f_0^*$ in Monte Carlo calculations of $I(t)$ is derived from this inequality.

It is not difficult to show that, for the absorption estimate, the infinum of the average variance is equal to $I_0^2 - \sum_t I^2(t)\mu(t)$ (see Sect. 3.4). The minimizing function $g$ for $\mathscr{D}_1$ (if $p \equiv 0$) can be shown to satisfy the nonlinear equation

$$g = [(K^*g)^2 + \rho^2]^{1/2}, \quad \text{where} \quad \rho^2 = \sum_t \varphi_t(2f_t^* - \varphi_t)\mu(t).$$

The solution of this equation exists, is unique, and satisfies the inequality $0 \leqslant g \leqslant f^*$.

## 3.7 Utilization of Asymptotic Solutions of the Milne Problem

Let the three-dimensional half-space $R_+^3 = \{z \geqslant 0\}$ be filled with a homogeneous absorption and scattering medium. Denote by $\mu$ the cosine of the angle between the direction of the photon motion and the $z$ axis. We use the following notation.

$\psi(z, \mu)$: density of the photon source,

$l$: free length,

$q$: probability of survival of the photon undergoing a collision,

$\omega(\mu, \mu')$: scattering function satisfying

$$\omega(\mu, \mu') = \omega(\mu', \mu), \qquad \int_{-1}^{1} \omega(\mu', \mu)\, d\mu' = 1.$$

The density $f(z, \mu)$ of the number of the scattering photons (including the source) satisfies the integral equation $f = Kf + \psi$. We suppose that

(i) $K \in [L^1 \to L^1]$ and hence $K^* \in [L^\infty \to L^\infty]$,

(ii) $\|K\| = \|K^*\| < 1$.

The operator $K^*$ is defined as

$$[K^*h](z, \mu) = \begin{cases} \dfrac{q}{l|\mu|} \displaystyle\int_0^z \exp\left\{-\dfrac{z - z'}{l|\mu|}\right\} \int_{-1}^{+1} \omega(\mu, \mu')h(z', \mu')\, d\mu' dz', & \mu < 0 \\ \dfrac{q}{l} \displaystyle\int_{-1}^{+1} \omega(0, \mu')h(z, \mu')\, d\mu', & \mu = 0 \\ \dfrac{q}{l\mu} \displaystyle\int_z^{\infty} \exp\left\{-\dfrac{z' - z}{l\mu}\right\} \int_{-1}^{+1} \omega(\mu, \mu')h(z', \mu')\, d\mu' dz', & \mu > 0, \end{cases} \tag{3.14}$$

where $z \geqslant 0$, $-1 \leqslant \mu \leqslant +1$.

Let us assume that the function $g(z, \mu)$ satisfies the equation $g = K_1^* g$, where $-\infty < z < +\infty$ and $K_1^*$ is given by (3.14) provided that the integration in (3.14) for the case $\mu < 0$ is carried out over $(-\infty, z)$.

It is known [14] that we may write

$$g(z, \mu) = \exp\left(-\frac{z}{L}\right) a(\mu),$$

where the diffusion length $L > l$ and the function $a(\mu)$ satisfy the equation

$$\left(\mu \cdot \frac{l}{L} + 1\right) a(\mu) = q \int_{-1}^{+1} \omega(\mu, \mu') a(\mu')\, d\mu'. \tag{3.15}$$

Suppose that there exist constants $M_1, M_2$ so that $0 < M_1 \leqslant a(\mu) \leqslant M_2 < \infty$. These inequalities are satisfied if, for instance, the function $\omega(\mu, \mu')$ satisfies the similar restriction. Let us consider the problem of finding the probability of photon escape from $R_+^3$. This probability is given by $I_\varphi = (f, \varphi)$, where

$$\varphi(z, \mu) = \begin{cases} \exp\left(-\dfrac{z}{|\mu| l}\right) & \text{if } \mu < 0, \\ 0 & \text{if } \mu \geqslant 0. \end{cases}$$

From the definition of $K^*$, $K_1^*$, $g$, $\varphi$, it follows immediately that

$$[K^*g](z, \mu) = [Kg](z, \mu) - \varphi(z, \mu)g(0, \mu) = g(z, \mu) - \varphi(z, \mu)a(\mu). \tag{3.16}$$

Thus, $K^*g/g \leqslant 1$. Combining (3.16) with $f^* = K^*f^* + \varphi$, we have $g = \alpha \cdot f^*$, $M_1 \leqslant \alpha(z, \mu) \leqslant M_2$. From this, putting $c = 2/(M_1 + M_2)$, we have

$$cg = (1 + \varepsilon)f^*, \qquad |\varepsilon(z, \mu)| \leqslant \frac{M_2 - M_1}{M_2 + M_1}.$$

Notice that for isotropic scattering, i.e., when

$$\omega(\mu, \mu') \equiv 1/2, \quad \text{we have} \quad a(\mu) = \frac{1}{1 + \mu l/L}, \qquad \frac{M_2 - M_1}{M_2 + M_1} = \frac{l}{L}.$$

Define a Markov chain $\{x_n\}$ as follows:

$$r(x', x) = \frac{k(x', x)g(x)}{[K^*g](x)'}, \qquad r_0(x) = \frac{\psi(x)g(x)}{(\psi, g)},$$

$p(x_n) = 0$ if $n < m$ and $p(x_n) < \delta_p$ if $n \geqslant m$.

Let $\xi = \sum_{n=0}^{N} Q_n\varphi(x_n)$; then $E\xi = I_\varphi = (f, \varphi)$.

*Theorem.* Let

$$\|K\| = q < 1, \; K^*g/g \leqslant 1 \text{ and } g = \text{const}\,(1 + \varepsilon)f^*, \; |\varepsilon(x)| \leqslant \delta < 1.$$

Then if $q' = q/(1 - \delta_p) < 1$, we have

$$\mathscr{D}\xi \leqslant 2\frac{1 + \delta}{1 - \delta} \cdot I_\varphi\left[I_\varphi + \frac{q'}{1 - q'}\|\varphi\| \cdot \|\psi\| \cdot q^m\right]. \tag{3.17}$$

*Proof.* We have

$$Q_n = \frac{(\psi, g)}{g_0}\left(\prod_{k=1}^{n} \frac{K^*g_{k-1}}{g_k}\right) \cdot \left(\prod_{k=m}^{n-1} \frac{1}{1 - p_k}\right),$$

writing $\psi_0 = \psi(x_0)$, $\varphi_n = \varphi(x_n)$, $f_n^* = f^*(x_n)$, $K^*f_n^* = [K^*f^*](x_n)$, $p_n = p(x_n)$ for convenience.
Hence

$$Q_n \leqslant \begin{cases} \dfrac{(\psi, g)}{g_n(1 - \delta_p^{n-m})}, & n > m, \\[2ex] \dfrac{(\psi, g)}{g_n}, & n \leqslant m. \end{cases}$$

Then, by (3.11) and latter inequality, making use of the iterated averaging and keeping in mind the assumptions of the theorem, we have

$$E\left(\sum_{n=0}^{N} Q_n \cdot \varphi_n\right)^2 = E\left[\sum_{n=0}^{N} Q_n^2 \cdot \varphi_n \cdot \left(\sum_{s=n+1}^{N} 2\frac{Q_s}{Q_n}\varphi_s + \varphi_n\right)\right]$$

$$= E\left[\sum_{n=0}^{m} Q_n^2 \cdot \varphi_n \cdot (f_n^* + K^* f_n^*) + \sum_{n=m+1}^{N} Q_\eta^2 \varphi_n (f_n^* + K^* f_n^*)\right]$$

$$\leqslant 2E\left[\sum_{n=0}^{m} Q_n^2 \cdot \varphi_n \cdot f_n^* + \sum_{n=m+1}^{N} Q_n^2 \cdot \varphi_n \cdot f_n^*\right]$$

$$\leqslant 2(\psi, g)\left[\sum_{n=0}^{\infty}\left(K^n\psi, \varphi \cdot \frac{f^*}{g}\right) + \sum_{n=m+1}^{\infty} \frac{\left(K^n\psi, \varphi \cdot \dfrac{f^*}{g}\right)}{(1-\delta_p)^{n-m}}\right]$$

$$\leqslant 2 \cdot I_\varphi(1+\delta)\left[\frac{I_\varphi}{1-\delta} + \frac{q' \|\psi\| \cdot \|\varphi\|}{(1-\delta)\cdot(1-q')} \cdot q^m\right]. \qquad \text{Q.E.D.}$$

If we take the number $m$ of pure-scattering collisions equal to $\ln I_\varphi / \ln q$, we obtain from (3.17) that

$$\mathscr{D}\xi \leqslant C_1 \cdot I_\varphi^2.$$

Thus we have obtained an estimate of the relative probability error not depending on the quantity $I_\varphi$ with an amount of labor proportional to $\sigma \approx |\ln I_\varphi|$.

The labor expended in obtaining this estimate when the direct simulation method is used is proportional to $1/I_\varphi$.

Let us now discuss the practical realization of the algorithm under study. The expression $k(z_0, \mu_0, z, \mu) \cdot g(z, \mu) = \omega(\mu_0, \mu) a(\mu) k_1(z_0, \mu_0, z) \exp\{-z/L\}$ shows that the free-path length is simulated according to the corresponding sampling procedure discussed in Sect. 2.4 (with parameter $c = 1/L$ and without escape). Scattering is simulated by use of $\omega(\mu_0, \mu) \cdot a(\mu)$. The integral of this may be found from (3.15), provided that $L$ and $a(\mu)$ are estimated sufficiently precisely. For example, a transport approximation may be used if $q \sim 1$, that is $L$ and $a(\mu)$ are as in isotropic scattering, with

$$l' = \frac{l}{q(1-\nu) + 1 - q},$$

where $\nu$ is an average cosine of the scattering angle (see [8]).

There are Monte Carlo algorithms (see Sect. 2.4) based on the representation $\Phi(z, \mu) = \Phi_1(z, \mu) \exp(-cz)$, where $c < \sigma$ is a constant, $\sigma = 1/l$. The integro-differential equation for $\Phi_1$ can be regarded as the equation of photon transfer in a fictitious medium with extinction cross section $\sigma - c\mu$ and survival probability $\sigma_s/(\sigma - c\mu)$, where $\sigma_s = q \cdot \sigma$.

We have investigated in Sect. 2.4, such a transformation for the case of spherical geometry (i.e., when the photon flux is estimated at a point). If $q_1 = \sigma_s/(\sigma - c\mu) > 1$, it is necessary either to simulate a fission, which complicates the algorithm, or to multiply the weight by $q_1$, which often leads to increased variance. An attempt may be made to correct this weight fluctuation by special modification of the scattering, i.e., $\omega_1(\mu', \mu) = \omega(\mu', \mu)\cdot a(\mu)$, where $a(\mu)$ is a function to be found.

After simulation (without absorption) of such scattering, the previous weight $Q'$ is multiplied by the factor:

$$Q = Q'\cdot \frac{q \int_{-1}^{+1} \omega(\mu', \mu)a(\mu)\, d\mu}{(1 - c\cdot l\cdot \mu')a(\mu)} .$$

It follows from this that, if (3.5) is satisfied, the weight fluctuations do not accumulate.

Thus, we have obtained the characteristic equation of the Milne problem without utilizing the asymptotic solution of the adjoint equation.

## 3.8 Local Estimates

Let $\Phi$ be the photon flux to be estimated at the desired point $\boldsymbol{x}^* = (\boldsymbol{r}^*, \omega^*)$. For simplicity, let $\psi(\boldsymbol{x}^*) = 0$ and $\sigma = \sigma_s$, i.e., $\sigma_c = 0$. Substituting $\boldsymbol{x}^*$ into (2.14) and dividing both sides of this equation by $\sigma(\boldsymbol{r})$, we get

$$\Phi(\boldsymbol{x}^*) = \int_X \frac{k(\boldsymbol{x}', \boldsymbol{x}^*)}{\sigma(\boldsymbol{r}^*)} \cdot f(\boldsymbol{x}')\, d\boldsymbol{x}'. \tag{3.18}$$

Thus the quantity $\Phi(\boldsymbol{r}^*)$ is formally represented as the linear functional of the collision density. However, the kernel $k(\boldsymbol{x}', \boldsymbol{x}^*)$ involves a $\delta$ function. To eliminate this, we integrate (3.18) over an arbitrary subset $\Omega_i$ of the domain of directions $\Omega$,

$$\int_{\Omega_i} \Phi(\boldsymbol{r}^*, \omega^*)\, d\omega^* = \int_X l_i(\boldsymbol{x}', \boldsymbol{x}^*) f(\boldsymbol{x}')\, d\boldsymbol{x}' = E \sum_{n=0}^{N} Q_n l_i(\boldsymbol{x}_n, \boldsymbol{x}^*),$$

where

$$l_i(\boldsymbol{x}, \boldsymbol{x}^*) = \frac{\exp[-\tau(\boldsymbol{r}, \boldsymbol{r}^*)] g(\mu^*)}{2\pi|\boldsymbol{r} - \boldsymbol{r}^*|^2} \cdot \Delta_i(\boldsymbol{s}^*). \tag{3.19}$$

Here,

$$s^* = \frac{r^* - r}{|r^* - r|}, \qquad \mu^* = \omega \cdot s^*,$$

and $\Delta_i(S)$ is the characterisric function of the domain $\Delta_i$.

Formula (3.19) defines the well-known local estimate of the photon flux. Unfortunately, this estimate is not applicable when the flux at a desired point in a given direction must be estimated. Besides, it has infinite variance.

Equation (2.14) may be written as $f = K^2 f + K\psi + \psi$. The local estimate for this equation may be called a double local estimate. Let $\psi$ be a density of artificial collisions, analogous to the flux of particles incident on the medium. Then the collision density $K\psi$ corresponds to the nonscattering flux in the medium. Therefore,

$$\Phi(x^*) = \int_X \left[\frac{\sigma(r^*)}{k_1(x', x^*)}\right]^{-1} f(x')\, dx',$$

defines the double local estimate of the scattering flux. Here

$$\frac{k_1(x', x^*)}{\sigma(r^*)} = \frac{1}{\sigma(r^*)} \int_X k(x', x'')k(x'', x^*)\, dx'', \tag{3.20}$$

where the integration is performed over $r''(t) = r^* - \omega^* t$, $t > 0$. The double local estimate may be used to calculate the intensity at a desired point $x^*$ of the phase space. Its variance, though infinite, diverges more slowly than does the variance of the local estimator. The integral in (3.20) may be estimated by using a single sample of $\rho''$.

In the most simple sampling, we put

$$\rho'' = r^* - \omega^* l^*,$$

where $l^*$ is the free-path length measured in the direction $-\omega^*$ from the point $r^*$. The estimate of (3.20) in polar coordinates with center at the point $r^*$ is

$$\varphi_1(r, \omega', l^*) = g\left[\left(\omega' \cdot \frac{\rho'' - r'}{|\rho'' - r'|}\right)\right] g\left[\left(\frac{\rho'' - r'}{|\rho'' - r'|} \cdot \omega^*\right)\right] \cdot \frac{\exp\{-\tau(r', \rho'')\}}{2\pi|\rho'' - r'|^2}.$$

It is easy to verify that

$$\xi_1 = \sum_{n=0}^{N} Q_n \varphi_1(r_n, \omega_n, l_n^*), \tag{3.21}$$

is an unbiased estimate.

The phase function with a very sharp peak is two valued for $\mu = \varphi_1$. Therefore, in this case, the estimate (3.12) gives inaccurate results. However, similar remark applies also to the ordinary local estimate.

## 3.9 Universal Modification of the Local Estimate with Logarithmically Diverging Variance

In this section, we consider an easily realized modification of the local estimate based on a special upper bound for the extinction coefficient.

Suppose that it is desired to calculate at the point $\boldsymbol{r} = 0$ the total (i.e., integrated over $\boldsymbol{\omega}$) flux of photons that have undergone at least one scattering in the medium. This quantity is given (see Sect. 3.8) by

$$\Phi_0(0) = (f, \varphi_0) = \int\int f(\boldsymbol{r}, \boldsymbol{\omega})\varphi_0(\boldsymbol{r}, \boldsymbol{\omega})\, d\boldsymbol{r}\, d\boldsymbol{\omega},$$

where

$$\varphi_0(\boldsymbol{r}', \boldsymbol{\omega}) = \frac{g(-v)\exp\left\{-\int\limits_0^1 \sigma(\boldsymbol{r}\cdot t)\, dt\right\}}{2\pi \cdot r^2}, \qquad v = \frac{\boldsymbol{r}'\cdot\boldsymbol{\omega}}{r}. \tag{3.22}$$

The Monte Carlo algorithm then consists of calculating the quantities $\varphi_0(\boldsymbol{r}, \boldsymbol{\omega})$ and recording them for each collision of the photon path. The expectation of the sum of these quantities is equal to $(f, \varphi_0)$. The expectation of the square of the local estimate is not less than $(f, \varphi_0)$, which diverges as $1/r$ when $\boldsymbol{r} \to 0$.

As is easy to see from (3.12), this rate of divergence is exact. *Kalos* [15] has shown that the rate of probability convergence of the local estimate averaged over $N$ paths is given by $N^{-1/3}$ (instead of $N^{-1/2}$ for finite variance).

Various modifications of the local estimate reduce the rate of divergence and even make it finite. However, these modifications either essentially complicate the algorithm, or make use of some symmetry of the medium. In this section, we consider a simple modification of the local estimate with rate of divergence $|\ln r|$ as $r \to 0$. It is shown that this estimate may be used in practical calculations. Recall that the ordinary local estimate is applicable only after it is biassed, for instance, by putting $\sigma = 0$ if $r < \varepsilon$, $\varepsilon > 0$. To construct this modification, we shall essentially use the idea of the method of maximal cross section, which simplifies the simulation of the free-path length in an inhomogeneous medium.

We may generalize the justification of this method proposed in Sect. 2.3 as follows. Let $\sigma(\boldsymbol{r}) \leqslant \sigma_m(\boldsymbol{r})$. The photon flux $\Phi(\boldsymbol{r}, \boldsymbol{\omega})$ will not change if we replace $\sigma(\boldsymbol{r})$ with $\sigma_m(\boldsymbol{r})$ and introduce an artificial straight-forward scattering (i.e., in which $\boldsymbol{\omega}$ is unchanged) with probability $[\sigma_m(\boldsymbol{r}) - \sigma(\boldsymbol{r})]/\sigma_m(\boldsymbol{r})$. It follows from (2.1) that the collision density is then multiplied by $\sigma_m(\boldsymbol{r})/\sigma(\boldsymbol{r})$.

The above statement shows that the local estimate remains unbiased if the quantity (3.22) is multiplied by $\sigma(\boldsymbol{r})/\sigma_{\mathrm{m}}(\boldsymbol{r})$. Thus when the cross section $\sigma_{\mathrm{m}}(\boldsymbol{r})$ is used, the quantity

$$\varphi_{\mathrm{m}}(\boldsymbol{r}, \omega) = \frac{g(-\nu) \exp\left\{-\int\limits_0^1 \sigma(\boldsymbol{r}t)\, dt\right\}\sigma(\boldsymbol{r})}{2\pi r^2 \sigma_{\mathrm{m}}(\boldsymbol{r})}, \tag{3.23}$$

must be calculated at each collision point.
Let $\sigma(\boldsymbol{r}) \leqslant \sigma_0$ if $r \leqslant r_0$, $r_0 > 0$, and let

$$\sigma_{\mathrm{m}}(\boldsymbol{r}) = \sigma_{\mathrm{m}}^{(n)}(\boldsymbol{r}) = \begin{cases} \sigma(\boldsymbol{r}) & \text{if} \quad r > r_0, \\ \left(\dfrac{r_0}{r}\right)^n \sigma_0 & \text{if} \quad r \leqslant r_0. \end{cases}$$

Then, (3.23) for $r < r_0$ takes the form,

$$\varphi_{\mathrm{m}}^{(n)}(\boldsymbol{r}, \omega) = \frac{g(-\nu) \exp\left\{-\int\limits_0^1 \sigma(\boldsymbol{r}t)\, dt\right\}\sigma(\boldsymbol{r})}{2\pi r^{2-n} r_0^n \sigma_0}. \tag{3.24}$$

It will be shown in Sect. 3.9 that for $n = 1$, $n = 2$ the rate of divergence of the modified local estimate is equal to $|\ln r|$ and cannot be improved by choice of $n$. In addition, it is easy to simulate the free-path length in accordance with $\sigma^{(1)}$, $\sigma^{(2)}$ if, respectively, $n = 1, 2$. We discuss in detail algorithms for $n = 2$ and $n = 1$.

**$n = 2$**

The case $n = 2$ seems to be attractive because the quantity $\varphi_{\mathrm{m}}^{(n)}$ is, clearly, bounded. However, the variance of this estimate is infinite, because the distribution of the number of collisions is modified (see Sect. 3.9). Consider the algorithm of simulation of the free-path length. Let us consider a photon that is at the point $\boldsymbol{r}$ traveling in the direction $\omega$. It is sufficient to examine the case $r < r_0$, because if $r > r_0$, the run is simulated in the physical medium until it interesects the sphere $r = r_0$; further, the run may be considered to originate at the point of intersection and to simulate the trajectory followed in the case $r < r_0$.

The free-path length is obtained from

$$\int\limits_0^l \sigma_{\mathrm{m}}^{(n)}(\boldsymbol{r} + t\omega)\, dt = -\ln \alpha, \tag{3.25}$$

(see Sect. 2.2).

Let $l_0$ be the distance from the point $\boldsymbol{r}$ to the sphere $r = r_0$ along the direction $\boldsymbol{\omega}$. For $l < l_0$, (3.25) takes the form

$$\int_0^l \frac{\sigma_0 r_0^2 \, dt}{r^2 + t^2 + 2vrt} = -\ln \alpha. \tag{3.26}$$

Hence,

$$\arctan \frac{l + vr}{r(1 - v^2)^{1/2}} = M,$$

where

$$M = |\ln \alpha| \frac{r(1 - v^2)^{1/2}}{\sigma_0 r_0^2} + \frac{\pi}{2} - \arccos v.$$

The equation (3.26) has no solution if $M > \pi/2$, which implies that the photon escapes from the sphere $r = r_0$. In this case, the run may be considered to originate at the point $\boldsymbol{r} + \boldsymbol{\omega} l_0$ and to simulate a new run in the physical medium.

The resultant expression may be written

$$l = r[-v + (1 - v^2)^{1/2} \tan M].$$

This procedure is repeated if $l > l_0$. If the new point of collision $|\boldsymbol{r}_1| = |\boldsymbol{r} + l\boldsymbol{\omega}| < r_0$, then the photon moves in the same direction $\boldsymbol{\omega}$ with probability

$$\frac{\sigma_m^{(2)}(\boldsymbol{r}_1) - \sigma(\boldsymbol{r}_1)}{\sigma_m^{(2)}(\boldsymbol{r}_1)}.$$

In this case, the new values of $v$ and $\arccos v$ may be written

$$v_1 = \frac{l + rv}{r_1}, \qquad \arccos v_1 = \frac{\pi}{2} - M.$$

The probability of scattering is

$$\frac{\sigma(\boldsymbol{r}_1)}{\sigma_m^{(2)}(\boldsymbol{r}_1)} = \frac{\sigma(\boldsymbol{r}_1)}{\sigma_0} \cdot \left(\frac{r_1}{r_0}\right)^2.$$

The local estimate for the given trajectory equals the sum of $\varphi_m^{(2)}$ over all collisions.

$\boldsymbol{n} = \mathbf{1}$

In the case $n = 1$, (3.26) takes the form

$$\int_0^l \frac{\sigma_0 r_0 dt}{(r^2 + t^2 + 2rt\nu)^{1/2}} = -\ln \alpha. \tag{3.27}$$

Integration of (3.27) yields

$$\ln \frac{(r^2 + l^2 + 2rl\nu)^{1/2} + l + r\nu}{r(1 + \nu)} = -\frac{\ln \alpha}{\sigma_0 r_0}. \tag{3.28}$$

By an elementary calculation, we obtain

$$l = \frac{1}{2} br[(1/b - \nu)^2 - 1], \tag{3.29}$$

where

$$b = \exp [\ln \alpha/(\sigma_0 r_0)]/(1 + \nu).$$

If $l > l_0$ then the photon escapes from the sphere $r = r_0$.

The formulas just obtained show that it is convenient to put $r_0 = (\sigma_0)^{-1}$, because then $b = \alpha/(1 + \nu)$ and the free-path-length simulation (3.29) is more efficient than the physical simulation in an homogeneous medium according to the formula $l = -\ln \alpha/\sigma$. When we compare these two modifications of the local estimate, we must take into account the average number of collisions given by

$$S_n = \int\int \sigma_m^{(n)}(r)\Phi(\mathbf{r}, \omega)\, d\mathbf{r}\, d\omega.$$

If $r_0$ is fixed then, obviously, $S_1 < S_2$.

Let $\{(\mathbf{r}_k, \omega_k)\}$ be a sequence of photon collisions, $k = 0, \ldots, N$; $N$ is the random number of the last collision.

Consider a random variable

$$\xi = \sum_{k=0}^{N} \varphi(\mathbf{r}_k, \omega_k), \tag{3.30}$$

where $\varphi \geqslant 0$. It is known that $M\xi = I_\varphi = (f, \varphi)$. *Ermakov* and *Zolotukhin* [9] have shown that

$$E\xi^2 = (f, \varphi[2f^* - \varphi]), \tag{3.31}$$

when $f^*$ is the importance function, defined by

$$f^*(\mathbf{r}, \omega) = \varphi(\mathbf{r}, \omega) + E_{(\mathbf{r}, \omega)} \sum_{k=1}^{N} \varphi(\mathbf{r}_k, \omega_k). \tag{3.32}$$

The subscript $(\boldsymbol{r}, \omega)$ denotes that the average is over all trajectories that originate from the collision point $(\boldsymbol{r}, \omega)$. In view of (3.19, 30), (3.31, 32) may be used to evaluate the variance of the local estimate.

Suppose that the functions $\Phi(\boldsymbol{r}, \omega)$ and $g(\mu)$ are bounded and separated from zero, so that they may be ignored when considering the order of magnitude of (3.31). Then, using the expression for $f$, $\sigma_{\mathrm{m}}^{(n)}$, $\varphi_{\mathrm{m}}^{(n)}$, we obtain

$$E\xi_n^2 \sim R_n = \int_{\{r<1\}} f_0^{*(n)}(r) r^{-2} dr. \tag{3.33}$$

Here $\xi_n$ is defined by (3.30) provided that $\varphi = \varphi_{\mathrm{m}}^{(n)}$; $f_0^{*(n)}(r) = \int f^{*(n)}(\boldsymbol{r}, \omega) d\omega$, where $f^{*(n)}$ is the importance function that corresponds to $\varphi_{\mathrm{m}}^{(n)}$.

The quantities $f_0^{*(n)}(r)$ are evaluated on the basis of (3.32). Clearly,

$$\int \varphi_{\mathrm{m}}^{(n)}(\boldsymbol{r}, \omega)\, d\omega \sim r^{n-2}.$$

Further, from the definition of $\varphi_{\mathrm{m}}^{(n)}$,

$$I_1^{(n)}(r) = \int \left[ E_{(\boldsymbol{r},\omega)} \sum_{k=1}^{N} \varphi_{\mathrm{m}}^{(n)}(\boldsymbol{r}_k, \omega_k) \right] d\omega$$

is the flux (at the point $r = 0$) of scattering photons for an isotropic source at the point $\boldsymbol{r}$.

It is known (see, for example, [14]) that $I_1^{(n)}(r) \sim r^{-1}$, and so $f_0^{*(n)}(r) \sim c_1 r^{n-2} + c_2 r^{-1}$.

Consequently, using in (3.33) polar coordinates, we obtain

$$R_n \sim \lim_{r\to 0} \int_r^1 (c_1 t^{n-2} + c_2 t^{-1})\, dt.$$

Thus, the variance of the local estimate diverges as $|\ln r|$ when $r \to 0$, if $n \geqslant 1$.

However, if $n \geqslant 3$, the average number of collisions $(\sigma_{\mathrm{m}}^{(n)}, \Phi)$, which determines the computing time, becomes infinite. Thus we must take $1 \leqslant n < 3$.

Let $\xi$ be the local estimate with variance diverging as $|\ln r|$, and

$$\zeta_M = \frac{\sum_{i=1}^{M} \xi_i}{M},$$

where $\{\xi_i\}_{i=1}$ is the sequence of the independent samples of this estimate.

In order to study the probability convergence of the estimate $\zeta_M$ let us evaluate

$$U(x) = \int_0^x y^2 dF_\xi(y), \quad \text{where} \quad F_\xi(y) = P(\xi < y)$$

when $x \to \infty$.

Let $\delta_p(\mathbf{r})$ be the characteristic function of the domain $\{\mathbf{r}|r < \rho\}$, and let

$$\xi^{(1)}(\rho) = \sum_{k=0}^{N} \varphi(\mathbf{r}_k, \omega_k)\delta_p(\mathbf{r}_n),$$

$$\xi^{(2)}(\rho) = \sum_{k=0}^{N} \varphi(\mathbf{r}_k, \omega_k)[1 - \delta_p(\mathbf{r}_k)],$$

where $\xi = \xi^{(1)}(\rho) + \xi^{(2)}(\rho)$.

Obviously, $U(x) = E\xi_x$,

where

$$\xi_x = \begin{cases} \xi & \text{if } \xi \leqslant x, \\ 0 & \text{if } \xi > x. \end{cases}$$

Further $E\xi_x^2 \leqslant 2\{E[\xi_x^{(1)}(\rho)]^2 + E[\xi^{(2)}(\rho)]^2\}$,

because $\xi_x \leqslant \xi_x^{(1)}(\rho) + \xi^{(2)}(\rho)$.

Thus $E\xi^{(2)2}(\rho) \sim |\ln \rho|$ as $\rho \to 0$.

On the other hand, $E\xi_x^{(1)2}(\rho) \leqslant xE\xi^{(1)}(\rho) \leqslant c_1 x\rho$,

because $E\xi^{(1)}(\rho) = (f, \varphi\delta_p)$;

therefore $U(x) \leqslant c_1 x\rho + c_2|\ln x|$,

where $\rho > 0$ is an arbitrary small number. Putting $\rho = 1/x$, we obtain $U(x) \leqslant c \ln x$ when $x \to \infty$. This estimate cannot be improved.

Let $U(x) \sim c|\ln x|$. The random variable [16]

$$\gamma_M = \left(\frac{M}{\ln M}\right)^{1/2} (\xi_M - M\xi_M),$$

is asymptotically normal with mean 0, and standard deviation $d$, which we write: asymptotically normal $(0, d)$.

To estimate the parameter $d$ all samples of $\xi$ must be divided into $s$ parts (each having $N$ samples) and the standard error (i.e., the standard deviation of the sampling distribution) of the random variable $\gamma_M$ must be calculated.

The standard error of the resultant estimate is given by

$$d^*(\zeta_{Ms}) = d^*\left(\frac{\ln (Ms)}{Ms}\right)^{1/2}, \tag{3.34}$$

which differs from the analogous quantity for the case of bounded variance by the factor $[\ln (Ms)/\ln M]^{1/2}$. In spite of the fact that the variance of $\xi_{Ms}$ is infinite, it is possible to construct (approximately) a confidence interval for $E\xi$, as was obtained for the case of a normal distribution with $d = d^*(\zeta_{Ms})$ by use of the limit theorem, mentioned previously. Thus, the rate of the probability convergence of the estimate $\zeta_N$ is equal to $[(\ln N)/N]^{1/2}$, where $N$ is the number of samplings.

Sometimes, it is necessary to calculate the local estimate of the photon flux at a point that is selected at random, for each trajectory or even for each collision, from a certain domain. This method is applied also to estimate the integral of the flux over a given domain, and to estimate the flux at a given point $(\boldsymbol{r}^*, \boldsymbol{\omega}^*)$ of the phase space (see Sect. 3.8); the origin of coordinates must be put at a point preselected at random, for the next trajectory. Simple speculations show that the variance of this estimate diverges as $\ln x$.

In order to compare efficiencies of various local estimates, the integral scattered flux $I_s(0)$ at the center of the sphere $r \leqslant 1$ was calculated for isotropic scattering with

$$\sigma(r) = \begin{cases} 1 & \text{if} \quad r \leqslant 1, \\ 0 & \text{if} \quad r > 1, \end{cases}$$

and for a surface source with the Lambert density $2\mu$ $(\mu > 0)$, where $\mu$ is the cosine of angle between the direction of emission and the radius vector of the source position. This source corresponds to the isotropic flux in a space.

The calculations were carried out for the following scattering cross sections in the sphere $r < 1$:

A. $\sigma_{\mathrm{m}}(\boldsymbol{r}) = \sigma = 1$ (ordinary estimate)
B. $\sigma_{\mathrm{m}}(\boldsymbol{r}) = \sigma^{(1)}(r) = 1/r$
C. $\sigma_{\mathrm{m}}(\boldsymbol{r}) = \sigma^{(2)}(r) = 1/r^2$

Statistical estimates of mathematical expectations and of standard deviations of the curtailed local estimates $\xi^{(2)}(\rho)$ (i.e., local estimates obtained by decreasing $\varphi(\boldsymbol{r}, \boldsymbol{\omega})$ to zero in the neighborhood $r < \rho$) were obtained for all these variants.

Variances of the random variables $\xi^{(2)}$ are bounded if $\rho > 0$; $\xi^{(2)}(0) = \xi$. The results for the flux $I_s(0)$ and the corresponding standard errors are shown in Tables 3.1, 2. Each variant was obtained in less than 5 minutes computer time of BESM-6; $N$ is the number of trajectories per variant. In order to estimate the errors of the main results (if $\rho = 0$) the trajectories were divided into groups each having $M$ trajectories (Table 3.1) and subsequently formula (3.34) was applied. Computational results and the exact expression (2.21) for $I_s(0)$,

**Table. 3.1.** Estimates of the flux $I_s(0)$

| Variant \ $\rho$ | 0 | 0.01 | 0.03 | 0.06 | 0.1 | $M$ | $N$ |
|---|---|---|---|---|---|---|---|
| $A$ | 0.219 | 0.198 | 0.191 | 0.183 | 0.172 | — | $230 \times 10^3$ |
| $B$ | 0.201 | 0.199 | 0.193 | 0.183 | 0.172 | 2000 | $220 \times 10^3$ |
| $C$ | 0.196 | 0.194 | 0.190 | 0.182 | 0.171 | 1000 | $72 \times 10^3$ |

**Table 3.2.** Standard Errors

| Variant \ $\rho$ | 0 | 0.01 | 0.03 | 0.06 | 0.1 |
|---|---|---|---|---|---|
| $A$ | — | 0.0037 | 0.0019 | 0.0013 | 0.0010 |
| $B$ | 0.0022 | 0.0014 | 0.0011 | 0.0009 | 0.0007 |
| $C$ | 0.0029 | 0.0021 | 0.0017 | 0.0014 | 0.0012 |

$$I_s(0) = \frac{1 - e^{-1}}{\pi} = 0.20121\ldots$$

agree fairly well.

Tables 3.1, 2 show that the modification B is more efficient than the ordinary local estimate and far better than modification C. It is interesting to note that the coefficients $d^*$ in (3.34) were actually equal in both variants (0.27 and 0.26). The time of simulation of one trajectory in variant $C$ is approximately three times as large as in variant $B$ (or in variant $A$).

To examine the hypothesis of normality of $\gamma_M$, we divide the real axis into 10 equal (in the probabilistic sense) parts according to the hypothetical normal law with parameters that have been evaluated statistically for the case $B$ and in 8 analogous parts for the case $C$. The corresponding numbers of the observations are given in Table 3.3. Values of the chi-squared test for these data are: $\tilde{\chi}_7^2 = 5.3$, $\tilde{\chi}_5^2 = 9.1$; the corresponding fiducial probabilities are 0.75, 0.10 respectively.

Thus the departures from normality are sufficiently small and corroborate the correctness of the calculated errors (Table 3.2).

**Table 3.3.** Values of $\gamma_M$ for various intervals

| Variant | Number of observations | | | | | | | | | | M | Mean number of observations |
|---|---|---|---|---|---|---|---|---|---|---|---|---|
| $B$ | 9 | 14 | 13 | 12 | 11 | 10 | 12 | 8 | 10 | 11 | 1000 | 11 |
| $C$ | 6 | 10 | 13 | 14 | 11 | 6 | 6 | 6 | — | — | 1000 | 9 |

## 3.10 Other Universal Modifications of Local Estimates

Suppose that we want to estimate the quantity

$$I(r^*) = \int_{\Omega} \Phi(x^*)\, d\omega^*.$$

Integrating (3.20) with respect to $\omega^*$, we obtain

$$I(r^*) = \frac{1}{\sigma(r^*)} E \sum_{n=0}^{N} Q_n \int_X k(x_n, x'') \int_{\Omega} k(x'', x^*)\, d\omega^* dx''. \tag{3.35}$$

It is known that the variance of such an estimate for $I$ is finite. However, the double integral in (3.35) is not computable in practice. The estimate of this integral may be randomized as follows. The direction of an auxiliary run that originates at $r_n$ is drawn from a given indicatrix; next the integral

$$I(r_n, \omega) = \int_0^{\infty} \frac{\sigma[r(t)] \exp\{-\tau[r_n, r(t)] - \tau[r(t), r^*]\} g\left[\omega \cdot \dfrac{r^* - r(t)}{|r^* - r(t)|}\right]}{2\pi |r(t) - r^*|^2}, \tag{3.36}$$

is calculated. Here $r(t) = r_n + \omega t$. This estimate was investigated by *Podlivaev* and *Ruzu* [17]. Its variance diverges as $\ln x$, because $I(r_n, \omega) \sim (\sin\theta)^{-1}$ if the angle $\theta$ between $\omega$ and $r^* - r_n$ is small. The variance would become finite if the singularity could be included in the distribution density for $\omega$. However it is difficult to include it in a realistic indicatrix.

*Kalos* has suggested the following modification of the local estimate with finite variance. The integral in (3.35) is calculated by use of a single sample from the density

$$c \frac{|r_n - r^*|}{|r'' - r_n|^2 |r'' - r^*|^2},$$

for $r''$. The proof of finiteness of the variance for this modification is given, for instance, in [5]. The complexity of $r''$ sampling makes it difficult to realize *Kalos*'s estimate (a more convenient modification of this technique is given in [18]). Besides, the estimate of the integral includes two values of the indicatrix. That lowers the quality of the estimate if the indicatrix has a forward peak (such as often appears in atmospheric- and ocean-optics problems).

An original method of improvement of the local estimates was suggested in [19]. As a special, simple example of this method, consider a symmetric system such that the photon source and the detector are interchangeable but do not coincide (for example, an isotropic source in an infinite medium) and the quantity $I(\boldsymbol{r}^*)$ is to be estimated. Let $\boldsymbol{r}_0, \boldsymbol{r}_1, \ldots, \boldsymbol{r}_N, \boldsymbol{r}_{N+1}$ be an arbitrary trajectory that terminates at the detector.

Introduce the notation $l_0 = |\boldsymbol{r}_1 - \boldsymbol{r}_0|$, the length of the first run, $l_N = |\boldsymbol{r}_{N+1} - \boldsymbol{r}_N|$, the length of the last run. By virtue of the symmetry of the system, one half of all the trajectories satisfy the in equality $l_0 < l_N$. This fact may be used to estimate the quantity $I(\boldsymbol{r}^*)$, as follows. The local contribution from the collision $\boldsymbol{x}_n$ is counted if and only if $|\boldsymbol{r}_n - \boldsymbol{r}^*| > l$; the result is multiplied by 2.

The variance of this estimate diverges as $\ln x$, but it becomes finite if $l_0$ is drawn from the density $C_m r^{-m} \exp\{-\sigma r\}$, $0 < m < 1$ and if an appropriate weight is introduced. An analogous technique may be applied to an arbitrary system; trajectories must then also be simulated for the adjoint equation of transfer (adjoint trajectories). In the general case, the results for the direct and adjoint equations are added. When the adjoint equation is completely defined by the function $\varphi(\boldsymbol{x})$, the above method is directly applied to estimating only a single functional of the form $I_\varphi = (f, \varphi)$. Notice that this technique improves the local estimate if the detector may be separated from the source.

If all characteristics of the system depend on a single coordinate $z$, the local estimates may be obtained by use of the transfer equation for plane geometry (see, for instance, Sect. 3.7). We shall consider the more general problem of estimating the integral $I_{z_0}(\omega^*)$ of the function $\Phi(\boldsymbol{r}, \boldsymbol{\omega})$ over the plane $z = z_0$ in an arbitrary medium.

This integral is determined by $(f, \varphi^*_{z0})$, where $\varphi^*_{z0}(\boldsymbol{x})$ is the flux of photons that reach the plane $z = z_0$ in the direction $\boldsymbol{\omega}^*$, having left the previous collision at $\boldsymbol{x}$.

Obviously,

$$\varphi^*_{z0}(\boldsymbol{r}, \boldsymbol{\omega}) = \begin{cases} \dfrac{g(\mu^*)}{2\pi c^*} \exp(-\tau^*), & (z_0 - z)\cdot c^* > 0, \\ 0, & (z_0 - z)\cdot c^* < 0, \end{cases}$$

where $\mu^* = \boldsymbol{\omega}\cdot\boldsymbol{\omega}^*$, $c^* = \omega^*_z$, $\tau^*$ = optical distance from $\boldsymbol{r}$ to the plane $z = z_0$ along the direction $\boldsymbol{\omega}^*$, hence

$$I_{z0}(\boldsymbol{\omega}^*) = \sum_{n=0}^{N} Q_n \cdot \varphi^*_{z0}(\boldsymbol{r}_n, \boldsymbol{\omega}_n).$$

Analogous arguments show that the number of photons crossing the plane $z = z_0$ can be estimated by calculating at each scattering the quantity

$$\varphi_{z0}(\boldsymbol{r}, \boldsymbol{\omega}_1) = \begin{cases} \exp(-\tau_1), & (z_0 - z)\cdot c_1 > 0, \\ 0, & (z_0 - z)\cdot c_1 < 0. \end{cases}$$

Here $\boldsymbol{\omega}_1$ is the direction after scattering at the point $\boldsymbol{r}$, $c_1 = \omega_{1z}$, $\tau_1$ is the optical distance from $\boldsymbol{r}$ to the plane $z = z_0$ along the direction $\boldsymbol{\omega}_1$.

Local estimates may be applied to calculating the integral of the photon flux over a certain domain of the phase space.

Let

$$\Phi(\boldsymbol{x}) = E \sum_{n=0}^{N} Q_n \cdot \varphi(\boldsymbol{x}_n, \boldsymbol{x});$$

hence

$$\int_{\mathscr{D}} \Phi(\boldsymbol{x})\, d\boldsymbol{x} = E \sum_{n=0}^{N} Q_n \int_{\mathscr{D}} \varphi(\boldsymbol{x}_n, \boldsymbol{x})\, d\boldsymbol{x}.$$

The latter integral may be evaluated by use of a single sample that is drawn from domain $\mathscr{D}$ according to a given density $p(\boldsymbol{x}_n, \boldsymbol{y}_n)$. It is easy to show that the estimate

$$\xi = \sum_{n=0}^{N} Q_n \frac{\varphi(\boldsymbol{x}_n, \boldsymbol{y}_n)}{p(\boldsymbol{x}_n, \boldsymbol{y}_n)},$$

is unbiased. For variance reduction, it is desirable to include in the density $p(\boldsymbol{x}_n, \boldsymbol{y}_n)$ the singularity of the function $\varphi(\boldsymbol{x}_n, \boldsymbol{y}_n)$. In this way, integrals of the flux may be evaluated over the surface of the detector, over the aperture angle of the detector, etc.

## 3.11 Method of Dependent Sampling and Calculation of Derivatives of Linear Functionals

With the help of the weight method problems of transfer theory may be solved simultaneously for several systems by use of only one and the same appropriate Markov chain. This is possible if the kernel functions of the corresponding transfer equations and a given Markov chain satisfy (3.7). This method is especially effective when a dependence of results on small variations of the parameters of the system is investigated. Such a technique is called the method of dependent sampling.

Suppose that $\varphi$, $\psi$, the kernels of the integral equation, are functions that depend on some parameter $\lambda$:

$$\psi_\lambda(\boldsymbol{x}) = \psi(\boldsymbol{x}, \lambda), \qquad k_\lambda(\boldsymbol{x}', \boldsymbol{x}) = k(\boldsymbol{x}', \boldsymbol{x}, \lambda), \qquad \varphi_\lambda(\boldsymbol{x}) = \varphi(\boldsymbol{x}, \lambda),$$

Consequently,

$$I(\lambda) = (f_\lambda, \varphi_\lambda) = E \sum_{n=1}^{N} Q_n(\lambda) \cdot \varphi(\boldsymbol{x}_n, \lambda). \tag{3.37}$$

In the dependent-sampling algorithm, direct simulation is frequently realized for some particular value of the parameter $\lambda$, $\lambda_0$ say, i.e.,

$$r(\boldsymbol{x}', \boldsymbol{x})[1 - p(\boldsymbol{x}')] = k(\boldsymbol{x}', \boldsymbol{x}, \lambda_0), \qquad r_0(\boldsymbol{x}) = \psi(\boldsymbol{x}, \lambda_0).$$

After the transition from $\boldsymbol{x}'$ to $\boldsymbol{x}$ a new weight is obtained by

$$Q(\lambda) = Q'(\lambda)\cdot\frac{k(\boldsymbol{x}', \boldsymbol{x}, \lambda)}{k(\boldsymbol{x}', \boldsymbol{x}, \lambda_0)}.$$

In practice, to simulate transitions $\boldsymbol{x}' \to \boldsymbol{x}$, successive simulations of elementary random variables—the direction of scattering, the free-path length, the collision-type number which is sometimes introduced into the phase space—are used. The product of the conditional-probability densities of these random variables determines $k(\boldsymbol{x}', \boldsymbol{x}, \lambda)$ to within a constant factor that does not depend on $\lambda$. Thus, we obtain a simple realization of the algorithm under study: we must multiply the photon weight by the ratio of the corresponding probability densities for the values $\lambda$ and $\lambda_0$ after each elementary sampling.

The similar-trajectory method is a special case of technique described. It is sometimes necessary to carry out calculations for systems that are similar in the following sense. The multiplication of a unit vector by $\lambda$ for one system is equivalent to multiplication of the effective cross section by $\lambda$ for another system. Assume, for example, that it is desired to estimate the probability of the passage of a particle through homogeneous layer of medium of thickness $H$. Obviously, calculations for the thickness $\lambda H$ are equivalent to calculations for the thickness $H$ with cross section $\sigma\cdot\lambda$ (this follows, for example, from the formula for free-path-length simulation $l = -\ln\alpha/\sigma$).

We have, for the similar-trajectories method,

$$\frac{k(\boldsymbol{x}', \boldsymbol{x}, \lambda)}{k(\boldsymbol{x}', \boldsymbol{x}, \lambda_0)} = \frac{\lambda}{\lambda_0}\exp[-\sigma(\lambda - \lambda_0)l].$$

If the particle escapes from the medium its weight is multiplied by $\exp\{-\sigma\Delta l(\lambda - \lambda_0)\}$, where $\Delta l$ is the length of the last run in the medium.

Suppose that it is possible to differentiate the expression (3.37) under the expectation sign; then

$$\frac{dI(\lambda)}{d\lambda} = E\sum_{n=0}^{N}\left[\frac{d\ln Q_n(\lambda)}{d\lambda} + \frac{d\ln\varphi(\boldsymbol{x}_n, \lambda)}{d\lambda}\right]Q_n(\lambda)\varphi(\boldsymbol{x}_n, \lambda). \tag{3.38}$$

The interchange of the averaging and differentiation is permissible here if there exists a function $\Psi(\boldsymbol{x}_n)$ such that

$$\left|\frac{d\ln Q_n(\lambda)}{d\lambda} + \frac{d\ln\varphi(\boldsymbol{x}_n, \lambda)}{d\lambda}\right|\cdot|Q_n(\lambda)\varphi(\boldsymbol{x}_n, \lambda)| \leqslant \Psi(\boldsymbol{x}_n)|\cdot Q_n(\lambda_0)\varphi(\boldsymbol{x}_n, \lambda_0)|,$$

for $\lambda_0 - \varepsilon < \lambda < \lambda_0 + \varepsilon$ and if

$$E \sum_{n=0}^{N} |Q_n(\lambda_0)\varphi(\mathbf{x}_n, \lambda_0)| \Psi(\mathbf{x}_n) < +\infty.$$

Notice that the inequality holds if, for example, $\Psi(\mathbf{x}_n) < cn$ and $\|K_1\| < 1$. Use of the logarithmic derivative greatly simplifies the construction of algorithms for calculating derivatives of linear functionals of the transfer theory. Let $\sigma = \sum_{i=1}^{k} \sigma_i + \sigma_0 t$ ($t \geqslant 0$ a parameter, $t_0 = 1$) be the effective total cross section for some zone $\mathscr{D}$ of the system.

For simplicity, let us assume that the functions $\varphi$, $\psi$ do not depend on $t$ and consider the direct simulation for $t = t_0 = 1$. After the free-path length $l$ is sampled, the weight must be multiplied by

$$\frac{\sigma(t, l) \exp[-\tau(t, l)]}{\sigma(t_0, l) \exp[-\tau(t_0, l)]} = \frac{\sigma(t, l)}{\sigma(t_0, l)} \exp[-\sigma_0 \Delta l(t - 1)],$$

($\Delta l$ = length of the photon trajectory in zone $\mathscr{D}$), and after the $i$th type of collision is sampled, the weight must be multiplied by

$$\frac{\sigma_i(t, l)}{\sigma(t, l)} \Big/ \frac{\sigma_i(t_0, l)}{\sigma(t_0, l)} = \frac{\sigma(t_0, l)}{\sigma(t, l)} \cdot \frac{\sigma_i(t, l)}{\sigma_i(t_0, l)}.$$

Consequently, the resulting weight is given by

$$Q(t) = t^m \exp\{-(t - 1)\Sigma\sigma_0 \Delta l\},$$

where $\Sigma$ denotes a sum over all photon runs, $m$ = number of collisions of the type that corresponds to the cross section $\sigma_0$. Further, $d \ln Q(t)/(dt)|_{t=t_0=1} = m - \Sigma\sigma_0\Delta l$, $Q(t_0) = 1$. Thus, in this case, we may write

$$\left.\frac{dI_\varphi}{dt}\right|_{t=t_0} = E \sum_{n=0}^{N} (m - \Sigma\sigma_0\Delta l)_n \cdot \varphi(\mathbf{x}_n),$$

if it is possible to differentiate under the expectation sign in the expression

$$I_\varphi(t) = E \sum_{n=0}^{N} Q_n(t)\varphi(\mathbf{x}_n).$$

This differentiation is possible if $t$ belongs to an interval such that, for $\forall t \in (t_1, t_2)$,

$$\left|\frac{dQ_n(t)}{dt}\right| \leqslant q_n, \qquad E \sum_{n=0}^{N} q_n \varphi(\mathbf{x}_n) < +\infty.$$

In our case, we may take $t_1 = t_0 - \varepsilon_1$, $t_2 = t_0 + \varepsilon$ and

$$q_n = [n(1+\varepsilon)^{n-1} + (\Sigma\sigma_0\Delta l)(1+\varepsilon)^n] \exp\{\varepsilon(\Sigma\sigma_0\Delta l)\}.$$

Suppose that the medium is bounded. Then $\|K\| < 1$, $\Delta l < c$. If $\|K\|(1+\varepsilon)\exp(\varepsilon c\sigma_0) < 1$, then

$$E\sum_{n=0}^{N} q_n\varphi(x_n) < \sum_{n=0}^{\infty} n(1+\varepsilon)^n(K^n\psi, \varphi)\, e^{n\varepsilon c\sigma_0}$$

$$+ \sigma_0 c \sum_{n=0}^{\infty} (1+\varepsilon)^n(K^n\psi, \varphi)\exp\{n\varepsilon c\sigma_0\} < +\infty.$$

## 3.12 Method of Expected Values in the Theory of Radiative Transfer

First, let us discuss a method that is very widely used, takes many forms, and is most frequently called the method of expected values. Suppose we wish to estimate, by use of the Monte Carlo method, the integral

$$I = \int_X \int_Y f(x, y) g(x, y)\, dx\, dy,$$

where $f(x, y)$ is the joint density function of random variables (or vectors) $\xi$, $\eta$; and $X$, $Y$ are spaces of arbitrary dimensions. Introduce the notation

$$f_1(x) = \int f(x, y)\, dy,$$

$$f_2(y|x) = f(x, y)/f_1(x), \qquad \zeta = g(\xi, \eta),$$

$$E_\eta[\zeta|x] = \int f_2(y|x) g(x, y)\, dy.$$

Here $f_1(x)$ is the marginal density function of $\xi$; $f_2(y|x)$ is the conditional density of $\eta$ relative to the hypothesis that $\xi = x$; $E_\eta[\zeta|x]$ is the conditional expected value of $\zeta$ relative to the hypothesis that $\xi = x$. The formula for the total expected value,

$$I = E\zeta = \int f_1(x)\, dx \int f_2(y|x) g(x, y)\, dy = E_\xi E_\eta[\zeta|\xi], \tag{3.39}$$

follows immediately from Fubini's theorem.

Thus, the problem may be solved by reducing it to the Monte Carlo calculation of the one-dimensional integral $\int f_1(x) E_\eta[\zeta|x] dx$, provided that there is a simple method of evaluating the quantity $E_\eta[\zeta|x]$ for each value of $x$. Besides,

$$\mathcal{D}_\xi E_\eta[\zeta|\xi] \leqslant \mathcal{D}\zeta,$$

because

$$\mathscr{D}\zeta = E_\xi \mathscr{D}_\eta[\zeta|\xi] + \mathscr{D}_\xi E_\eta[\zeta|\xi]. \tag{3.40}$$

Let us consider now a problem of Monte Carlo calculation of the functional $I_\varphi = (f, \varphi)$ in the theory of radiative transfer. Suppose that the importance function (i.e., the solution $\Phi^*(x)$ of the adjoint equation) is known for a certain domain $\mathscr{D}$ (or at least a good approximation of this function is known). Relation (3.11) shows that $\Phi^*(x)$ is the conditional expectation value of the contribution, relative to the hypothesis that the trajectory originates at the point $\boldsymbol{x}(\boldsymbol{r}, \omega)$. Notice that the method of expected values can be easily generalized to arbitrary path integrals. Let $\xi$ denote a part of the simulated trajectory from the source to the first collision in the domain $\mathscr{D}$, and $\eta$ denotes the part of the trajectory after this collision (if the trajectory does not interesect the domain $\mathscr{D}$, the contribution to $g(\xi, \eta)$ doesnot depend on $\eta$).

As follows from the method of expected values, the trajectory may be terminated after its first collision in the domain $\mathscr{D}$ (at a point $\boldsymbol{x}$); the quantity $\Phi^*(\boldsymbol{x})$ should be stored. Obviously, this algorithm would be effective if it is necessary to calculate one functional $I_\varphi$ for several photon sources. However, more frequently, it is desired to calculate various functionals for a single source (for instance, an angular-spatial distribution of solar radiation in a scattering atmosphere).

For natural sources, many approximate methods have been developed that give satisfactory solutions of the direct equation of transfer in certain domains of phase space. To utilize approximate solutions, the theorem of optical mutuality (see Sect. 2.7) may be applied. Then the direct and adjoint equations exchange roles and the algorithm of expected values is formulated as follows.

The photons are emitted with density $\sigma\varphi$ (see Sect. 2.7). After the first collision in the domain $\mathscr{D}$, the trajectory is terminated and the known solution of the transfer equation is stored. We give an example of effective utilization of this technique in Chap. 6.

## 3.13 The Splitting Technique

Sometimes, it is convenient to use several samplings of $\eta$ for a single value of $\xi$ (in the notation of Sect. 3.12). We shall now give a precise formulation of this method and the optimization of such a method, which is called the splitting technique.

Let $\xi$ be a random variable with probability density $f_1(x)$, $n(\xi) \geqslant 1$ is an integer number that depends on $\xi$, and let $\eta_1, \ldots, \eta_{n(\xi)}$ be a sequence of independent and identically (on condition) distributed random variables with density $f_2(y|x)$, provided that $\xi = x$.

The splitting method is based on the random estimate

$$\zeta_n = \frac{1}{n(\xi)} \sum_{k=1}^{n(\xi)} g(\xi, \eta_k).$$

Let $\boldsymbol{\eta} = (\eta_1, \ldots, \eta_{n(\xi)})$. Making use of the formula for the total expectation value, we obtain

$$E\zeta_n = E_\xi E_{\boldsymbol{\eta}}[\zeta_n|\xi] = E_\xi\left\{E_{\boldsymbol{\eta}}\left[\sum_{k=1}^{n(\xi)} g(\xi, \eta_k)|\xi\right]\Big/ n(\xi)\right\}$$

$$= E_\xi\left\{\frac{E_{\boldsymbol{\eta}}[\zeta|\xi]\cdot n(\xi)}{n(\xi)}\right\} = E_\xi E_{\boldsymbol{\eta}}[\zeta|\xi] = E\zeta = \int\int f(x, y)g(x, y)\,dx\,dy;$$

i.e., $\zeta_n$ is an unbiased estimate. Further

$$\mathscr{D}\zeta_n = \mathscr{D}_\xi E_{\boldsymbol{\eta}}[\zeta_n|\xi] + E_\xi \mathscr{D}_{\boldsymbol{\eta}}[\zeta_n|\xi] = \mathscr{D}_\xi E_{\boldsymbol{\eta}}[\zeta|\xi] + E_\xi\{\mathscr{D}_{\boldsymbol{\eta}}[\zeta|\xi]/n(\xi)\}.$$

Let $t_1$ be the average computation time per sample of $\xi$, $t_2(x)$ = the average computation time per sample of $\eta$, provided that $\xi = x$. Then the average computation time per sample of $\zeta_n$ is given by $t_n = t_1 + E_\xi[n(\xi)t_2(\xi)]$.

To optimize the splitting method, it is necessary to find a function $n(\xi)$ that minimizes $S_n = t_n \mathscr{D}\zeta_n$. For simplicity, let us assume that $n(\xi) = \text{const} = n$. In this simple case, $t_n = t_1 + nt_2$, $\mathscr{D}\zeta_n = A_1 + A_2/n$, $t_2 = E_\xi t_2(\xi)$. Clearly, the value

$$n = \left[\frac{A_2 \cdot t_1}{A_1 \cdot t_2}\right]^{1/2}, \tag{3.41}$$

$$A_1 = \mathscr{D}_\xi E_\eta[\zeta|\xi], \qquad A_2 = E_\xi \mathscr{D}_\eta[\zeta|\xi],$$

minimizes the quantity $S_n = (t_1 + nt_2)(A_1 + A_2/n)$.

The quantities $t_1$, $t_2$, $A_1$, $A_2$ may be evaluated by use of results of a preliminary special calculation.

*Example.* The splitting method is very useful for solving some problems of radiative-transfer theory. Consider the problem of estimating the probability $p$ that a photon traverses a layer of medium $0 \leqslant z \leqslant H$. Assume that the paths between photon collisions are straight lines. When a collision takes place, the photon is either absorbed in the medium or scattered according to a certain probability law. The photon source is placed on the plane $z = 0$. If the photon escapes from the layer through the plane $z = H$, we put for this trajectory $\zeta = 1$, otherwise $\zeta = 0$; $M\zeta = p$. Consider a simplest variant of the splitting method. The phase point $Q$ of the first intersection with the splitting plane

$z = z_1$ is determined. Next, the trajectory is splited into $n$ independent branches that stem from $Q$, each having a weight that is $n^{-1}$ times the weight of the parent trajectory. The problem is to find the optimal values of the parameters $z_1$ and $n$. Introduce the notation: $p_1$ is the probability that the photon reaches the splitting plane, $p_2$ is the conditional probability that the photon escapes through the plane $z = H$, provided that it has intersected the splitting plane ($p_1 p_2 = p$).

Define discrete random variables $\xi$ and $\eta$ as follows. If the photon intersects the splitting plane, then $\xi = 1$, otherwise $\xi = 0$; if the photon has traversed the layer, then $\eta = 1$; otherwise, $\eta = 0$.

Let

$$\zeta = g(\xi, \eta) = \begin{cases} 0 & \eta = 0, \\ 1 & \eta = 1. \end{cases}$$

The distributions of $\xi$ and $\eta$ are defined by

$$P(\xi = 1) = p_1, \qquad p(\xi = 0) = 1 - p_1,$$

$$P(\eta = 1|\xi = 0) = 0, \qquad P(\eta = 0|\xi = 0) = 1,$$

$$P(\eta = 1|\xi = 1) = p_2, \qquad P(\eta = 0|\xi = 1) = 1 - p_2.$$

From this, it can easily be shown that

$$A_1 = \mathscr{D}_\zeta M_\eta[\zeta|\xi] = p_2^2 p_1(1 - p_1) = pp_2(1 - p_1),$$

$$A_2 = E_\xi \mathscr{D}_\eta[\zeta|\xi] = p_2(1 - p_2)p_1 = p(1 - p_2).$$

Assume now that $t_2 = t_1 \cdot p_1$, i.e., the average time of simulation of one trajectory after the splitting is equal to the average time of simulation before the splitting. Then

$$S_n = t_1(1 + np_1)p\left[p_2(1 - p_1) + \frac{1 - p_2}{n}\right],$$

and the optimal $n$ is given by

$$n = \left[\frac{1}{p}\cdot\frac{1 - p_2}{1 - p_1}\right]^{1/2}.$$

Thus $S_n$ is given by

$$S_n = t_1 p[\sqrt{p_2(1 - p_1)} + \sqrt{p_1(1 - p_2)}]^2.$$

It has a minimum if $p_1 = p_2 = \sqrt{p}$; consequently

$$S_n^{\text{opt.}} = 4t_1 p\sqrt{p}(1 - \sqrt{p}).$$

The computer-time score is given by

$$\frac{S_1}{S_n^{\text{opt.}}} = \frac{t_1(1 + \sqrt{p})p(1 - p)}{4t_1 p(1 - \sqrt{p})} = \frac{1}{4} \cdot \frac{(1 + \sqrt{p})^2}{\sqrt{p}}.$$

If $p \ll 1$ we have

$$n \approx p^{-1/2}, \qquad \frac{S_1}{S_n^{\text{opt.}}} \approx \frac{1}{4\sqrt{p}}. \tag{3.42}$$

*Ogibin* [20] has investigated in detail multiple splitting of the trajectories. The practical computations show that often the single splitting greatly improves the speed of calculations, whereas the multiple-splitting improvement does not compensate for the complications that then arise in computational programs and in determining the parameters. This is particularly the case for problems in which the probability is not too small ($\sim 10^{-4}$).

## 3.14 Statistics of Unbiased Estimates with Finite and Infinite Variances

To estimate the expected value of $\xi$, one usually uses a sequence $x_1, \ldots, x_N$ of independent samples of $\xi$ obtained by Monte Carlo. Then, by the law of large numbers,

$$\begin{aligned} E\xi &\approx \bar{x}_N = \frac{1}{N}\sum_{k=1}^{N} x_k, \\ \mathscr{D}\xi &= \sigma^2 \approx \bar{\sigma}^2 = \frac{N}{N-1}\left(\frac{1}{N}\sum_{k=1}^{N} x_k^2 - \bar{x}_N^2\right). \end{aligned} \tag{3.43}$$

To evaluate the error of estimate (3.43), a confidence interval is usually calculated which contains $E\xi$ with probability $1 - \varepsilon$, where $0 < \varepsilon \ll 1$. The central-limit theorem can be used to construct the confidence interval approximately, provided that $\mathscr{D}\xi < +\infty$. For $N$ sufficiently large, the following asymptotic equality holds

$$P(|\bar{x}_N - I| < \beta\sigma/N^{1/2}) = P\left(\frac{\sqrt{N}}{\sigma}|\bar{x}_N - I| < \beta\right) \approx \frac{1}{\sqrt{2\pi}}\int_{-\beta}^{\beta} e^{-(x^2/2)}dx = p(\beta).$$

Thus, setting $p(\beta) = 1 - \varepsilon$, we obtain the desired confidence interval: $\bar{x}_N \pm \beta\sigma/(\Psi N)$. In practice, $\sigma$ is usually unknown as well, and sample estimates $\bar{\sigma}$ must be used for it. In practice, the actual values of $\beta$ are taken to be 1, 2 or 3, which correspond to the confidence levels $p(\beta) = 0.68, 0.95, 0.997$, respectively.

As a rule, the same trajectories are used to estimate a number of functionals for which the sample estimates are obtained as total contributions from various collisions. Clearly, it is not efficient in such calculations, to estimate $\tilde{\sigma}$ by summation of $x^2$ over all trajectories. Therefore, the trajectories are divided up into a few groups each of which has $M$ trajectories; from those data, the standard deviation of the random variable is estimated

$$\zeta_M = \frac{1}{M} \sum_{i=1}^{M} \xi_i,$$

where $\xi_i$ are independent samplings of $\xi$.

Let $\sigma_{M,S}$ be a statistical estimate of the quantity $\sigma_M = (D\zeta_M)^{1/2}$ calculated for $S$ groups ($N = MS$). Then the standard error of the quantity $\tilde{x}_N$ can be estimated as

$$\sigma(\tilde{x}_N) = \frac{\sigma}{\sqrt{N}} \approx \frac{\bar{\sigma}_{M,S}}{\sqrt{S}}. \tag{3.44}$$

Sometimes it is necessary to use, in the Monte Carlo method, estimates that have infinite variance (see Sects. 3.8–10). Let $F_\xi(y)$ be a distribution function of the random variable $\xi$ and let

$$U(x) = \int_0^x y^2 dF_\xi(y).$$

If the variance $D\xi$ diverges as $\ln x$ (i.e., $U(x) \sim c \cdot \ln x$), then the random variable

$$\gamma_M = \left[\frac{M}{\ln M}\right]^{1/2} (\zeta_M - E\xi),$$

is asymptotically normal $(0, d)$, where $d$ is a standard deviation (see [16], p. 374). The standard error (see Sect. 3.9) of the resultant estimate is given by

$$d^*(\zeta_{MS}) = d^* \left[\frac{\ln (MS)}{M \cdot S}\right]^{1/2},$$

which differs from the analogous quantity for the case of bounded variance by the factor $[\ln (MS)/\ln M]^{1/2}$. Although the variance of $\zeta_{MS}$ is infinite the foregoing limit theorem enables us to construct approximately a confidence interval for $E\xi$, as was explained for the normal distribution with $d = d(\zeta_{MS})$. Indeed, for large $N = M \cdot S$, the divergence of $D\zeta_M$ is due to the tail of distribution which does not essentially affect the observations.

This supposition agrees with the results of calculations cited in Sect. 3.9. Therefore, the rate of probability convergence of estimate $\zeta_N$ is given by $t = [\ln N/(N)]^{1/2}$.

# 4. Monte Carlo Methods for Solving Direct and Inverse Problems of the Theory of Radiative Transfer in a Spherical Atmosphere

In this chapter, we shall consider Monte Carlo algorithms for solving problems of the solar-radiation propagation in a scattering and absorbing planetary atmosphere, for example, that of the Earth. In the last few years, the importance of these problems, in which multiple scattering, sphericity of the atmosphere, and the light polarization for a detailed radiation model of the atmosphere are taken into account, has grown because of the need to interpret optical observations.

Establishment of connections between characteristics of the scattered-radiation field and the parameters of a meteorological model of the atmosphere makes it possible to formulate and solve inverse problems of atmospheric optics. In this chapter, axial symmetry of the planet-atmosphere-sun system will be assumed. This assumption does not essentially distort the actual physical process, while it simplifies the computations and enables us to construct a series of effective Monte Carlo algorithms.

We now give a general formulation of these problems. Let $A$ (see Fig. 4.1) be an observation point with coordinates $x^*$, $y^* = 0$, $z^* > 0$. We assume that the solar radiation incident at the top of the atmosphere (considered here as a sphere of radius $R = R_2 > R_0$, where $R_0$ is the Earth's radius) is collimated and has the direction $(-\boldsymbol{x})$. The angular distribution of the radiation as observed at the point $A$ is measured in polar coordinates with center $A$, axis $OA$ and plane of reference of the azimuthal angle $XOZ$. Figure 4.1 shows the connection between the latitude angle $\theta$ and the altitude $h$ of the sight line.

The most important characteristic of radiation is its intensity (radiance) which can be calculated from the equation of transfer without taking into account the polarization. If there only the intensity is needed, this approach will

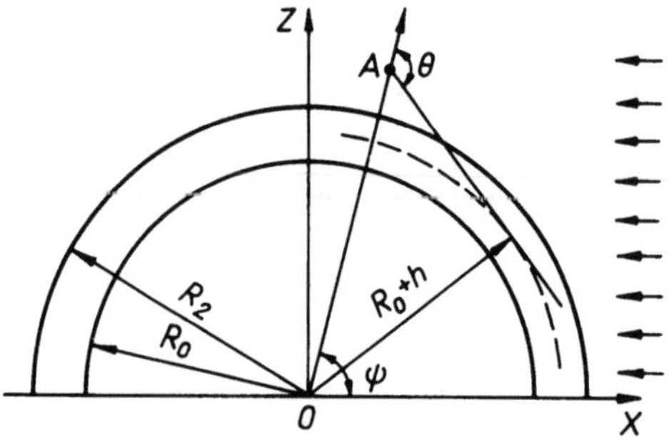


**Fig. 4.1.** Schematic diagram of observation of solar radiation at point $A$

be fully acceptable because, as calculations show, the errors in the evaluation of intensity due to neglect of polarization are very small (see Sect. 4.12).

However, a light beam is more completely described by the Stokes vector, taken here as $\boldsymbol{I} = (I, Q, U, V)$. It is assumed that the incident light is initially unpolarized, i.e. $\boldsymbol{I}_c = (I, 0, 0, 0)$. Programs based on the algorithms of this chapter enable us to obtain the Stokes vector in two forms:

1) the $\theta$ distribution of $\boldsymbol{I}$ at each observation point $\boldsymbol{r}_i^* = (x_i^*, 0, z_i^*)$, $i = 1, \ldots, n_r$, is obtained as a histogram, i.e., as a sequence of estimates of integrals

$$\boldsymbol{I}_k(\boldsymbol{r}_i^*, \varphi_j) = \int\limits_{\theta(h_{k-1})}^{\theta(h_k)} \boldsymbol{I}(\boldsymbol{r}_i^*, \varphi_j, \theta) \sin\theta \, d\theta, \quad k = 1, \ldots, n_\theta,$$

for desired values of the azimuthal angle $\varphi_j$, $j = 1, \ldots, n_\varphi$;

2) the values of the vector function $\boldsymbol{I}$ are evaluated along strictly determined directions $\omega_{jk} = (\varphi_j, \theta_k)$: $\boldsymbol{I}_{ijk} = \boldsymbol{I}(\boldsymbol{r}_i^*, \varphi_j, \theta_k)$.

Simultaneously, scattering multiplicity distributions of some characteristics of the scattered radiation can be obtained. Moreover, by use of the method of dependent sampling, these quantities can be calculated simultaneously for several wavelengths that are of primary interest, when the dependence of the radiation-field characteristics on the medium parameters is investigated.

*Atmospheric Model*

The atmospheric model is here specified by the parameters:

1) $\sigma_a(h, \lambda)$: cross section for aerosol scattering;
$\sigma_M(h, \lambda)$: cross section for molecular scattering;
$\sigma(h, \lambda) = \sigma_a(h, \lambda) + \sigma_M(h, \lambda)$: total scattering cross section.

Here $h$ is the altitude, $\lambda$ the wavelength. The absorption is characterized by either the transmission function or the absorption coefficients $\sigma_c^i(h, \lambda)$ for various gas components. It is assumed that the coefficients $\sigma_a$, $\sigma_M$ and $\sigma_c^i$ are step functions; i.e., the spherical-shell atmosphere is divided into concentric spherical-shell layers with radii $r_i$ $(i = 1, \ldots, N + 1,\ r_1 = R_0,\ r_{N+1} = R_2)$, each having constant coefficients.

2) $R_a(h, \mu, \lambda)$ and $R_M(\mu)$ [or the indicatrices $g_a(h, \mu, \lambda)$, $g_M(\mu)$, when the transfer equation is treated without taking polarization into account] are matrices for aerosol (respectively, molecular) scattering; $\mu = \omega' \cdot \omega$ – cosine of the scattering angle. In order to define the matrix $R_a$ (respectively, indicatrix $g_a$) the atmosphere is divided into layers each having constant characteristics (i.e., which do not depend on altitude $h$) $R_a$, $(g_a)$. The components of the matrix $R_a$ (indicatrix $g_a$) are assumed to be linear within the intervals $[\mu_{i-1}, \mu_i]$, $i = 1, \ldots, n$, $\mu_0 = 1$, $\mu_n = -1$. The matrix of total scattering of light (by air) is defined as a weighted average of the matrices for molecular and aerosol scattering:

$$R(h, \mu, \lambda) = \frac{R_a(h, \mu, \lambda)\sigma_a(h, \lambda) + R_M(\mu)\sigma_M(h, \lambda)}{\sigma_a(h, \lambda) + \sigma_M(h, \lambda)}.$$

Notice that the matrix $R$ is normalized so that $\int_{-1}^{1} R_{11}(h, \mu, \lambda)d\mu = 1$ for fixed $h$ and $\lambda$.
3) $P_a(\lambda)$, the albedo of the ground. The reflection law may be quite arbitrary. In particular, a Lambert surface: $P(\mu) = 2\mu$, $0 \leqslant \mu \leqslant 1$, may be used. In this case, when polarization is calculated, the incident light is assumed to be unpolarized, i.e., the vector of the incident radiation is transformed by a matrix $R$ such that $R_{11} = 2\mu$, with the other components zero.
4) Transmission functions: $0 \leqslant P_1[\lambda, \tau_\chi(\lambda)] \leqslant 1$ of the photon path $(L)$ with the following optical-path length with respect to absorption,

$$\tau_\chi(\lambda) = \int_{(L)} \chi(h(l), \lambda)\, dl,$$

which is the effective mass of the absorbing material in the path $(L)$. The quantity $P_1$ is equal to the survival probability of a photon while traveling along the path $(L)$; therefore $P_1(\lambda, 0) = 1$; and $P_1(\lambda, \infty) = 0$. $\chi(h, \lambda)$ is a step function; $\chi(h, \lambda) = \chi_i(\lambda)$ for $r_i < h \leqslant r_{i+1}$, $i = 1, \ldots, N$. The function $P_1[\lambda, \tau_\chi(\lambda)]$ may be tabulated and linear interpolation may be used between points of the table $\tau_{\chi k}$ $(k = 1, \ldots, n_\chi)$. Empirical formulas may also be used (see Sect. 4.9).

*Flow Diagram*

To simulate the free-path length and evaluate the optical thicknesses $\tau(\lambda)$ and $\tau_0$, it is necessary to determine the indexes of layers $A_i$, $i = 1, \ldots, N_l + 1$ that are intersected by the photon path, and to calculate the path length in every layer.

Let $(x, y, z)$ be a point of collision, and let $A_1$ be the index of the layer that contains this point. The direction of the path of the photon after it is scattered is characterized by a unit vector $(a, b, c)$. The distance from the collision point to the sphere of radius $r(A)$ satisfies the system

$$x' = x + al, \quad y' = y + bl, \quad z' = z + cl, \quad x'^2 + y'^2 + z'^2 = r^2(A).$$

Writing $x^2 + y^2 + z^2 = R$, $sx + by + cz = S$, $S^2 - R = S_1$, we obtain from this system that

$$l = -S \pm [S_1 + r^2(A)]^{1/2}.$$

The expression for $S$ shows that the distance to the upper boundary of the layer is given by last expression with the positive sign if $S > 0$. In this case there is no need to check for an intersection with the lower boundary.

If $S < 0$, it is necessary to check, first, the intersection with the lower boundary of the layer given by taking the minus sign. Next, the distance to the upper boundary should be checked if necessary. The indexes of the layers (Fig. 4.2) are redetermined according to the scheme: $A_{i+1} = A_i + 1$ if the photon intersects the upper boundary (at the distance $l_i$) having the index $A_i$, and $A_{i+1} =$

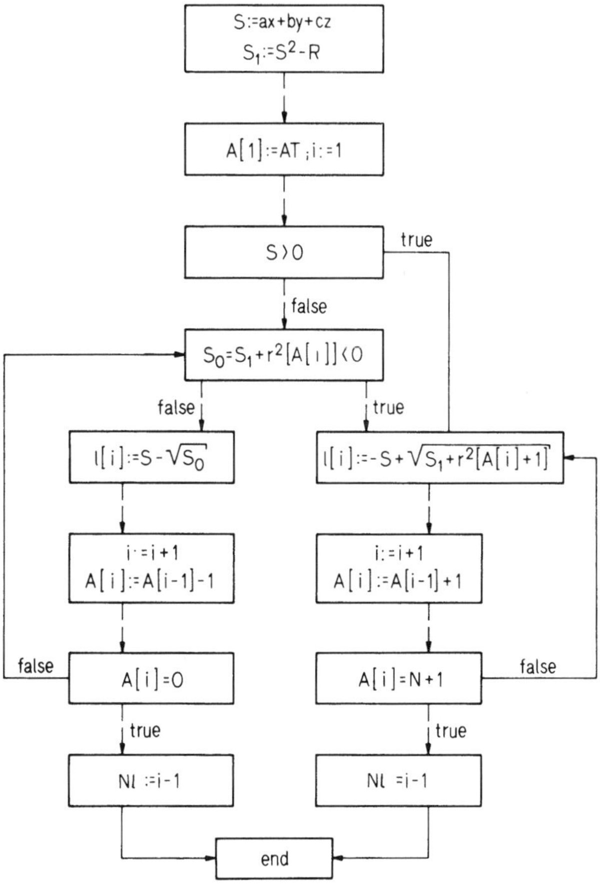


Fig. 4.2. Flow diagram of the geometrical block

$A_i - 1$ if the particle intersects the lower boundary. Thus, all distances $l_i$ may be obtained and the indexes of the layers traversed can be determined until the photon either escapes from the atmosphere or reaches the surface of the planet.

## 4.1 Modifications of the Local Estimates for Axial Symmetry

The general form of the local estimates for calculating a photon flux $\Phi$ at a given point of the phase space $X = R \times \Omega$ is given in Sect. 3.8. In this section, we shall develop two modifications of the Monte Carlo method designed to estimate characteristics of the scattered radiation at a given point for the axially symmetric planet-atmosphere-sun system.

Let us consider a detector placed at a point $\boldsymbol{r}^* = (x^*, 0, z^*)$ (see Fig. 4.1). We wish to calculate the intensity (i.e., the density of the photon flux) $\Phi$ at a given point $\boldsymbol{x}^* = (\boldsymbol{r}^*, \omega^*)$ of the phase space.

The estimates presented in this section are based on preliminary averaging of (3.19, 20) over a rotation angle (with the line joining the planet to the sun as the reference axis). An improved local estimate allows us to calculate the intensity and other characteristics of the scattered light at a point $\boldsymbol{r}^*$ and for a desired value of the azimuthal angle, by averaging over some intervals of the latitude angle. The variance of this estimate diverges as $\ln x$. An improved double local estimate makes it possible to calculate these quantities at a point $\boldsymbol{r}^*$ and for a given direction $\omega^*$. Its variance diverges (as $\ln x$) but is finite for the case of single scattering. The first estimate is preferable, however, if calculation of the integral of intensity over a sufficiently large interval of the angle $\theta$ is desired.

*Modification of the Local Estimate*

In view of the symmetry of the planet-atmosphere-sun system, it may be assumed that the collision point is uniformly distributed on a circle of radius $\rho = \sqrt{y^2 + z^2}$. The problem treated in this section is the evaluation of scattered radiation (due to collisions within a region $\varphi_0 \leqslant \varphi \leqslant \varphi_0 + d\varphi$) incident at the receiver position $\boldsymbol{r}^*$.

The next collision point in turn is rotated into the plane $\varphi = \varphi_0$, and the local estimate is multiplied by an appropriate weight, proportional to the length of the arc of circle of radius $\rho$ contained in the region $\varphi_0 \leqslant \varphi \leqslant \varphi_0 + d\varphi$. Introduce the following notation:

| | |
|---|---|
| $(x, y, z)$: | collision point; |
| $(x_1, y_1, z_1)$: | collision point after the rotation; |
| $(a, b, c)$: | unit vector giving the photon direction before scattering; |
| $(a_1, b_1, c_1)$: | same direction, after rotation; |
| $\boldsymbol{r}^* = (x^*, 0, z^*)$: | receiver position, $x^{*2} + z^{*2} = \rho_c^2$; |

$(l, m, n)$: unit normal to the plane $\varphi = \varphi_0$;
$(a_2, b_2, c_2)$: unit direction vector from $(x_1, y_1, z_1)$ to $(x^*, 0, z^*)$;
$r_1$ = distance between $(x_1, y_1, z_1)$ and $(x^*, 0, z^*)$:

$$r_1 = [(x_1 - x^*)^2 + y_1^2 + (z_1 - z^*)^2]^{1/2}.$$

First, we obtain the quantities $l, m, n$. If the angle $\varphi$ is measured from the plane $Y = 0$ and if each plane $\varphi = \varphi_0$ contains the axis $Or^*$, we have

$$\frac{m}{\sqrt{l^2 + m^2 + n^2}} = \cos\varphi, \quad l^2 + m^2 + n^2 = 1, \quad lx^* + nz^* = 0.$$

Hence,

$$l = \frac{z^*}{\rho_c}\sin\varphi_0, \quad m = \cos\varphi_0, \quad n = -\frac{x^*}{\rho_c}\sin\varphi_0.$$

The plane $\varphi = \varphi_0$ intersects the circle

$$y_1^2 + z_1^2 = \rho^2, \quad x_1 = x, \tag{4.1}$$

at a point $(x_1, y_1, z_1)$. Therefore,

$$lx_1 + my_1 + nz_1 = 0, \quad y_1^2 + z_1^2 = \rho^2, \quad x_1 = x. \tag{4.2}$$

Hence

$$x_1 = x, \quad y_1 = \frac{-B \pm (B^2 - AC)^{1/2}}{A}, \quad z_1 = -\frac{lx_1 + my_1}{n},$$

where

$$A = m^2 + n^2, \quad B = lmx, \quad C = l^2 \cdot x^2 - n^2\rho^2.$$

From this, it appears that the point $(x_1, y_1, z_1)$ may belong to the plane $\varphi = \varphi_0 + \pi$. To exclude this possibility, only points $\boldsymbol{r}_1 = (x_1, y_1, z_1)$ such that $\cos\varphi_0 \cdot \cos\psi > 0$, where $\psi$ is the angle between $\boldsymbol{r}_1 \times \boldsymbol{r}^*$ and $(0, 1, 0)$, are chosen. Because

$$\cos\psi = \frac{z_1 x^* - z^* x_1}{[y_1^2 z^{*2} + (z_1 x^* - z^* x_1)^2 + x^{*2} y_1^2]^{1/2}},$$

we rewrite the foregoing condition for choice of $\boldsymbol{r}_1$ as

$$(z_1 \cdot x^* - z^* \cdot x_1)\cos\varphi_0 > 0. \tag{4.3}$$

If both solutions of (4.2) satisfy the condition (4.3), the contributions from both points must be calculated.
It is easy to see from geometric consideration that

$$a_1 = a, \quad b_1 = b\cos\beta - c\sin\beta, \quad c_1 = c\cos\beta + b\sin\beta,$$

where

$$\sin\beta = \frac{y_1 z - z_1 y}{\rho^2}, \quad \cos\beta = \frac{y y_1 + z z_1}{\rho^2}.$$

Further, we have

$$a_2 = \frac{x^* - x_1}{r_1}, \quad b_2 = -\frac{y_1}{r_1}, \quad c_2 = \frac{z^* - z_1}{r_1},$$

$$\mu_2 = a_1 a_2 + b_1 b_2 + c_1 c_2.$$

It is easy to show that the arc length $\delta_1$ of the circle (4.1) contained in the domain $(\varphi_0, \varphi_0 + d\varphi)$ is

$$\delta_1 = \frac{r_1 \cdot |\sin\theta|}{|\cos\gamma|}\, d\varphi,$$

where $\gamma$ is the angle between the normal to the plane $\varphi = \varphi_0$ and the tangent to the circle (4.1) at the point $(x_1, y_1, z_1)$, $\theta$ is the angle between $\boldsymbol{r}^*$ and the vector $(a_2, b_2, c_2)$. Because the unit tangent vector is given by $\boldsymbol{m} = (0, z_1/\rho, -y_1/\rho)$, we have

$$\cos\gamma = \frac{z_1}{\rho} \cdot \cos\varphi_0 + \frac{y_1 \cdot x^*}{\rho \cdot \rho_c} \cdot \sin\varphi_0.$$

Clearly, $\cos\theta = (a_2 x^* + c_2 z^*)/\rho_c$; therefore, after rotation of the collision point into the plane $\varphi = \varphi_0$ (3.19) takes the form

$$F = \frac{\exp\{-\tau(\boldsymbol{r}, \boldsymbol{r}^*)\} g(\mu_2) \cdot |\sin\theta|}{(2\pi)^2 \cdot r_1 \cdot \rho |\cos\gamma|} \cdot \Delta_i(\theta), \tag{4.4}$$

where $\Delta_i(\theta)$ is the indicator function of the latitude angle.

Thus, the modified local estimate enables us to calculate integrals over $\theta$ for any desired value of $\varphi$. If $\sigma(\boldsymbol{r}^*) \neq 0$, the variance of the estimate is infinite, owing to the divisor $r_1$ in (4.4). To make it finite, it is necessary to bias the estimate by putting $F = 0$ if $r_1 < \varepsilon$. For the collision density $f(\boldsymbol{x}) \sim r_1$ in the neighborhood of the point $r^*$ (in the plane $\varphi = \varphi_0$), we have $D\xi \sim |\ln \varepsilon|$ and the standard error of this estimate might been estimated as it was in Sect. 3.13.

*Modification of the Double Local Estimate*

The above estimate enables us to calculate an average of the intensity over a solid angle that contains the line of sight. The disadvantages of this algorithm are that either (i) if increases the angle then the error of estimate increases too; or (ii) if the angle decreases, then the variance increases.

We wish now to construct a modified double local estimate that will enable us to calculate, for a spherical-shell atmosphere, characteristics of the scattered light along strictly determined directions at several observation points. In addition, the contribution from each collision (scattering) is calculated for all observation variants that essentially improve the evaluation of the spatial-angle dependence of the quantities under study. The double local calculation presented here uses the axial symmetry of the planet-atmosphere-sun system.

Thus, the problem is to calculate the intensity of the scattered solar radiation at the detector position $r^*$, and in a desired direction $\omega^*$. Let $\Phi(r, \omega)$ and $f(r, \omega)$ be the density of the photon flux and the collision density at the point $x = (r, \omega)$, respectively.

Then

$$\Phi(x^*) = \int_R \int_\Omega f(r, \omega)\varphi^*(r, \omega)\, dr\, d\omega + (\psi, \varphi^*), \tag{4.5}$$

where

$$\varphi^*(r, \omega) = \frac{g(\mu^*) \exp[-\tau(r, r^*)]}{2\pi |r - r^*|^2} \delta(s - \omega), \quad s = \frac{r - r^*}{|r^* - r|}.$$

Integrals of the form (4.5) are usually evaluated, in a Monte Carlo technique, by computational simulation of photon trajectories and averaging the sums of contributions $\varphi^*(x)$ from the collisions of various orders. In our case, this method is not applicable because of the $\delta$ function in the expression for $\varphi^*$.

Let $L = \{r = r^* - \omega^*\lambda, \lambda \geqslant 0\}$ be a direction of observation contained in a small solid angle $\Omega_0$ at the point $r^*$. The desired density $\Phi(x^*)$ can be taken approximately as scattered radiation at the point $r^*$ for photons that have undergone collisions (scattering) within the angle $\Omega_0$ divided by the size of this solid angle. For each point $r_0 = (x_0, y_0, z_0) \in \Omega_0$, consider a circle of radius $\rho = \sqrt{y_0^2 + z_0^2}$, lying in the plane $x = x_0$, with center at $(x_0, 0, 0)$.

Thanks to the symmetry of the planet-atmosphere-sun system, the photon collisions are expected to be uniformly distributed on these circles. Consequently, instead of estimating the radiation due to the collisions within an angle, it is possible to consider the radiation due to the collisions within the space obtained by rotation of the solid angle $\Omega_0$ about the symmetry axis $OX$. Furthermore, the weight (i.e., the contribution to the estimate) is multiplied by the area of the cross section $S$ obtained by intersecting the angle $\Omega_0$ with the plane $x = x_0$, and divided by the area of the ring obtained by rotation of the cross section $S$ about the axis in the plane $x = x_0$ [here $(x_0, y_0, z_0)$ is the collision point].

Finally, we introduce the idea of an artificial collision: if the direction of the photon motion intersects the boundary of the body, we shall consider the photon to have undergone a collision within the body with a weight equal to the probability of this collision. Taking the limit as $|\Omega_0| \to 0$ we obtain the following algorithm for calculating the quantity $\Phi(\boldsymbol{x})$.

Let $\boldsymbol{\omega}$ be the direction of the photon motion after the collision at a point $\boldsymbol{r} = (x, y, z)$ and let $\boldsymbol{r}_1 = (x_1, y_1, z_1)$ be the point of intersection of the surface obtained by rotating $L$ about the axis $X$, with this direction.

The point $\boldsymbol{r}_1$ is rotated along the arc of a circle of radius $\rho_1 = \sqrt{y_1^2 + z_1^2}$, with its center at $(x_1, 0, 0)$, into the azimuthal observation plane $\varphi = \varphi'$; the point $\boldsymbol{r}_1 = (x_1, y_1, z_1)$ turns then into the point $\boldsymbol{r}_2 = (x_2, y_2, z_2) \in \Gamma$, and the direction $\boldsymbol{\omega} = (a, b, c)$ turns into the direction $\boldsymbol{\omega}_1 = (a, b_1, c_1)$.

Taking into account these weights, we find that, to estimate the desired quantity, it is necessary to average (over all collisions) the quantity $j(x) = 0$ if the photon direction after a collision does not intersect the surface $\Gamma$, and

$$j(x) = \frac{\sigma(\boldsymbol{r}_1)\exp[-\tau(\boldsymbol{r}, \boldsymbol{r}_1) - \tau(\boldsymbol{r}_2, \boldsymbol{r}^*)]g(\mu^*)(1 - n_1^2)^{1/2}}{4\pi^2|a^*|\cdot\rho_1\cdot|\boldsymbol{n}\cdot\boldsymbol{\omega}|}, \tag{4.6}$$

otherwise. Here $\boldsymbol{\omega}^* = (a^*, b^*, c^*)$ is the direction of observation, $\boldsymbol{n} = (n_1, n_2, n_3)$ is the normal to the surface $\Gamma$ at a point $\boldsymbol{r}_1$, $\sigma(\boldsymbol{r})$ = extinction coefficient.

In order to justify the foregoing speculations, consider transformations of the integral equation of transfer (2.14) for the collision density $f(\boldsymbol{x})$. Recall that $f(\boldsymbol{x}) = \Phi(\boldsymbol{x})\cdot\sigma(\boldsymbol{r})$; it is sufficient, therefore, to estimate the quantity $f(x^*)$. Let us rewrite the kernel of (2.14) in the form

$$k(\boldsymbol{x}, \boldsymbol{x}^*) = l(\boldsymbol{x}, \boldsymbol{x}^*)\delta(\boldsymbol{s} - \boldsymbol{\omega}),$$

where

$$\boldsymbol{s} = \frac{\boldsymbol{r}^* - \boldsymbol{r}}{|\boldsymbol{r}^* - \boldsymbol{r}|} \in \Omega, \quad l(\boldsymbol{x}, \boldsymbol{x}^*) = \frac{g(\mu^*)\sigma(\boldsymbol{r}^*)\exp[-\tau(\boldsymbol{r}, \boldsymbol{r}^*)]}{2\pi|\boldsymbol{r} - \boldsymbol{r}^*|^2}.$$

Fix $\boldsymbol{\omega}$ and consider the first term on the right-hand side of (2.14), expressed in polar coordinates with center at $\boldsymbol{r}^*$. The point $\boldsymbol{r} = (x, y, z)$ then turns into the point $(r, \boldsymbol{s})$ such that $\boldsymbol{r} = \boldsymbol{r}^* - \boldsymbol{s}r$, where the Jacobian of the transformation is equal to $r^2$.

Using this transformation, it is possible to eliminate the $\delta$ function from the integrand of

$$\int_R f(\boldsymbol{r}, \omega) l(\boldsymbol{x}, \boldsymbol{x}^*) \delta(s - \omega^*) d\boldsymbol{r} = \int_S \int_0^\infty f((r, \boldsymbol{s}), \omega)\, l(((r, \boldsymbol{s}), \boldsymbol{\omega}), \boldsymbol{x}^*)$$

$$\times\, \delta(s - \omega^*) r^2 dr\, ds = \int_0^\infty f((r, \omega^*), \omega) l(((r, \omega^*), \omega), x^*) r^2 dr. \quad (4.7)$$

Here, the integration is performed over $L$. Writing the latter expression in cartesian coordinates,

$$\int_0^\infty f(\boldsymbol{r}^* - r\omega^*, \omega) \cdot l((\boldsymbol{r}^* - r\omega^*, \omega), \boldsymbol{x}^*) r^2 dr$$

$$= \int_{x^*}^\infty f\left(\boldsymbol{r}^* - \frac{x - x^*}{a^*}\omega^*, \omega\right) l\left(\left(\boldsymbol{r}^* - \frac{x - x^*}{a^*}\omega^*, \omega\right), \boldsymbol{x}^*\right) \frac{r^2}{|a^*|} dx. \quad (4.8)$$

Consider the transformation $X_\psi$, rotation of the space $R$ around the x axis by the angle $\psi$. Due to the axial symmetry of the earth–atmosphere–sun system we have for any $\psi$

$$f(\boldsymbol{r}, \omega) = f(X_\psi(\boldsymbol{r}), X_\psi(\omega)), \quad l(\boldsymbol{x}, \boldsymbol{x}^*) = l(X_\psi(\boldsymbol{x}), X_\psi(\boldsymbol{x}^*)).$$

Here the following notation is used:

$$X_\psi(\boldsymbol{x}) = (X_\psi(\boldsymbol{r}), X_\psi(\omega)).$$

Now, let

$$\boldsymbol{r} = \boldsymbol{r}^* - \frac{x - x^*}{a^*}\omega^*.$$

Using cylindrical coordinates in (4.8), and keeping in mind the last relations, we obtain

$$\int_\Omega \int_{x^*}^\infty f(\boldsymbol{r}, \omega) l((\boldsymbol{r}, \omega), \boldsymbol{x}^*) \frac{r^2}{|a^*|} dx\, d\omega$$

$$= \frac{1}{2\pi} \int_0^{2\pi} \left[ \int_\Omega \int_{x^*}^\infty f(\boldsymbol{r}, \omega) l((\boldsymbol{r}, \omega), \boldsymbol{x}^*) \frac{r^2}{|a^*|} dx\, d\omega \right] d\psi$$

$$= \frac{1}{2\pi} \int_0^{2\pi} \left[ \int_\Omega \int_{x^*}^\infty f(X_\psi(\boldsymbol{r}), X_\psi(\omega)) l((X_\psi(\boldsymbol{r}), X_\psi(\omega)), X_\psi(\boldsymbol{x}^*)) \frac{r^2}{|a^*|} dx\, d\omega \right] d\psi. \quad (4.9)$$

For an arbitrary function $h$, consider the integral

$$\int_0^{2\pi} \left[ \int_\Omega h(X_\psi(\omega))\, d\omega \right] d\psi.$$

Now

$$\int_{\Omega} h(X_\psi(\omega))\,d\omega = \int_{\Omega} h(X_0(\omega))\,d\omega = \int_{\Omega} h(\omega)\,d\omega.$$

Consequently, (4.9) can be written in the form

$$\frac{1}{2\pi}\int_0^{2\pi}\int_{\Omega}\int_{x^*}^{\infty} f(X_\psi(\boldsymbol{r}), \omega) I((X_\psi(\boldsymbol{r}), \omega), X_\psi(\boldsymbol{x}^*)) \frac{r^2}{|a^*|}\,dx\,d\omega\,d\psi$$

$$= \frac{1}{2\pi}\int_{\Omega}\left[\int_0^{2\pi}\int_{x^*}^{\infty} f(X_\psi(\boldsymbol{r}), \omega) I((X_\psi(\boldsymbol{r}), \omega), X_\psi(\boldsymbol{x}^*)) \frac{r^2}{|a^*|}\,dx\,d\psi\right] d\omega. \qquad (4.10)$$

The expression in square brackets is an integral over the surface $\Gamma$ obtained by rotating $L$ around the axis $X$. Taking, in (4.10), a surface measure, we obtain the desired quantity:

$$\frac{1}{2\pi}\int_{\Omega}\int_{\Gamma} f(\Sigma, \omega) I((\Sigma, \omega), \boldsymbol{x}_{\Sigma}^*) \frac{r^2(1 - n_1^2)^{1/2}}{|a^*|\rho_{\Sigma}}\,d\Sigma\,d\omega, \qquad (4.11)$$

where $\boldsymbol{n} = (n_1, n_2, n_3)$ is a normal to the surface $\Gamma$ at the point $\Sigma$, and $\rho_{\Sigma}$ is the distance from $\Sigma$ to the axis $X$. Thus, the desired quantity is represented as an integral of the product of the collision density and a certain function over the surface $\Gamma$. It is known that the quantity $\Phi(\boldsymbol{r}, \omega)|\boldsymbol{n}\cdot\omega|d\Sigma$ is equal to the number of photons that intersect an area element $d\Sigma$, placed at the point $\boldsymbol{r}$, in the direction $\omega$. Therefore, keeping in mind

$$\int_{\Gamma} f(\boldsymbol{x})\chi(\boldsymbol{x})\,d\Sigma = \int_{\Gamma} \Phi(\boldsymbol{x})|\boldsymbol{n}\cdot\omega| \frac{\sigma(\boldsymbol{r})}{|\boldsymbol{n}\cdot\omega|}\chi(\boldsymbol{x})\,d\Sigma,$$

we obtain the following algorithm for calculating the desired quantity.

$$\frac{\sigma(\Sigma)\exp[-\tau(\boldsymbol{r}, \Sigma) - \tau(\Sigma, \boldsymbol{r}_{\Sigma}^*)]g(\mu^*)(1 - n_1^2)^{1/2}}{4\pi^2|a^*|\rho_{\Sigma}|\boldsymbol{n}\cdot\omega|}, \qquad (4.12)$$

is calculated for each intersection of the surface $\Gamma$. The sum of those contributions is averaged over all trajectories. If a photon intersects the surface at two points that are observable from the point $\boldsymbol{r}_{\Sigma}$, the contributions are calculated from both of those points. Instead of a real intersection, it is possible to consider an intersection with a weight equal to the probability that a photon reaches the surface. The described algorithm can be obtained also by transforming the integral that defines the double local estimate. Thus, replace $f(\Sigma, \omega)$ in (4.11) with the expression

$$f(\Sigma, \omega) = \int_X f(\boldsymbol{x})k(\boldsymbol{x}, (\Sigma, \omega))\,d\boldsymbol{x} + \psi(\Sigma, \omega),$$

and consider the first term when polar coordinates with the center at $\boldsymbol{r}$; $[\boldsymbol{x} = (\boldsymbol{r}, \boldsymbol{v})]$ are used. The jacobian of this transformation is

$$\frac{|\Sigma - \boldsymbol{r}|^2}{|\boldsymbol{n}\cdot\boldsymbol{s}|},$$

where

$$\boldsymbol{s} = \frac{(\Sigma - \boldsymbol{r})}{|\Sigma - \boldsymbol{r}|}, \quad \Sigma = \boldsymbol{r} + \boldsymbol{s}|\Sigma - \boldsymbol{r}| = \boldsymbol{r} + \boldsymbol{s}\lambda.$$

Consequently (4.11) may be rewritten as

$$\begin{aligned}
&\int_X f(\boldsymbol{x}) \int_\Omega \int_\Gamma l(\boldsymbol{x}, (\Sigma, \omega))\delta(\boldsymbol{s} - \omega) l((\Sigma, \omega), \boldsymbol{x}_\Sigma^*)\cdot \frac{r^2\sqrt{1 - n_1^2}}{2\pi|a|^*\rho_\Sigma}\, d\boldsymbol{x}\, d\Sigma\, d\omega \\
&= \int_X f(\boldsymbol{x}) \int_\Omega \int_\Omega l(\boldsymbol{x}, (\boldsymbol{r} + \boldsymbol{s}\lambda, \omega)) l((\boldsymbol{r} + \boldsymbol{s}\cdot\lambda, \omega), \boldsymbol{x}_\Sigma^*) \\
&\times \delta(\boldsymbol{s} - \omega)\frac{r^2\sqrt{1 - n_1^2}}{2\pi|a^*|\rho_\Sigma}\cdot\frac{|\Sigma - \boldsymbol{r}|^2}{|\boldsymbol{n}\cdot\boldsymbol{s}|}\, d\omega\, d\boldsymbol{s}\, d\boldsymbol{x} = \int_X f(\boldsymbol{x}) \int_\Omega l(\boldsymbol{x}, (\boldsymbol{r} + \boldsymbol{s}\cdot\lambda, \boldsymbol{s})) \\
&\times l((\boldsymbol{r} + \boldsymbol{s}\lambda, \boldsymbol{s}), \boldsymbol{x}_\Sigma^*)\frac{r^2\sqrt{1 - n_1^2}}{2\pi|a^*|\rho_\Sigma}\cdot\frac{|\Sigma - \boldsymbol{r}|}{|\boldsymbol{n}\cdot\boldsymbol{s}|}\, d\boldsymbol{x}\, d\boldsymbol{s} \\
&= \int_X \int_\Omega f(\boldsymbol{x})\cdot\frac{g(\mu)}{2\pi}\cdot\frac{\sigma(\Sigma)\exp[-\tau(\Sigma, \boldsymbol{r}) - \tau(\boldsymbol{r}^*, \Sigma)]g(\mu^*)\sqrt{1 - n_1^2}}{4\pi^2|a^*|\cdot\rho_\Sigma\cdot|\boldsymbol{n}\cdot\omega|}\, d\boldsymbol{x}\, d\omega.
\end{aligned}$$

The function $f(\boldsymbol{x})\cdot[g(\mu)]/2\pi$ is the density of the probability distribution of the scattered photon, and $\exp\{-\tau(\boldsymbol{r}, \Sigma)\}$ is equal to the probability that a photon reaches the surface $\Gamma$, provided that it originates at a point $\boldsymbol{r}$ and moves along the scattering direction. Thus, we have obtained a new proof of the algorithm (4.12).

To evaluate the variance of the estimate (4.12), it is sufficient to assume the intensity that corresponds to the density $2x$ $(0 < x < 1)$ to be isotropic for the $\mu$ distribution. Hence, the variance diverges as $\ln x$. However, the variance for the case of single scattering is finite. For multiple scattering this contribution to the result may be ignored if $|\mu| < \varepsilon \ll 1$.

For a realistic atmosphere whose phase function has a forward peak, this does not perceptibly bias the result, while the variance becomes finite. Consider now how the estimate just described compares with the ordinary local estimate obtained above. One may characterize the observation point with two angles, $\varphi_0$ and $\theta_0$. To obtain the density of the flux in a desired direction by use of the method given in Sect. 4.2.1, it is necessary to calculate the contributions from collisions within a body obtained by rotating the angle $\theta_0 \leqslant \theta \leqslant \theta_0 + \Delta\theta$ contained in the plane $\varphi_0$ around the axis $X$. In order to obtain the estimate for a unit solid angle, divide (4.4) by $|\sin \theta_0|\Delta\theta$; it then takes the form

$$\frac{\exp[-\tau(\boldsymbol{r}, \boldsymbol{r}^*)]g(\mu^*)}{4\pi^2 \cdot \rho \cdot r_1 |\cos \gamma| \Delta\theta}. \tag{4.13}$$

Again, as mentioned above, instead of calculating the contribution from collisions within a rotated space, the contributions from fictitious collisions can be estimated after each scattering. Furthermore, (4.13) must be multiplied by a weight equal to the probability that a photon will undergo a collision on the path $\Delta l$ contained in the rotated space. Because $\Delta l$ is small, this probability may be replaced by $h(\boldsymbol{r}_0)\Delta l$, where $h(\boldsymbol{r} + \lambda\boldsymbol{\omega})$ is the free-path-length distribution density defined for $\boldsymbol{r} + \lambda\boldsymbol{\omega}$. Here $\boldsymbol{\omega}$ is the scattering direction, and $\boldsymbol{r}_0$ is the point of intersection of the boundary of the rotated space with the ray $\boldsymbol{r} + \lambda\boldsymbol{\omega}$.

Multiplying (4.13) by

$$h(\boldsymbol{r}_0) = \sigma(\boldsymbol{r}_0) \exp[-\tau(\boldsymbol{r}, \boldsymbol{r}_0)], \quad \Delta l = |\boldsymbol{n} \cdot \boldsymbol{\omega}|^{-1} \cdot r_1 \Delta\theta,$$

yields

$$\frac{\sigma(\boldsymbol{r}_0) \exp[-\tau(\boldsymbol{r}, \boldsymbol{r}_0) - \tau(\boldsymbol{r}_1, \boldsymbol{r}^*)] \cdot g(\mu^*)}{4\pi^2 \rho |\boldsymbol{n} \cdot \boldsymbol{\omega}| \cdot |\cos \gamma|}.$$

This is equivalent to (4.12), because

$$|\cos \gamma|^{-1} = (1 - n_1^2)^{1/2}/|a^*|. \tag{4.14}$$

To see this, let $A$ be a plane that contains a point $\boldsymbol{r}_0$ of the surface $\Gamma$ and axis $X$ and let $B$ be a plane tangent to the surface $\Gamma$ at the point $\boldsymbol{r}_0$. Denote the vector tangent to the circle $x = x_0$, $y^2 + z^2 = \rho^2$ ($\rho^2 = y_0^2 + z_0^2$) at the point $\boldsymbol{r}_0$ by $\mathbf{H}$, and denote the normal to the surface $\Gamma$ at the point $\boldsymbol{r}_0$ by $\boldsymbol{n}$. The plane $B$ is perpendicular to the plane $A$ because of the orthogonality of the normals $\mathbf{H}$ and $\boldsymbol{n}$ to these planes (because the plane $B$ contains the vector $\mathbf{H}$). Consequently, a projection $\boldsymbol{\omega}_A^*$ of the vector $\boldsymbol{\omega}^* \in B$ on the plane $A$ is equal to the projection of $\boldsymbol{\omega}^*$ on the line of intersection of planes $A$ and $B$, whereas the projection of $\boldsymbol{\omega}_A^* \in A$ on the axis $X$ is equal to the projection of $\boldsymbol{\omega}^*$ on the axis $X$.

Because the angle between the plane $\varphi = \varphi_0$ that contains the vector $\boldsymbol{\omega}^*$ and the projection plane $A$ is equal to $\gamma$, we have $|\boldsymbol{\omega}^*| = \cos \gamma$, whereas the length $a^*$ of the projection $\boldsymbol{\omega}_A^*$ on the axis $X$ is equal to $|\boldsymbol{\omega}_A^*| \cdot (1 - n_1^2)^{1/2} = \cos \gamma (1 - n_1^2)^{1/2}$. Thus (4.14) is proved.

Consider now the problem of finding the point $\boldsymbol{r}_0$ at which the ray $\boldsymbol{r} + \lambda\boldsymbol{\omega}$ ($\lambda \geqslant 0$) intersects the surface $\Gamma$. Let $\boldsymbol{r} = (x, y, z)$ be a collision point and let $\boldsymbol{\omega} = (a, b, c)$ be the direction after the collision. The problem of finding the point of intersection $\boldsymbol{r}_0 = (x_0, y_0, z_0)$ reduces to finding $\lambda$ such that $\boldsymbol{r}_0 = \boldsymbol{r} + \lambda\boldsymbol{\omega} \in \Gamma$. Now, $x_0 = x + \lambda a$. Since the surface $\Gamma$ is obtained by rotating the ray $L = \{\boldsymbol{r} = \boldsymbol{r}_0 + v\boldsymbol{\omega}^*, v \geqslant 0\}$ around the axis $X$, the cross section of the surface $\Gamma$ by the plane $x = x_0$ is a circle that contains two points: $\boldsymbol{r}_0$; and $\boldsymbol{r}_1$, which is the point at which the ray intersects the plane $x = x_0$. The second point is easily found from

$$x^* + va^* = x_0 = x + \lambda a \Rightarrow v = (x + \lambda a - x^*)/a^*; \quad \boldsymbol{r}_1 = \boldsymbol{r}^* + v\boldsymbol{\omega}^*.$$

Because the distances from the points $\boldsymbol{r}_0$ and $\boldsymbol{r}_1$ to the axis $X$ are equal, we have $y_1^2 + z_1^2 = y_0^2 + z_0^2$. Hence

$$\left[y^* + \frac{x + \lambda a - x^*}{a^*} b^*\right]^2 + \left[z^* + \frac{x + \lambda a - x^*}{a^*} c^*\right]^2 = (y + \lambda b)^2 + (z + \lambda c)^2,$$

or $U\lambda^2 + 2V\lambda + W = 0$, where (because $y^* = 0$)

$$U = \left(\frac{a}{a^*}\right)^2 - 1,$$

$$V = \frac{ac^*}{a^*} z^* + a\left(\frac{1}{a^{*2}} - 1\right)(x - x^*) - by - cz,$$

$$W = z^{*2} + 2(x - x^*)z^* \frac{c^*}{a^*} + \left(\frac{1}{a^{*2}} - 1\right)(x - x^*)^2 - y^2 - z^2.$$

If the observation direction $\boldsymbol{\omega}^*$ intersects the surface of a planet at a point $\boldsymbol{r}_1$, it is necessary to calculate the contribution directly from the planetary surface. Consider a cone of directions $\Omega_0$, containing $\boldsymbol{\omega}^*$, which cuts out an area $s$ of the Earth's surface. Denote a direct contribution from Earth to the point $\boldsymbol{r}^*$ in direction $\boldsymbol{\omega}^*$ by $j(\boldsymbol{x}^*)$.
Clearly,

$$j(\boldsymbol{x}^*) = f(\boldsymbol{r}_1) \frac{\exp[-\tau(\boldsymbol{r}_1, \boldsymbol{r}^*)]P(\mu) \cdot P_a}{2\pi|\boldsymbol{r}_1 - \boldsymbol{r}^*|^2}, \tag{4.15}$$

where $\mu = -\boldsymbol{\omega}^* \cdot \boldsymbol{r}_1/|\boldsymbol{r}_1|$. The surface of the ground is assumed to be a Lambert reflector: $P(\mu) = 2\mu$. Denote the density of photon collisions with the Earth's surface at a point $\boldsymbol{r}_1$ by

$$f(\boldsymbol{r}_1) = \int_X \mu_1 f(\boldsymbol{x}) k(\boldsymbol{x}_1, \boldsymbol{x}_1)\, d\boldsymbol{x} + \psi(\boldsymbol{r}_1),$$

where $\mu_1 = \boldsymbol{\omega} \cdot \boldsymbol{r}_1/|\boldsymbol{r}_1|$, and $\boldsymbol{\omega}$ is the photon direction before the collision with the Earth's surface.

Let $P_a$ be the albedo of the ground surface. To calculate $f(\boldsymbol{r}_1)$, the quantity

$$\frac{\mu_1 \cdot \sigma(\boldsymbol{r}) \exp\left[-\tau(\boldsymbol{r}, \boldsymbol{r}_1)\right] g(\nu)}{2\pi|\boldsymbol{r} - \boldsymbol{r}_1|^2},$$

must be averaged over all collisions. Here

$$\mu_1 = \frac{\boldsymbol{r}_1 \cdot (\boldsymbol{r} - \boldsymbol{r}_1)}{|\boldsymbol{r}_1||\boldsymbol{r} - \boldsymbol{r}_1|}, \qquad \nu = \omega \cdot \frac{(\boldsymbol{r} - \boldsymbol{r}_1)}{|\boldsymbol{r} - \boldsymbol{r}_1|},$$

where $\omega$ is the direction of motion of the photon before the collision at the point $\boldsymbol{r}$.

In practice, the contribution from the albedo is calculated as follows. The mean number $p$ of photons hitting the domain $s$ is calculated, next the sum of these numbers is averaged over all trajectories. Keeping in mind the axial symmetry of the system we calculate the quantity $p$ as follows: sum the collisions of the photon with the Earth's surface within a spherical zone $S$ obtained by rotating an area element $s$ that contains the point $\boldsymbol{r}_1$ around the axis $X$; next average the estimate $\bar{p}$ and multiply it by the ratio $s/S$, i.e., $p = \bar{p}s/S$. This modification improves the statistics because of the increase of the number of particles that hit $s$. Denote the second factor in (4.15) by $j_1(\boldsymbol{r}_1)$. Then, calculating $j(\boldsymbol{x}^*)$ for a unit solid angle, we obtain

$$j(\boldsymbol{x}^*) = \frac{j_1(\boldsymbol{r}_1)\bar{p}}{|\Omega_0|}\frac{s}{S}. \tag{4.16}$$

Here

$$s = \frac{|\Omega_0||\boldsymbol{r}_1 - \boldsymbol{r}^*|^2}{\mu}, \tag{4.17}$$

where

$$\mu = \frac{-\omega^* \cdot \boldsymbol{r}_1}{|\boldsymbol{r}_1|}, \qquad S = 2\pi(\mu_2 - \mu_1)R_0^2, \tag{4.18}$$

$R_0$ is the Earth's radius, $\mu_1 = c_1/R_0$, $\mu_2 = c_2/R_0$, and the interval $[c_2, c_1]$ is the projection of the zone $S$ on the axis $X$.

Substituting (4.17), (4.18) and the expression for $j_1(\boldsymbol{r}_1)$ in (4.16) yields an expression for the contribution to $\Phi(\boldsymbol{x}^*)$ directly from Earth,

$$j(\boldsymbol{x}^*) = \frac{\exp\left[-\tau(\boldsymbol{r}_1, \boldsymbol{r}^*)\right]\bar{p}P_a}{(2\pi)^2(\mu_2 - \mu_1)R_0^2}.$$

The contribution $j(\boldsymbol{x}^*)$ is calculated and summed with the basic contribution in the final part of the calculation.

## 4.2 System of Integral Transfer Equations that Takes Polarization into Account

The following notation will be used from here on:

$\boldsymbol{r}$: spatial vector;
$\boldsymbol{\omega}$: unit direction vector;
$\sigma_s(\boldsymbol{r}), \sigma_c(\boldsymbol{r})$: coefficients of scattering and absorption, respectively;

$$\sigma(\boldsymbol{r}) = \sigma_s(\boldsymbol{r}) + \sigma_c(\boldsymbol{r}); \quad q(\boldsymbol{r}) = \sigma_s(\boldsymbol{r})/\sigma(\boldsymbol{r});$$

$P(\boldsymbol{\omega}', \boldsymbol{\omega}, \boldsymbol{r}) = \|p_{ij}(\boldsymbol{\omega}', \boldsymbol{\omega}, \boldsymbol{r})\|_{i,j=1,\ldots,4}$: the total scattering matrix;
$\boldsymbol{I}(\boldsymbol{r}, \boldsymbol{\omega}) = [I(\boldsymbol{r}, \boldsymbol{\omega}), Q(\boldsymbol{r}, \boldsymbol{\omega}), U(\boldsymbol{r}, \boldsymbol{\omega}), V(\boldsymbol{r}, \boldsymbol{\omega})]$: a vector function representing the light intensity.

It is known that this vector function satisfies a certain system of integro-differential equations (see Chap. 1). Perhaps the most proper constructions and justifications of the Monte Carlo methods are based on the integral transfer equation for the collision density $\boldsymbol{f}(\boldsymbol{r}, \boldsymbol{\omega}) = \boldsymbol{I}(\boldsymbol{r}, \boldsymbol{\omega})\sigma(\boldsymbol{r})$; here the name "integral" is used artificially, because of the $\delta$ function which appears in the kernel.

However, this does not matter when Monte Carlo methods are used, because the distribution, defined by the $\delta$ function is easily simulated. The system of integral transfer equations that takes polarization into account can be obtained by complicated transformations of the system of integro-differential equations described above.

However, we shall derive this system from a physical point of view. We introduce the notation: $\boldsymbol{f} = (f_1, f_2, f_3, f_4)$, a vector function for the collision density; $\boldsymbol{\psi} = (\psi_1, \psi_2, \psi_3, \psi_4)$, the source distribution density; $\boldsymbol{x} = (\boldsymbol{r}, \boldsymbol{\omega})$, a point of the phase space $X = R \times \Omega$ (with coordinates $\boldsymbol{r} \in R$ and directions $\boldsymbol{\omega} \in \Omega$); $K(\boldsymbol{x}', \boldsymbol{x})$, the transition matrix for the Markov chain that characterizes the radiative-transfer process taking polarization into account; $K(\boldsymbol{x}', \boldsymbol{x})\boldsymbol{I}$ is a conditional vectorial distribution density of the secondary collisions on the hypothesis that the photon underwent the previous collision at a point $\boldsymbol{x}'$, and that the Stokes vector was $\boldsymbol{I}$.

The expression for the transition matrix $K(\boldsymbol{x}', \boldsymbol{x})$ can be obtained from the well-known representation for the transition density $k(\boldsymbol{x}', \boldsymbol{x})$ of the transfer process without polarization by replacing the indicatrix $g(\mu)$ with a scattering matrix $P(\mu, \boldsymbol{r})$:

$$K(x', x) = \frac{q(r') \exp[-\tau(r', r)]\sigma(r)P(\mu, r)}{2\pi|r - r'|^2}\delta(\omega - s)$$

$$= \begin{pmatrix} k_{11}(x', x) \cdots\cdots\cdots k_{14}(x', x) \\ \cdots\cdots\cdots\cdots\cdots\cdots \\ k_{41}(x', x) \cdots\cdots\cdots k_{44}(x', x) \end{pmatrix}. \tag{4.19}$$

Here, $q$ is the survival probability, $\tau(r', r)$ is the optical length from $r'$ to $r$, $s = (r - r')/|r - r'|$, $\sigma$: total extinction coefficient, $P(\mu, r)$ is the matrix with elements $p_{ij}$, $i, j = 1, \ldots, 4$. The collision in a volume element $dx$ (for a given photon) is either its first collision, or it immediately follows a collision that the photon has undergone, for instance, in $(x', x + dx')$, having a vector distribution density $f(x)$. Consider the vectors $\psi(x)dx$ and

$$dx \int_X K(x', x)f(x')\,dx'.$$

Then we obtain the integral equation of transfer for the collision density $f$,

$$f(x) = \int_X K(x', x)f(x')\,dx' + \psi(x), \tag{4.20}$$

or

$$f_i(x) = \sum_{j=1}^{4} \int_X k_{ij}(x', x)f_{ij}(x')\,dx' + \psi_i(x), \quad i = 1, \ldots, 4, \tag{4.21}$$

where polarization is taken into account and $f$ is a vector function.

We now show that the Neumann series for the system of integral equations of transfer converges. Write the system (4.21) in the form

$$f_i(x) = [Kf]_i(x) + \psi_i(x), \quad i = 1, \ldots, 4, \tag{4.22}$$

where $X$ is a finite-dimensional Euclidean space, $f, \psi \in L$, the function space. Define

$$\|f\| = \sum_{i=1}^{4} \int_X |f_i(x)|\,dx.$$

It is supposed that $K \in [L \to L]$. It can be easily shown that

$$\|K\|_L \leqslant \sup_{j,x'} \sum_{i=1}^{4} \int_X |k_{ij}(x', x)|\,dx.$$

It is known (see, e.g., [21]) that the solution of (4.22) is given by the Neumann series

$$f = \sum_{n=0}^{\infty} K^n \psi, \quad K^0 \psi = \psi,$$

if $\|K^{n_0}\|_L < 1$ for some $n_0$.

Let $F$ be a set of Stokes-vector functions,

$$f = (I, Q, U, V) = (I[f], Q[f], U[f], V[f]).$$

It is known [2] that $I \geqslant 0$,

$$I^2 \geqslant Q^2 + U^2 + V^2. \quad \text{Hence} \quad |Q| + |U| + |V| \leqslant \sqrt{3} I$$

and

$$\|f\|_L \leqslant (1 + \sqrt{3}) \int_X I dx = (1 + \sqrt{3}) \|I[f]\|_L.$$

There is a representation $K = DS$ where $S$ is a scattering operator and $D$ is an extinction operator. From the physical significance of these operators, it follows that $K, D, S \in [F \rightarrow F]$, $\|I[Sf]\|_L = \|I[f]\|_L$ (i.e., the integral intensity after scattering is the same), and $\|I[Df]\|_L \leqslant q\|I[f]\|_L (q < 1)$ for either a bounded system or in the presence of absorption. Combining the latter relations, we obtain

$$\|K^n \psi\|_L \leqslant (1 + \sqrt{3}) \|I[K^n \psi]\|_L \leqslant (1 + \sqrt{3}) q^n \|I(\psi)\|_L;$$

this completes the proof, if $\psi \in F$.

## 4.3 Monte Carlo Solution of a System of Integral Equations

Usually the Monte Carlo method is applied to estimating the linear functionals of the solution of the integral equation under study. Consider now a general Monte Carlo technique for estimating these functionals in the case of a system of integral equations of the second kind.

Suppose that it is required to calculate the functional

$$I_\varphi = (f, \varphi) = \sum_{i=1}^{m} \int_X f_i(x) \varphi_i(x)\, dx = \sum_{n=0}^{\infty} (K^n \psi, \varphi),$$

of the solution of a system of integral equations

$$f_i(x) = [Kf]_i(x) + \psi_i(x) = \sum_{j=1}^{m} \int_X k_{ij}(x', x) f_j(x')\, dx' + \psi_i(x).$$

Here $\boldsymbol{\varphi}$ is a vector function that has bounded components. Define in phase space $X$ a stationary Markov chain $\{\boldsymbol{x}_n\}$ by the probability density $r_0(\boldsymbol{x})$ of the initial state $\boldsymbol{x}_0$, the probability density $r(\boldsymbol{x}', \boldsymbol{x})$ of transition from $\boldsymbol{x}'$ to $\boldsymbol{x}$ and the probability $p(\boldsymbol{x})$ of termination at the point $\boldsymbol{x}$. Introduce also a random weight vector by

$$Q_0^{(i)} = \frac{\psi_i(\boldsymbol{x}_0)}{r_0(\boldsymbol{x}_0)}, \quad Q_n^{(i)} = \sum_{j=1}^{m} Q_{n-1}^{(j)} \frac{k_{ij}(\boldsymbol{x}_{n-1}, \boldsymbol{x}_n)}{r(\boldsymbol{x}_{n-1}, \boldsymbol{x}_n)} \cdot \frac{1}{1 - p(\boldsymbol{x}_{n-1})}.$$

It can be shown, as in Sect. 3.3, that

$$E \sum_{n=0}^{N} \left[ \sum_{i=1}^{m} Q_n^{(i)} \varphi_i(\boldsymbol{x}_n) \right] = (\boldsymbol{f}, \boldsymbol{\varphi}) = I_\varphi, \tag{4.23}$$

where $N$ is the random number of the last state. The Monte Carlo method for estimating the quantity $I_\varphi$ is based on (4.23). In this section, we describe in detail the Monte Carlo algorithm for estimating the intensity and polarization of the multiply scattered radiation in a spherical-shell atmosphere. The most physical transition density $r(\boldsymbol{x}', \boldsymbol{x})$ for this problem is the one defined by the kernel $k_{11}(\boldsymbol{x}', \boldsymbol{x})$ that corresponds to the radiative-transfer process without polarization. Obviously, in order to simulate such a process the weight vector must be transformed after the scattering by a matrix that has the elements $p_{ij}(\omega', \omega, \boldsymbol{r})/p_{11}(\omega', \omega, \boldsymbol{r})$.

This scheme produces the least weight fluctuations, although minimal standard error of the estimate is not ensured.

Small-variance estimates may be obtained by using information about the solution of the adjoint problem. In particular, the following modifications can result in very effective reduction of variance: use of an importance distribution of the initial states of trajectories; simulation of a few first collisions without escape; multiplication of the probability that the photon reaches the Earth, which is proportional to the integral albedo, by an approximate importance of the albedo. The incident light is supposed to be initially unpolarized, i.e., $\boldsymbol{I}_c = (I_c, 0, 0, 0)$. The Stokes vector after scattering is then given by

$$\boldsymbol{I}_1(\boldsymbol{r}, \omega) = P(\omega', \omega, \boldsymbol{r}) \cdot \boldsymbol{I}(\boldsymbol{r}, \omega'), \tag{4.24}$$

where

$$P(\omega', \omega, \boldsymbol{r}) = L(\pi - i_2) R(\omega', \omega, \boldsymbol{r}) L(-i_1),$$

with

$$L(i) = \begin{bmatrix} 1 & 0 & 0 & 0 \\ 0 & \cos 2i & \sin 2i & 0 \\ 0 & -\sin 2i & \cos 2i & 0 \\ 0 & 0 & 0 & 1 \end{bmatrix}.$$

Here $i_1$ and $i_2$ are the angles between the scattering plane and the planes that contain the axis of the coordinate system and the vectors $\boldsymbol{\omega}'$ and $\boldsymbol{\omega}$, respectively, and $L$ is the rotation matrix [2]. The matrix of total scattering of light (by air) is defined as a weighted average of the matrices for molecular and aerosol scattering,

$$R(\omega', \omega, \boldsymbol{r}) = \frac{R_a(\omega', \omega, \boldsymbol{r})\sigma_a(\boldsymbol{r}) + R_M(\omega', \omega)\sigma_M(\boldsymbol{r})}{\sigma_a(\boldsymbol{r}) + \sigma_M(\boldsymbol{r})}.$$

In general, for an anisotropic medium, all matrix components are different. For an isotropic medium, the scattering matrix takes the form [3]

$$\begin{bmatrix} r_{11} & r_{12} & 0 & 0 \\ r_{21} & r_{22} & 0 & 0 \\ 0 & 0 & r_{33} & r_{34} \\ 0 & 0 & -r_{43} & r_{44} \end{bmatrix}. \tag{4.25}$$

If the scatterers are homogeneous spherical particles, then $r_{11} = r_{22}$, $r_{12} = r_{21}$, $r_{33} = r_{44}$, $r_{34} = r_{43}$. As numerous experiments show, for aerosol-photon scattering $R_a$ also has the form (4.25). For molecular scattering, $R_M$ is given by

$$R_M(\omega', \omega) = \frac{3}{4}\begin{bmatrix} \frac{1}{2}(1+\mu^2) & -\frac{1}{2}(\mu^2-1) & 0 & 0 \\ -\frac{1}{2}(1-\mu^2) & \frac{1}{2}(1+\mu^2) & 0 & 0 \\ 0 & 0 & \mu & 0 \\ 0 & 0 & 0 & \mu \end{bmatrix},$$

where $\mu = \omega' \cdot \omega$ is the cosine of scattering angle. The matrix $R_a$ is normalized so that

$$\int_{-1}^{1} r_{11}(\mu)\, d\mu = 1.$$

$\mu$ is sampled from $R_{11}$, i.e., from the indicatrix (see Sect. 4.5); the azimuthal angle $\varphi$ is assumed to be isotropic. The new photon direction (after scattering) is characterized by angle $\theta$ and $\varphi$. In order to describe the quantities $I$, $Q$, $U$, $V$, we introduce a coordinate system with axis coinciding with the radius vector of the point of scattering. In these coordinates, the vector function $\boldsymbol{I}$ is symmetric with respect to an axis parallel to the direction of solar radiation, which makes it possible to apply the modifications of local calculation method that are described in Sect. 4.1. Furthermore, this coordinate system is most useful when comparisons with the experimental data are investigated.

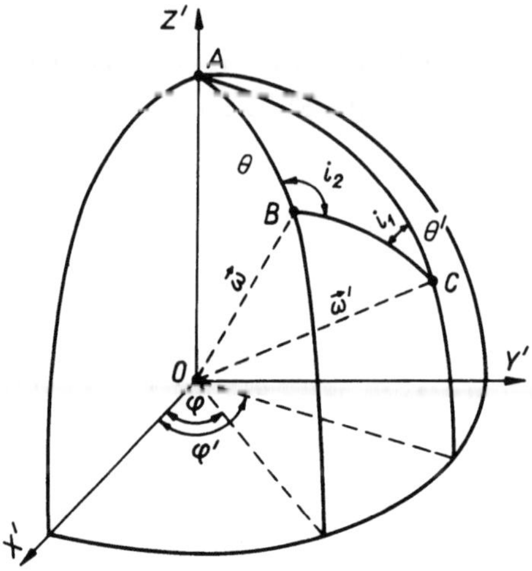


Fig. 4.3. Geometry of the scattering problem

When a new direction is chosen, the angles $i_1$ and $i_2$ can be obtained by making use of the formulas of spherical trigonometry. (Fig. 4.3) Introduce the notation: $x, y, z$ are coordinates of the scattering point $O$; $r$ is the distance from the scattering point to the Earth's center. The direction of the $OZ'$ axis is defined by the unit vector $s = (x/r, y/r, z/r)$.

$$\mu = \cos\theta = \omega' \cdot \omega, \quad \mu_0 = \sin\theta,$$

$$\mu_1 = \cos v' = \omega' \cdot s, \quad \mu_2 = \sin v',$$

$$\mu_3 = \cos v = \omega \cdot s, \quad \mu_4 = \sin v.$$

From the spherical triangle $ABC$, we obtain

$$\mu_3 = \mu_1\mu + \mu_2\mu_0 \cos i_1, \quad \mu_1 = \mu_3\mu + \mu_4\mu_0 \cos i_2;$$

hence

$$\cos i_1 = \frac{\mu_3 - \mu_1\mu}{\mu_2\mu_0}, \quad \cos i_2 = \frac{\mu_1 - \mu_3\mu}{\mu_4\mu_0},$$

$$\sin i_1 = (1 - \cos^2 i_1)^{1/2} \operatorname{sgn}(q),$$

$$\sin i_2 = (1 - \cos^2 i_2)^{1/2} \operatorname{sgn}(q), \quad q = \omega' \cdot [\omega r].$$

In order to determine the direction of rotation of the coordinates of the Stokes vector from the plane $\omega', r$ to the plane $\omega, r$, the signs of the sines are defined by the sign of the vector triple product $q$.

An important point of the algorithm is the local-calculation method for estimating the photon flux at the observation point. Coordinates of the Stokes vector for the local estimate are recalculated for $\mu^* = \omega' \cdot \omega^*$, the cosine of the angle between the photon direction $\omega'$ before the collision and the direction $\omega^*$ from the scattering point $r(x, y, z)$ to the observation point $r^*(x^*, y^*, z^*)$:

$$I_1(r, \omega^*) = P(\omega', \omega^*, r) \cdot I(r, \omega').$$

The photon history after scattering is followed in the usual way by transforming parameters according to (4.24). The Stokes vector is recalculated by

$$I(r, \omega) = R_{11} \cdot I_0(r, \omega') + R_{12} \cdot A,$$
$$Q(r, \omega) = [R_{21} I_0(r, \omega') + A R_{22}] \cos i_2 - [R_{33} B - R_{34} V_0(r, \omega')] \sin 2i_2,$$
$$U(r, \omega) = [R_{21} I_0(r, \omega') + A R_{22}] \sin 2i_2 + [R_{33} B - R_{34} \cdot V_0(r, \omega')] \cos 2i_2,$$
$$V(r, \omega) = R_{43} B + R_{44} V_0(r, \omega'),$$

where

$$A = Q_0(r, \omega') \cos 2i_1 - U_0(r, \omega') \sin 2i_1,$$
$$B = Q_0(r, \omega') \sin 2i_1 + U_0(r, \omega') \cos 2i_1.$$

## 4.4 Estimates of the Importance Function and Their Applications to Modified Simulation of Trajectories

Suppose that it is desired to evaluate a linear functional $I_\varphi = (f, \varphi)$ of the solution $f$ of an integral equation of the second kind (2.14). Consider then an appropriate adjoint equation

$$f^*(x) = \int_X k(x, x') f^*(x') \, dx' + \varphi(x), \tag{4.26}$$

or $f^* = K^* f^* + \varphi$.

It is known (see Chap. 3) that $I_\varphi = (f, \varphi) = (f^*, \psi)$. In particular, putting $\psi(x) = \delta(x - x_0)$, we obtain $I_\varphi = f^*(x_0)$, i.e., $f^*(x_0)$ equals the value of the functional $I_\varphi$ for a source of unit power at a point $x_0$.

Therefore, with respect to a functional to be estimated, the function $f^*(x)$ is an importance function. If the trajectory is simulated so that the transition density is multiplied by the importance of the points of phase space, the variance of the estimation for the functional $I_\varphi$ can be made zero. Thus, using information about the solution of the adjoint equation effective Monte Carlo modifications can be constructed. This section is devoted to finding approximate importance functions and their applications to simulating some elements of the transfer process.

As a rule, in radiative-transfer problems, it is desired to estimate the net flux distribution, either in a given domain of the phase space, or in a desired direction. These distributions can be estimated by histograms that use (for a discrete parameter $\lambda$) a function $\varphi(\mathbf{x}, \lambda)$ of the form

$$\varphi(\mathbf{x}, \lambda) = \varphi_1(\mathbf{x})\delta[\omega(\mathbf{x}), \lambda], \tag{4.27}$$

where

$$\delta[\omega(\mathbf{x}), \lambda] = \begin{cases} 1, & \text{if } \omega(\mathbf{x}) = \lambda, \\ 0, & \text{otherwise,} \end{cases}$$

and $\lambda, \omega(\mathbf{x}) = 1, 2, \ldots, S.$

In this case, it is useful to take (see Sect. 3.6) an importance function $g = f_0^*$ satisfying the equation

$$f_0^* = K^* f_0^* + \varphi_0,$$

where

$$\varphi_0(\mathbf{x}) = \left[\sum_{\lambda=1}^{S} \varphi^2(\mathbf{x}, \lambda)\mu(\lambda)\right]^{1/2}$$
$$= \varphi_1(\mathbf{x})\left[\sum_{\lambda=1}^{S} \delta[\omega(\mathbf{x}), \lambda]\mu(\lambda)\right]^{1/2} = \varphi_1(\mathbf{x})\{\mu[\omega(\mathbf{x})]\}^{1/2}.$$

Thus, for $\mu(\lambda) = S^{-1}$, we have

$$\varphi_0(\mathbf{x}) = S^{-1/2} \cdot \varphi_1(\mathbf{x}), f_0^* = S^{-1/2} \cdot f_1^*,$$

where

$$f_1^* = \sum_{\lambda=1}^{S} f_\lambda^*, \quad \text{since} \quad \varphi_1 = \sum_{\lambda=1}^{S} \varphi_\lambda.$$

It is easy to see that the Monte Carlo algorithms under study do not depend on the constant factor of the modifying function $g$. Consequently, the relation $g = f_0^*$ is equivalent to $g = f_1^*$, provided that (4.27) holds and that

$$\mu(\lambda) = S^{-1}.$$

We present some algorithms for calculating the function $f_1^*$ for the atmospheric-optics problems that have been formulated at the beginning of this chapter. The problem of estimating the angle distribution of the net flux simultaneously at several observation points $r_i^*$, $i = 1, \ldots, n_x$ is treated. Thus, the parameter $\lambda$ of this problem is in fact the index of combination: observation-point position to the value of the azimuthal angle interval (value) of the latitude angle, for which the intensity is estimated. In this case, the function $\varphi(\boldsymbol{x}, \lambda)$ takes the form (4.27), where

$$\varphi_1(\boldsymbol{x}) = \frac{\exp[-\tau(\boldsymbol{r}, \boldsymbol{r}^*)]g(\mu_2)|\sin\theta|}{(2\pi)^2 \cdot r_1 \rho \cdot |\cos\gamma|},$$

or

$$\varphi_1(\boldsymbol{x}) = \frac{\sigma(\boldsymbol{r}_1)\exp[-\tau(\boldsymbol{r}, \boldsymbol{r}_1) - \tau(\boldsymbol{r}_1, \boldsymbol{r}^*)]g(\mu^*)(1 - n_1^2)^{1/2}}{(2\pi)^2 \rho |a^*||\boldsymbol{n} \cdot \boldsymbol{\omega}|},$$

for calculating the integral of flux over given intervals of the latitude angle or for estimating the flux in given directions, respectively. The function $f_1^*$ is calculated by use of a special program based on (3.11),

$$f_1^*(\boldsymbol{x}_0) = \varphi_1(\boldsymbol{x}_0) + E \sum_{n=1}^{N} Q_n \varphi_1(\boldsymbol{x}_n).$$

Estimation of the function $f_1^*$ does not require much computer time, because the averaging is over all observation variants. In an importance sampling, $K^*g$ must be calculated. It is impossible to calculate this quantity for realistic systems, because the space $X$ is, in the theory of photon transfer, a phase space of coordinates and velocities. However, the kernel function $k(\boldsymbol{x}', \boldsymbol{x})$ of the integral transfer equation is a combination of the conditional distribution densities of elementary random variables; namely, the components of the velocity of the particle, the free-path length $l$, and the initial coordinates $\boldsymbol{x}_0$ of the trajectory of the photon. As pointed out in [22], the variance of the estimate can be made zero if the trajectory is simulated in such a manner that the conditional probability densities of the elementary random variables are multiplied by an appropriate importance of the points of phase space (i.e., an importance sampling).

*A Modified Simulation of the Initial Coordinates*

The solar radiation incident at the top of the atmosphere may be assumed to be parallel. Therefore, the distribution density of $\rho_0 = (y_0^2 + z_0^2)^{1/2}$ is given by $2\rho/R_2^2$, $0 \leqslant \rho_0 \leqslant R_2$, where $R_2$ is the radius of the exterior of the atmosphere (Fig. 4.4). The following approximation is used for the importance function of the initial coordinates:

$$f_1^*(\rho) = c_i \geqslant 0 \quad \text{if} \quad \rho_{k-1} \leqslant \rho < \rho_k, \quad k = 1, \ldots, n_\rho, \rho_0 = 0, \quad c_k = \text{const.}$$

It is easy to see by geometric arguments that, for observation points lying in the

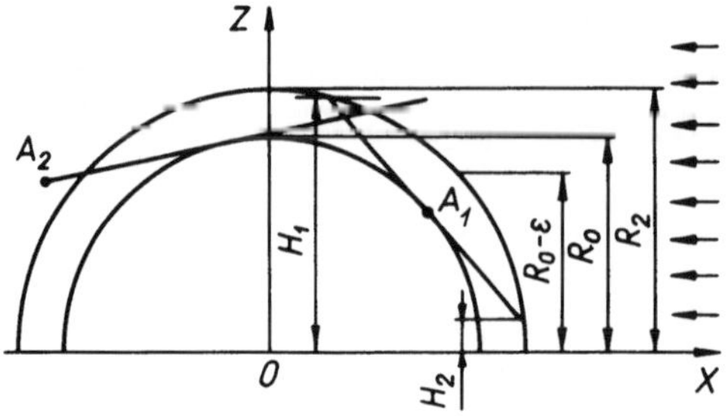


**Fig. 4.4.** Geometry of the problem of finding the importance of the initial coordinates of trajectories

illuminated part of an atmosphere (point $A_1$, Fig. 4.4), the importance of photons that start in the domains $\rho > H_1$ and $\rho < H_2$ is relatively small and vanishes for a phase function with a very sharp peak, because secondary scattering of photons that start in these domains does not make a noticeable contribution to the radiation at point $A_1$. Thus, for an appropriate choice of $\varepsilon_1$ and $\varepsilon_2$, it is possible to assume that $f^*(\rho) = 0$ if $\rho > H_1 + \varepsilon_1$ and $\rho < H_2 - \varepsilon_2$. If the observation point is situated in a twilight part (point $A_2$, Fig. 4.4), it is sufficient to simulate the first particle coordinates of the photons in the domain $\rho > R_0 - \varepsilon$ [i.e., $f^*(\rho) = 0$ if $\rho < R_0 - \varepsilon$], where $R_0$ is the Earth's radius. For domains where the $c_i$ cannot be assumed to be zero, these quantities are evaluated from certain geometrical arguments or can be calculated with a special program based on (3.11).

From the importance-sampling technique, we find that $\rho_0$ can be sampled from the density

$$f(\rho) = \frac{2\rho f_1^*(\rho)}{\int_0^{R_2} 2\rho f_1^*(\rho)\, d\rho}, \quad 0 \leqslant \rho \leqslant R_2.$$

We now present an algorithm for the simulation of $\rho^*$ based on the well-known relation

$$\xi = F_\xi^{-1}(\alpha), \tag{4.28}$$

where $F_\xi(x)$ is the distribution function of $\xi$, and $\alpha$ is the random variable uniformly distributed in [0, 1]. To realize (4.28), it is necessary to solve

$$\int_0^{\rho^*} f(\rho)\, d\rho = \alpha.$$

Let

$$\int_0^{\rho_{k-1}} f(\rho)\, d\rho \leqslant \alpha < \int_0^{\rho_k} f(\rho)\, d\rho.$$

Then

$$\int_0^{\rho_k} f(\rho)\,d\rho = \alpha + \int_{\rho^*}^{\rho_k} f(\rho)\,d\rho = 0.$$

Hence

$$\rho^* = \sqrt{\rho_k^2 + M\cdot c/c_k},$$

where

$$c = \int_0^{R_2} 2\rho f_1^*(\rho)\,d\rho = \sum_{i=1}^{n_\rho} c_i(\rho_i^2 - \rho_{i-1}^2), \quad M = \alpha - \int_0^{\rho_k} f(\rho)\,d\rho,$$

and the initial photon weight is given by

$$Q_0 = \frac{1}{R_2^2}\cdot\frac{\int_0^{R_2} 2\rho f_1^*(\rho)\,d\rho}{f_1^*(\rho)} = \frac{c}{R_2^2\cdot c_k}.$$

Numerical experiments were carried out to investigate the efficiency of this modification. A problem of simulating the observations in an illuminated part of atmosphere was considered. The quantity (see Sect. 3.6)

$$\mathscr{D}_1 = S^{-1}\sum_{t=1}^{S} \mathscr{D}\xi(t),$$

was evaluated for two cases.

First the quantities $c_i$ were taken to be 1 for all intervals $[\rho_{i-1}, \rho_i]$, $i = 1, \ldots, n_\rho$. In this case, $D_1^{(0)} = 0.1532$; simultaneously, importances $c_i \geqslant 0$ were calculated to be used in the following calculation for modified simulation of the initial coordinates of trajectory. In this case, $D_1^{(1)} = 0.0303$.

It is known that the efficiency of the Monte Carlo algorithms is defined by a number that is inversely proportional to

$$S = T\mathscr{D}\xi, \tag{4.29}$$

where $T$ is the labor expended in obtaining one sample of $\xi$. The number of histories was the same in the two cases. Because the modification of $\rho^*$ simulation does not require large amounts of computer time,

$$\frac{S_0}{S_1} = \frac{T_0\mathscr{D}_1^{(0)}}{T_1\mathscr{D}_1^{(1)}} \approx \frac{\mathscr{D}_1^{(0)}}{\mathscr{D}_1^{(1)}} = 5.13;$$

i.e., the computer time would be reduced by the factor 5.13.

*Simulation of the Cosine of Scattering Angle*

First, let us consider a simulation scheme in which this modification is used after the first scattering. Let $g(\mu)$ be the density of the distribution of the cosine of the scattering angle ($\mu = \cos\theta$) after scattering at a certain point in phase space; that is, the indicatrix is assumed to be linear on each of the intervals $\mu_i \leqslant \mu < \mu_{i-1}$, $i = 1, \ldots, n_\mu$, $\mu_0 = 1$, $\mu_{n_\mu} = -1$, $g(\mu_i) = g_i$. Denote the corresponding importance function by $g^*(\mu)$, which is chosen from prior physical arguments, and is more precisely defined by special calculations based on (3.11).

The function $g^*(\mu)$ is supposed to be a step function on the mentioned intervals and is equal to $g_k^*$ if $\mu \in [\mu_{k-1}, \mu_k]$. Thanks to the importance-sampling technique, $\mu$ is chosen from

$$\frac{g(\mu)\cdot g^*(\mu)}{(g, g^*)}; \quad \text{also,} \quad Q_n = Q_{n-1}\cdot\frac{(g, g^*)}{g^*(\mu)}.$$

Introduce the notation:

$$\Delta\mu_i = \mu_{i-1} - \mu_i, \qquad S_i = \frac{1}{2}\Delta\mu_i g_i^*(g_{i-1} + g_i),$$

$$c = (g, g^*) = \int_{-1}^{1} g(\mu)g^*(\mu)\,d\mu$$

$$= \sum_{i=1}^{n_\mu} g_i^* \int_{\mu_i}^{\mu_{i-1}} \left[g_i + \frac{\mu - \mu_i}{\Delta\mu_i}(g_{i-1} - g_i)\right] d\mu = \sum_{i=1}^{n_\mu} S_i.$$

By (4.28), we have

$$\int_{\mu}^{1} g(t)g^*(t)\,dt = \alpha c.$$

Let

$$M = \alpha c - \sum_{i=1}^{k} S_i \leqslant 0,\ M + S_k > 0, \quad \text{then} \quad M + S_k - g_k^* \int_{\mu}^{\mu_{k-1}} g(t)\,dt = 0.$$

Consequently

$$S_k + g_k^* \int_{\mu_{k-1}}^{\mu} g(t)\,dt = -M, \quad g_k^* \int_{\mu_k}^{\mu} g(t)\,dt = -M.$$

Because the function $g(t)$ is piecewise linear, we have

$$\mu = \mu_k - \frac{g_k\Delta\mu_k + [(g_k\Delta\mu_k)^2 - 2M\Delta\mu_k(g_k - g_{k-1})/g_k^*]^{1/2}}{g_k - g_{k-1}}. \tag{4.30}$$

After simulating $\mu$ by this formula, we recalculate the photon weight by

$$Q_n = Q_{n-1} c / g_k^*.$$

For the $n$th scattering, where $n > 1$, $\mu$ simulation is performed by use of the usual simulation formula for $\mu$ that coincides with (4.30) if $g^*(\mu) \equiv 1$. This modification seems to be useful when twilight effects are calculated.

We now consider a general modification of scattering simulation appropriate for calculating the intensity of different parts of atmosphere. Let $G(\omega)$ be the density of distribution of photon directions $\omega$ after scattering at a point of phase space, and let $G^*(\omega)$ be the corresponding approximate importance function.

According to the importance-sampling technique, we should sample $\boldsymbol{\omega}$ from the density

$$\frac{G(\omega)G^*(\omega)}{(G, G^*)}.$$

However, the direct realization of this simulation is too difficult, because it is necessary to evaluate a four-dimensional integral $(G, G^*)$. But under certain conditions (for example, when the observation point is situated in a twilight region and the albedo is small), $G^*$ can be represented approximately as follows,

$$G^*(\omega) \approx G_0^* + G_1^*(\omega),$$

where $G_0^*$ is a constant, and $G_1^*(\omega) \neq 0$ only in a domain $\mathscr{D}^*$ localized about the direction $\boldsymbol{\omega}^*$. Hence

$$G^*(\omega)G(\omega) \approx G_0^* G(\omega) + G_1^*(\omega)G(\omega) \approx G_0^* G(\omega) + G(\omega^*)G_1(\omega).$$

This is a special case of the expression

$$\text{const}\,[qG(\omega) + pG_1(\omega)], \tag{4.31}$$

where $q, p > 0$, $q + p = 1$. Sampling from (4.31) will be useful provided that the domain $\mathscr{D}^*$ is symmetric with respect to the radius vector $\boldsymbol{r}$ of the scattering point and is defined by inequality $-1 \leqslant \eta_{n\eta} \leqslant \eta \leqslant \eta_0 \leqslant 1$, where $\eta$ is the cosine of the angle between $\omega$ and $\boldsymbol{r}$. Then the density (4.31) can be rewritten in the form $qg(\mu) + pg^*(\eta)$ where $g(\mu)$ is the physical indicatrix of total scattering. The quantity $p$ and function $g^*(\eta)$ are chosen from prior physical arguments, and are tested in special computations. The function $g^*$ is taken to be $g_k^*$ if $\eta_k \leqslant \eta < \eta_{k-1}$, $k = 1, \ldots, n_\eta$, $\eta_0 = 1$, $\eta_{n_\eta} = -1$. The scattering is simulated as follows: a random number $\alpha_0$ is chosen, and if $\alpha_0 < p$, then a random variable $\eta$ is sampled from $g^*(\eta)$. By (4.28) and because $g^*(\eta)$ is a step function, we have

$$\eta = \eta_k - M/g_k^*,$$

where

$$M = \alpha - \sum_{i=1}^{k} g_i^*(\eta_{i-1} - \eta_i) = \int_{\eta}^{\eta_k} g^*(\lambda)\, d\lambda \leqslant 0.$$

The coordinates $(a, b, c)$ of a new photon direction are calculated using the fact that the previous direction coincides with $\boldsymbol{r}$, and cosine of the scattering angle is $\eta$. For $\alpha_0 > p$, the scattering is simulated according to the physical indicatrix $g(\mu)$, i.e., to formula (4.30), where $g_k^* = 1$, $k = 1, \ldots, n_\mu$. After the direction of scattering is simulated, the weight is multiplied by

$$\frac{g(\mu)}{qg(\mu) + pg^*(\eta)} \leqslant \frac{1}{q}.$$

In order to investigate the efficiency of a modified simulation of $\mu$ with the use of the approximate importance function $g^*(\eta)$, this modification was used to solve the problem discussed above. For $c_i \equiv 1$ and $p = 0$ we obtained $\mathscr{D}_1^{(0)} = 0.1552$, whereas for $c_i \equiv 1$, $p = 1/3$, and an appropriate approximate function $g^*$ previously evaluated,

$$\mathscr{D}_1^{(1)} = 0.0731. \quad \mathscr{D}_1^{(0)}/\mathscr{D}_1^{(1)} = 2.12.$$

*Modified Free-Path-Length Simulation*

The strong extinction of sunlight due to the great optical thickness of the atmosphere creates difficulties in Monte Carlo calculations of some twilight problems (e.g., evaluation of the intensity and polarization of light). To obtain sufficiently precise results by the direct simulation method, great amounts of computer time are required.

We now present some modified free-path-length simulations based on use of approximate information about the importance function. A modification of the free-path-length simulation will be useful, as simple physical arguments show, if the observation point is situated in (or near) a shadow region. Then, after a scattering, the photon escapes from the atmosphere with a large probability that its escape may be interpreted as absorption with a coefficient nearly equal to the total scattering cross section. Thus, the photon flux incident on the Earth's surface changes in a manner similar to that in the Milne problem, with a small survival coefficient. The corresponding importance function is approximately $\exp \tau$, where $\tau = \tau(l)$ is the optical-path length of the run $l$. The free-path-length distribution density is $\sigma(l) \exp[-\tau(l)]$; therefore $\tau$ is usefully sampled from

$$\frac{1}{\tau_0}, \quad 0 < \tau < \tau_0, \tag{4.32}$$

where $\tau_0$ is the optical distance from the origin of the run to the boundary of the atmosphere.

The photon weight is then multiplied by $\tau_0 \cdot \exp[-\tau(l)]$. In order to investigate the efficiencies of various modifications of free-path-length simulation, numerical experiments were carried out in which the intensity and polarization of scattered sunlight were calculated for the twilight region. The following modifications were considered.

a) The free-path-length $l$ is sampled from the density

$$1/L, \quad 0 < l < L, \tag{4.33}$$

where $L$ is a distance from the run origin to the boundary of atmosphere. After each sampling, the weight is multiplied by $L \cdot \sigma(l) \exp[-\tau(l)]$.

b) A physical simulation without escape, i.e., $l$ is sampled from the density

$$\frac{\sigma[\mathbf{r}(l)] \exp[-\tau(l)]}{1 - \exp(-\tau_0)}, \quad 0 < \tau < \tau_0,$$

and the weight is multiplied by $1 - \exp(-\tau_0)$.

Calculations were carried out also for sampling from the physical density $\sigma[\mathbf{r}(l)] \exp[-\tau(l)]$. Experiments show that, for the problem under study, the best modification is the second [i.e., in which the density (4.33) is used].

Further, we describe a modification of free-path-length simulation based on the use of approximate information about $f^*(l)$ for twilight problems. It is very difficult to calculate the function $f^*(l)$ for problems with complicated geometry and in which multiple scattering is taken into account. Therefore, the present modification is only for the free-path-length $l$ of the first-run, the importance of which for the present atmospheric-optics problems can be estimated sufficiently precisely (see Sect. 4.1). Consider an atmosphere (Fig. 4.5, in which $R_2$ is the exterior radius of atmosphere, $R_0$ is the Earth's radius, and $H = R_2 - R_0$ is the altitude of atmosphere) divided into $N$ layers of radii $r_i$, $i = 1, \ldots, N + 1$ with $r_1 = R_0$, $r_{N+1} = R_2$. Consider the following approximation to the importance function $f^*$ for the initial photon collisions.

The function $f^*$ is assumed to be constant ($= g_{ki}^* \geqslant 0$) in each domain $A_{ki}$, $k = 1, \ldots, n_\rho$, $i = 1, \ldots, N + 1$. The quantities $g_{ki}^*$ are obtained for given sets $\{\rho_k\}$ and $\{r_k\}$ by special calculations (here $\rho_k$ is a subdivision used in the simulation of initial coordinates of the photon trajectory). Notice that a finer subdivision may be used than that shown in Fig. 4.5, where $n_\rho = N + 1$. Moreover, the finer the subdivision, the better is a piecewise-constant approximation to the function $f^*(l)$. For reasons of convenience, the set $\{r_i\}$ $i = 1, \ldots, N + 1$ is involved in the set $\{\rho_k\}$, $k = 1, \ldots, n_\rho$, so that $N + 1 \leqslant n_\rho$, $\rho_1 = R_0$, $\rho_{n_\rho} = R_2$. In the importance-sampling technique, $l^*$ is sampled from the density

$$\frac{g(l) f^*(l)}{\int_0^\infty g(l) f^*(l)\, dl},$$

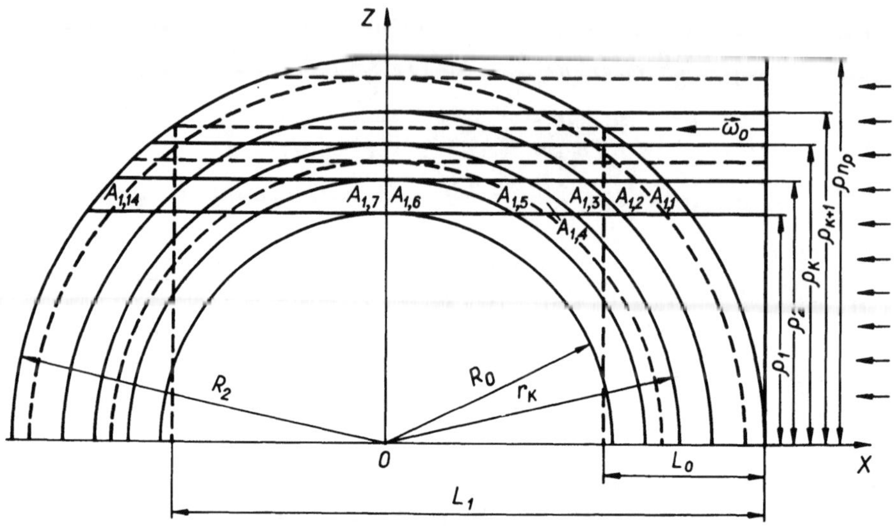


**Fig. 4.5.** Geometry of the importance sampling technique for free-path length simulation

and the weight is recalculated from

$$Q_n = Q_{n-1} \frac{\int_0^\infty g(l) f^*(l)\, dl}{f^*(l)} ,$$

Here $g(l) = \sigma(l) \cdot \exp[-\int_0^l \sigma(t) dt]$ is the distribution density of free-path lengths. But as is easily seen, the function $g(l)$ vanishes if $l < L_0$ and $l > L_1$, where $L_0$ and $L_1$ are distances from $\boldsymbol{r}_0$ to the upper bound (see Fig. 4.5). The point $\boldsymbol{r}_0$ is chosen in the circle $x = R_2$, $y^2 + z^2 = R_2^2$ as mentioned above. Thus, $l^*$ must be sampled from the density

$$\frac{g(l) f^*(l)}{\int_{L_0}^{L_1} g(l) f^*(l)\, dl} , \quad L_0 \leqslant l \leqslant L_1 .$$

Notice that the modification under study coincides with the physical simulation without escape, if $f^* = 1$. Keeping the foregoing arguments in mind and assuming that the photon starts in the region $[\rho_k, \rho_{k+1}]$, we shall formulate an algorithm for simulating $l^*$ and recalculating the weights:
Introduce the notation

$$a_i = \int_0^{l_i} \sigma(t)\,dt = \sum_{j=1}^{i} \sigma_j(l_j - l_{j-1}),$$

$$b_i = \exp(-a_i), \quad i = 1, \ldots, n_l,$$

$$\gamma = \int_{L_0}^{L_1} g(l) f^*(l)\,dl = \sum_{j=1}^{nl} g^*_{kj} \int_{l_{j-1}}^{l_j} g(l)\,dl = \sum_{j=1}^{nl} g^*_{kj}(b_{j-1} - b_j),$$

where $l_j$ are the distances from $\boldsymbol{r}_0$ to intersection points of direction $\boldsymbol{\omega}_0$ with the boundaries of the spherical layers, $l_0 = L_0$, $l_{n_l} = L_1$; $\sigma$ is the total extinction coefficient constant ($= \sigma_i$) in each atmospheric layer whose boundaries are at $r_i$ and $r_{i+1}$, $i = 1, \ldots, N$. The algorithm for simulating $l^*$ is similar to that used for simulating the cosine of scattering angle. By (4.28), we have

$$\int_{L_0}^{l^*} g(l) f^*(l)\,dl = \alpha\gamma,$$

where $\alpha$ is a random variable uniformly distributed on [0, 1].
Let

$$M = \alpha\gamma - \sum_{j=1}^{i} g^*_{kj}(b_{j-1} - b_j) \leqslant 0, \quad M + g^*_{ki}(b_{i-1} - b_i) > 0.$$

Then

$$M + g^*_{ki} \int_{l^*}^{l_i} g(l)\,dl = 0.$$

Hence

$$l^* = l_i - [\ln(b_i - M/g^*_{ki}) + a_i]/\sigma_i,$$

and

$$Q_n = Q_{n-1}\gamma/g^*_{ki}.$$

In order to investigate the efficiency of this algorithm for estimating the light intensity in the twilight region, numerical experiments were done. Observations at the top of an atmosphere, at the solar zenith angle $\psi = 100°$ were simulated by the Monte Carlo method. An atmosphere (altitude 100 km) was divided (over $\rho$ and $r$) into 30 layers. The scattering coefficients and indicatrix were taken correspond to an atmospheric model described in Sect. 4.10. Thus, the function $f^*(l)$ was represented as a $30 \times 60$ matrix. Notice that in preliminary calculations of quantities $g^*_{ki}$, it is advantageous to sample the free-path length $l^*$ of the first run from the uniform distribution on $[l_0, l_{n_l}]$, which provides uniform distribution of the first collision points in the domains $A_{ki}$. The next 2 (or 3) collisions were simulated without escape, or uniformly distributed [i.e., according to the density (4.32)]. For an angle of observation lying in the plane $\varphi = 0°$ we obtained: $D_1^{(1)} = 0.36$, and for $\varphi = 180°$, $D_1^{(2)} = 0.94$, where the 3 initial runs were sampled from the uniform $\tau$ distribution. When the foregoing modification of the first-run simulation was used, in which the 2 subsequent runs were sampled from the uniform $\tau$ distribution, the corresponding values of $D_1$ were

$$\mathscr{D}_1^{(3)} = 0.32 \quad \text{and} \quad \mathscr{D}_1^{(4)} = 0.16.$$

In both of these cases, the first coordinates of trajectories were sampled from the uniform distribution on $\{R_0 - \varepsilon \leqslant \rho \leqslant R_2\}$. The slight decrease of $D_1$ for $\varphi = 0°$ could probably be explained by the crude estimate of the importance function $f^*(l)$.

In conclusion, let us make some remarks about the importance-sampling technique for simulating the process by which a photon reaches the Earth's surface.

Let $P_a$ be the albedo of the Earth's surface (i.e. the probability of reflection) and let $f_0^*$ be the main importance of a reflected photon. Let $f_1^*$ be the main importance of a collision within an atmosphere and let $\tau_0$ be the optical length of the path of the photon to the Earth (it is supposed that the photon moves to the Earth). The physical probability that the photon reaches the Earth's surface is $\exp\{-\tau_0\}$. By the importance-sampling technique, we obtain that the process by which the photon reachs the Earth's surface must be simulated with probability

$$\frac{f_0^* \exp(-\tau_0)\cdot P_a}{f_1^*[1 - \exp(-\tau_0)] + f_0^* P_a \exp(-\tau_0)},$$

which can be easily calculated, provided that $f_0^* = f_1^*$. The weight is then multiplied by $[1 - (1 - P_a)\exp(-\tau_0)]P_a^{-1}$. This method significantly improves the estimate of the contribution of multiply scattered radiation. This modification seems to be useful when the observation point is situated in the illuminated region.

Notice that the importances of absorbed photons are zero. Therefore, the absorption must be simulated rarely. The photon weight is then multiplied by $\sigma_s/\sigma$, where $\sigma_s$ is the scattering coefficient, and $\sigma$ is the total cross section.

## 4.5 Simulation of Adjoint Trajectories

It follows from $I_\varphi = (f, \varphi) = (f^*, \psi)$ (see Sect. 3.1) that the functional $I_\varphi$ can be estimated by simulation of the adjoint transfer equation. In Sect. 2.7 it was shown that

$$I_\varphi = \int_R \int_\Omega f_1^*(\boldsymbol{r}, \boldsymbol{\omega}) \frac{\Phi_0(\boldsymbol{r}, -\boldsymbol{\omega})}{\sigma(\boldsymbol{r})} d\boldsymbol{r}\, d\boldsymbol{\omega}, \tag{4.34}$$

where $f_1^*$ is a collision density that corresponds to the transfer equation with the source density $p_1(\boldsymbol{r}, \boldsymbol{\omega}) = \varphi(\boldsymbol{r}, -\boldsymbol{\omega})\sigma(\boldsymbol{r})$, and $\Phi_0(\boldsymbol{r}, \boldsymbol{\omega})$ is the source distribution density in the direct transfer equation. The expression (4.34) gives an algorithm for estimating the functional

$$I_\varphi = E\xi, \quad \xi = \sum_{n=0}^{N} Q_n \psi_1(\boldsymbol{x}_n), \quad \psi_1(\boldsymbol{x}) = \frac{\Phi_0(\boldsymbol{r}, -\boldsymbol{\omega})}{\sigma(\boldsymbol{r})};$$

i.e., in order to estimate $I_\varphi$, it is necessary to simulate the transfer process from a source that has the density $p_1(\boldsymbol{r}, \boldsymbol{\omega})$, and to calculate $\psi_1(\boldsymbol{x}_n)$ at each collision. Consider unit flux of collimated solar radiation incident on the spherical-shell atmosphere in the direction $\boldsymbol{\omega}^{(0)}$. Then the distribution density of first-scattered photons is

$$\Phi_0(\boldsymbol{x}) = \sigma(\boldsymbol{r}) \frac{g(\mu)}{2\pi} \cdot \frac{1}{\pi R_2^2} \exp\{-\tau(l(\boldsymbol{r})\},$$

where $R_2$ is exterior radius of the atmosphere, $1/\pi R_2^2$ is the distribution density in the plane $YZ$ (it is supposed that a single photon is incident on the surface), $g(\mu)/2\pi$ is the distribution density of a new direction $\boldsymbol{\omega}$, $\mu = \boldsymbol{\omega}^{(0)} \cdot \boldsymbol{\omega}$. As is seen from (see also Sect. 2.7)

$$I_0(\boldsymbol{r}^*) = \int_\Omega \Phi(\boldsymbol{r}^*, \boldsymbol{\omega})\, d\boldsymbol{\omega} = \int_\Omega \int_R \Phi(\boldsymbol{r}, \boldsymbol{\omega})\delta(\boldsymbol{r} - \boldsymbol{r}^*)\, d\boldsymbol{r}\, d\boldsymbol{\omega},$$

the net flux at the observation point corresponds to the function $p_1(\boldsymbol{x}) = \delta(\boldsymbol{r} - \boldsymbol{r}^*)$. Consequently, in order to calculate the integral intensity $I_0(\boldsymbol{r}^*)$ of multiply scattered sunlight at a particular point $\boldsymbol{r}^*$, sample the random directions of trajectories (originating at a point $\boldsymbol{r}^*$) from the isotropic density $p_1(\boldsymbol{x})/4\pi$, and calculate, at each collision point $\boldsymbol{x}_n$,

$$\psi_1 = \frac{\Phi_0(\boldsymbol{x}_n)}{\sigma(\boldsymbol{r}_n)} \cdot 4\pi = \frac{2e^{-\tau_0} g(-\boldsymbol{\omega}^{(0)}, \boldsymbol{\omega})}{\pi R_2^2}, \tag{4.35}$$

where $\tau_0$ is the optical length of a straight path from the point $\boldsymbol{r}_n$ to the boundary of the atmosphere in the direction $-\boldsymbol{\omega}^0$. The relation

$$I(\boldsymbol{r}^*, \boldsymbol{\omega}^*) = \Phi(\boldsymbol{r}^*, \boldsymbol{\omega}^*) = \int_R \int_\Omega \Phi(\boldsymbol{r}, \boldsymbol{\omega})\delta(\boldsymbol{r} - \boldsymbol{r}^*)\delta(\boldsymbol{\omega} - \boldsymbol{\omega}^*)\, d\boldsymbol{r}\, d\boldsymbol{\omega},$$

shows that the calculation of photon flux at a point $x^* = (r^*, \omega^*)$ of phase space corresponds to the function $p_1(x) = \delta(r - r^*)\delta(\omega - (-\omega^*))$. This implies that to calculate the intensity of multiply scattered light at the point $(r^*, \omega^*)$, it is necessary to simulate the photon trajectories from $r^*$ in the direction $-\omega^*$ and, at each collision point, to calculate

$$\psi_1 = \frac{\Phi_0(x_n)}{\sigma(r_n)} = \frac{e^{-\tau_0} g(-\omega^{(0)}, \omega)}{2\pi^2 R_2^2}. \tag{4.36}$$

It is more difficult to obtain the corresponding estimate when polarization is taken into account, because there is no theorem of optical mutuality for this case. However an estimate can be obtained by of use the integral transfer equation in a coordinate system whose center is at the observation point $r^*$ [23]. As mentioned in Sect. 4.4, the new photon direction $\omega'$, after the scattering at point $r$ can be sampled from the density $p_{11}(\omega', \omega, r)$ that corresponds to the scattering function for unpolarized light; the photon weight is then divided by the value of this function, and the Stokes vector is multiplied by the scattering matrix $P(\omega', \omega, r)$. Thus, to estimate the Stokes vector at a point $(r^*, \omega^*)$, it is necessary to estimate

$$\psi_1 = \frac{e^{-\tau_0}}{2\pi^2 R_2^2} P(\omega_1, -\omega^*, r_1) \times P(\omega_2, \omega_1, r_2) \times P(\omega_{n-1}, \omega_{n-2}, r_{n-1}) \times P(\omega^{(0)}, -\omega_{n-1}, r_n) \cdot I_0,$$

at each collision point of the adjoint trajectory. Here $\omega_{n-1}$ is the direction of motion of the photon before the collision at point $r_n$, $P$ is the scattering matrix (see Sect. 4.3), $I_0 = (1, 0, 0, 0)$ is the Stokes vector for the incident radiation.

Consequently, it is necessary to multiply two matrices at each collision point. The advantages of the method of adjoint-walk simulation are:

1) there is no need for preliminary calculation of the importance of the initial states of trajectories;
2) trajectories leave the observation points in desired directions; and
3) the estimates (4.35, 36) have finite variances.

However, the simulated trajectory yields only a single functional, whereas local-calculation methods (see Sect. 4.1) yield the contribution from each collision point, for several observation points $r_i^*$, $i = 1, \ldots, n_x$ and several lines of sight $\omega_j^*$, $j = 1, \ldots, n_\theta$. In order to compare the efficiencies of the method of adjoint-walk simulation and the improved double local estimate (see Sect. 4.1), the intensities of multiply scattered solar radiation at different observation points with $\psi = 88°$, $60°$; $H = 0$, $250$ km ($\psi$ = solar zenith angle, $H$ = altitude) have been calculated. The atmosphere was assumed to be inhomogeneous with altitude, the optical thickness $\tau$ (measured vertically) of the atmosphere was taken to be 0.38. The phase function was assumed to be constant (i.e., not dependent on altitude), and the albedo of the Earth's surface was taken to be zero. Calculations show that, if only a few functionals are desired, the method of adjoint-walk simulation is more effective than the improved double local estimate. For example, calculations of the intensities for 5 receiver viewing directions at $\psi = 88°$, $H = 250$ km and azimuth $\varphi = 180°$ by using the method of adjoint-walk simulation and the double-local-estimate method, required 11 and 30 minutes of BESM-6 computer time, respectively. However, when a large number of functionals have to be evaluated, the computer time of the method of adjoint-walk simulation for calculating the intensity at the same point but for 40 viewing directions is approximately four times as great as with the double-local-estimate method. Therefore, the method of adjoint-walk simulation has a significant advantage when used to calculate a few functionals with great accuracy. A simple analysis shows that the adjoint-walk method is more effective for an illuminated region than for a twilight one. Numerical experiments show that in calculations of twilight problems, the first free-path length is preferably sampled from the density $1/L_0$, $0 \leqslant l \leqslant L_0$.

Here, $L_0$ is the distance from the point $\boldsymbol{r}^*$ to the boundary of the atmosphere in the direction $-\boldsymbol{\omega}^*$. This can be explained, probably, by the fact that in other modifications (see Sect. 4.5) the first-collision points are concentrated near the Earth's surface, and, therefore, do not contribute to the functional. In this modification, the weight is multiplied by $L_0 \exp[-\tau(\boldsymbol{r}^*, \boldsymbol{r})]\sigma(\boldsymbol{r})$. For directions of view situated in an illuminated part of the atmosphere, it is convenient to choose, for the free-path length, the physical simulation without escape (see Sect. 4.4). The photon weight is then multiplied by $1 - \exp(-\tau_0)$. Absence of the coefficient $\sigma(\boldsymbol{r})$ from these weight factors, which strongly vary with altitude $h$, explains why this modification is more effective than simulation according to the density $L_0^{-1}$. This was corroborated by calculations of the intensity for $\psi = 88°$, $\varphi = 0°$; and $\psi = 60°$, $\varphi = 0°$ and $180°$. For example, the computer time required for the sampling the density $1/L_0$, to calculate five functionals at the point $\psi = 88°$, $H = 250$ km and $\varphi = 0°$ is approximately 16 times as great as for physical simulation without escape.

## 4.6 Method of Dependent Sampling and Evaluation of Derivatives

For the sake of simplicity, let us consider the dependent-sampling method for explicit simulation of the physical process of transfer in a basic system (i.e., for $\lambda = \lambda_0$).

Various integral characteristics of the transfer process can be represented by linear functionals of the solution of the transfer equation

$$I_\varphi = (f, \varphi) = \int_X f(\boldsymbol{x})\varphi(\boldsymbol{x})\,d\boldsymbol{x} = \sum_{k=0}^{\infty} (K^n\psi, \varphi). \tag{4.37}$$

It follows from (4.37) that, to estimate the functional by the Monte Carlo method, it is necessary to average sums of the contributions $\varphi(\boldsymbol{x})$ from collisions of various orders. The dependent-sampling method for solving radiative-transfer problems consists of photon-path simulations that use the same trajectories for different systems; biases that arise are eliminated by introducing special weight factors.

Let the wavelength $\lambda$ be a parameter of a system, so that $k(\boldsymbol{x}, \boldsymbol{x}') = k(\boldsymbol{x}, \boldsymbol{x}', \lambda)$, $\varphi(\boldsymbol{x}) = \varphi(\boldsymbol{x}, \lambda) \equiv \varphi_\lambda$.
Then

$$I_\varphi = \sum_{n=0}^{\infty} (K_\lambda^n \psi, \varphi_\lambda).$$

Consider the relation

$$\begin{aligned}(K_\lambda^n \psi, \varphi_\lambda) &= \overbrace{\int \cdots \int}^{n+1} \psi(\boldsymbol{x}_0) k(\boldsymbol{x}_0, \boldsymbol{x}_1, \lambda), \ldots, k(\boldsymbol{x}_{n-1}, \boldsymbol{x}, \lambda) \\ &\quad \times \varphi(\boldsymbol{x}, \lambda) \cdot d\boldsymbol{x}_0 d\boldsymbol{x}_1, \ldots, d\boldsymbol{x}_{n-1} d\boldsymbol{x} \\ &= \overbrace{\int \cdots \int}^{n+1} \psi(\boldsymbol{x}_0) k(\boldsymbol{x}_0, \boldsymbol{x}_1, \lambda_0), \ldots, k(\boldsymbol{x}_{n-1}, \boldsymbol{x}, \lambda_0) \\ &\quad \times \frac{k(\boldsymbol{x}_0, \boldsymbol{x}_1, \lambda)}{k(\boldsymbol{x}_0, \boldsymbol{x}_1, \lambda_0)} \cdots \frac{k(\boldsymbol{x}_{n-1}, \boldsymbol{x}, \lambda)}{k(\boldsymbol{x}_{n-1}, \boldsymbol{x}, \lambda_0)} \varphi(\boldsymbol{x}, \lambda) d\boldsymbol{x}_0 d\boldsymbol{x}_1, \ldots, d\boldsymbol{x}_{n-1} d\boldsymbol{x}.\end{aligned}$$

The trajectories constructed for $\lambda = \lambda_0$ can be used to estimate $I_\varphi(\lambda)$, provided that after each transition $\boldsymbol{x} \to \boldsymbol{x}'$ the photon weight is multiplied by

$$\frac{k(\boldsymbol{x}, \boldsymbol{x}', \lambda)}{k(\boldsymbol{x}, \boldsymbol{x}', \lambda_0)}.$$

We assumed that there are no points $\boldsymbol{x}$, $\boldsymbol{x}'$ for which $k(\boldsymbol{x}, \boldsymbol{x}', \lambda) \neq 0$, when $k(\boldsymbol{x}, \boldsymbol{x}', \lambda_0) = 0$. In practice, $k(\boldsymbol{x}, \boldsymbol{x}', \lambda)$ is represented as a product of conditional probability densities of some elementary random variables (free-path length, direction of scattering, etc.), and, after each elementary sampling, the photon weight is multiplied by the ratio of the appropriate probability densities for $\lambda$ and $\lambda_0$. For example, after the next free-path length is sampled, the weight must be multiplied by

$$\frac{\sigma(\boldsymbol{r}', \lambda) \exp\left[-\int_0^{|\boldsymbol{r}'-\boldsymbol{r}|} \sigma(\boldsymbol{r} + \boldsymbol{\omega}' t, \lambda)\, dt\right]}{\sigma(\boldsymbol{r}', \lambda_0) \exp\left[-\int_0^{|\boldsymbol{r}'-\boldsymbol{r}|} \sigma(\boldsymbol{r} + \boldsymbol{\omega}' t, \lambda_0)\, dt\right]} = \frac{\sigma(\boldsymbol{r}', \lambda)}{\sigma(\boldsymbol{r}', \lambda_0)} \exp\{-[\tau(\boldsymbol{r}, \boldsymbol{r}', \lambda) - \tau(\boldsymbol{r}, \boldsymbol{r}', \lambda_0)]\};$$

i.e.

$$Q_n(\lambda) = Q_{n-1}(\lambda)\frac{\sigma(\boldsymbol{r}', \lambda)}{\sigma(\boldsymbol{r}', \lambda_0)} \exp\{-[\tau(\boldsymbol{r}, \boldsymbol{r}', \lambda) - \tau(\boldsymbol{r}, \boldsymbol{r}', \lambda_0)]\}.$$

After the cosine of the scattering angle ($\mu = \boldsymbol{\omega}\cdot\boldsymbol{\omega}'$) is chosen, we have

$$Q_n(\lambda) = Q_{n-1}(\lambda)\frac{g(\mu, \boldsymbol{r}, \lambda)}{g(\mu, \boldsymbol{r}, \lambda_0)}.$$

The method of dependent sampling makes it possible to evaluate also the sensitivity of the radiation field for small changes of (i) the coefficient of aerosol scattering, (ii) the albedo, or (iii) the phase function.

Consider a quantity $I_k = I_k(\lambda) = I_{\varphi_k}(\lambda)$ that depends on a parameter $t$. Then

$$I_k(t) = \sum_{n=0}^{\infty} \overbrace{\int \cdots \int}^{n+1} \psi(\boldsymbol{x}_0) \prod_{p=0}^{n-1} k(\boldsymbol{x}_p, \boldsymbol{x}_{p+1}, t_0, \lambda_0)$$
$$\times \prod_{p=0}^{n-1} \frac{k(\boldsymbol{x}_p, \boldsymbol{x}_{p+1}, t, \lambda)}{k(\boldsymbol{x}_p, \boldsymbol{x}_{p+1}, t_0, \lambda_0)} \cdot \varphi_k(\boldsymbol{x}_n, t, \lambda)\, d\boldsymbol{x}_0 d\boldsymbol{x}_1, \ldots, d\boldsymbol{x}_{n-1} d\boldsymbol{x}_n.$$

Suppose that it is desired to estimate $\partial I_k/\partial t|_{t=t_0}$. Assume that the series can be differentiated termwise, and that the differentiation can be performed under the integral signs. For $t = \sigma_a$ ($\sigma_a$ = coefficient of aerosol scattering), we now show that the derivative of the series, obtained in that manner, has an absolutely convergent dominating function that does not depend on $\sigma_a$.

Formally, we have, for the present

$$\frac{\partial}{\partial t}\left[\psi(\boldsymbol{x}_0) \prod_{p=0}^{n-1} k(\boldsymbol{x}_p, \boldsymbol{x}_{p+1}, t_0, \lambda_0) \prod_{p=0}^{n-1} \frac{k(\boldsymbol{x}_p, \boldsymbol{x}_{p+1}, t, \lambda)}{k(\boldsymbol{x}_p, \boldsymbol{x}_{p+1}, t_0, \lambda_0)} \cdot \varphi_k(\boldsymbol{x}_n, t, \lambda)\right]\Bigg|_{t=t_0}$$
$$= \psi(\boldsymbol{x}_0)\left[\prod_{p=0}^{n-1} k(\boldsymbol{x}_p, \boldsymbol{x}_{p+1}, t_0, \lambda_0)\right] Q_n(\lambda, t_0)\varphi_k(\boldsymbol{x}_n, t_0, \lambda)$$
$$\times \left[\frac{\partial \ln \varphi_k(\boldsymbol{x}_n, t, \lambda)}{\partial t} + \frac{\partial \ln [Q_n(\lambda, t)]}{\partial t}\right)\Bigg]_{t=t_0}$$
$$= \psi(\boldsymbol{x}_0)\left[\prod_{p=0}^{n-1} k(\boldsymbol{x}_p, \boldsymbol{x}_{p+1}, t_0, \lambda_0)\right] Q_n(\lambda, t_0)\varphi_k(\boldsymbol{x}_n, t_0, \lambda)\, \Psi_n(\lambda, t_0);$$

where

$$\Psi_n(\lambda, t) = \frac{\partial \ln \varphi_k(\boldsymbol{x}_n, t, \lambda)}{\partial t} + \sum_{p=0}^{n-1} \frac{\partial \ln k(\boldsymbol{x}_p, \boldsymbol{x}_{p+1}, t, \lambda)}{\partial t}. \tag{4.38}$$

Before calculating the derivatives of the local estimate, we introduce the notation:

$\sigma_a(\boldsymbol{r}, \lambda)$: coefficient of aerosol scattering with phase function $g_a(\mu, \boldsymbol{r}, \lambda)$,
$\sigma_M(\boldsymbol{r}, \lambda)$: coefficient of molecular scattering with phase function $g_M(\mu)$,
$\sigma_c(\boldsymbol{r}, \lambda)$: absorption coefficient,
$\mu = \omega \cdot \omega'$: cosine of the angle between the directions of the photon previous to, and after, the collision.
$\sigma(\boldsymbol{r}, \lambda) = \sigma_a + \sigma_M + \sigma_c$: total cross section,

$$g(\mu, \boldsymbol{r}, \lambda) = \frac{g_a(\mu, \boldsymbol{r}, \lambda)\sigma_a(\boldsymbol{r}, \lambda) + g_M(\mu)\sigma_M(\boldsymbol{r}, \lambda)}{\sigma(\boldsymbol{r}, \lambda)} \text{: total indicatrix,}$$

$$\tau(\boldsymbol{r}_n, \boldsymbol{r}_k, \lambda) = \int_0^{|\boldsymbol{r}_n - \boldsymbol{r}_k|} \sigma(\boldsymbol{r}_n + \omega_k l, \lambda)dl\text{: optical length of the path from } \boldsymbol{r}_n \text{ to } \boldsymbol{r}_k,$$

where

$$\omega_k = (\boldsymbol{r}_k - \boldsymbol{r}_n)/|\boldsymbol{r}_k - \boldsymbol{r}_n|.$$

Consider photon transfer in a medium that has these specified characteristics. Assume that an estimate of the integral of intensity over directions at a given point $\boldsymbol{r}^*$, is desired. It is known (see Sect. 3.8) that in this case,

$$\begin{aligned}\varphi_k^* &= \varphi_k^*(\boldsymbol{r}_n, \omega, \lambda)\\ &= c_1 \frac{\exp[-\tau(\boldsymbol{r}_n, \boldsymbol{r}^*, \lambda)][g_a(\mu^*, \boldsymbol{r}_n, \lambda)\sigma_a(\boldsymbol{r}_n, \lambda) + g_M(\mu^*)\sigma_M(\boldsymbol{r}_n, \lambda)]}{\sigma(\boldsymbol{r}_n, \lambda)},\end{aligned}$$

and

$$\begin{aligned}k(\boldsymbol{x}_p, \boldsymbol{x}_{p+1}, \lambda) = c_2 \cdot \frac{\sigma(\boldsymbol{r}_{p+1}, \lambda)}{\sigma(\boldsymbol{r}_p, \lambda)} \exp\{-\tau(\boldsymbol{r}_p, \boldsymbol{r}_{p+1}, \lambda)\}\\ \times [g_a(\mu_p, \boldsymbol{r}_p, \lambda)\sigma_a(\boldsymbol{r}_p, \lambda) + g_M(\mu_p)\sigma_M(\boldsymbol{r}_p, \lambda)],\end{aligned}$$

where

$$\mu_p = \omega_{p-1} \cdot \omega_p.$$

Further,

$$I_k(\lambda) = E \sum_{n=0}^{N} Q_n(\boldsymbol{r}_n, \lambda)\varphi_k^*(\boldsymbol{r}_n, \lambda).$$

Assume that the atmosphere is divided into $n_l$ layers each of which has a constant coefficient of aerosol scattering. Introduce the notation:

$$\sigma(\boldsymbol{r}, \lambda) = \sigma(m, \lambda), \quad g(\mu, \boldsymbol{r}, \lambda) = g(\mu, m, \lambda),$$

if $h_m < |r| \leqslant h_{m+1}$, $m = 0, 1, \ldots, n_i - 1$,

$$\psi_m^k = \begin{cases} 1 & \text{if} \quad m = k, \\ 0 & \text{if} \quad m \neq k, \end{cases}$$

$\mu_n^*$: cosine of the angle between the particle direction of the photon before the $n$th collision at the point $\boldsymbol{r}_n$ and the direction $\omega^* = (\boldsymbol{r}^* - \boldsymbol{r}_n)/|\boldsymbol{r}^* - \boldsymbol{r}_n|$;

$N(n)$: index of the layer that contains the point of collision $\boldsymbol{r}_n$;

$m(n, i)$: indexes of the layers intersected by the straight path of a photon between $\boldsymbol{r}_{n-1}$ and $\boldsymbol{r}_n$; $i(n)$ is the number of those intersections, and $l_{n,i}$ are the corresponding pieces of the straight path;

$m^*(n, i)$, $i^*(n)$, $l_{n,i}^*$ are the same quantities for the straight path between $\boldsymbol{r}_n$ and $\boldsymbol{r}^*$;

$L_{n,m}$: length of the trajectory (inside the $m$th layer) of a photon moving from $\boldsymbol{r}_{n-1}$ to $\boldsymbol{r}_n$; and

$L_{n,m}^*$: the same quantity for the pair $\boldsymbol{r}_n$, $\boldsymbol{r}^*$.

We now estimate the derivatives with respect to the coefficient of aerosol scattering. In this case, $t = \sigma_a(m, \lambda)$, $t_0 = \sigma_a^{(0)}(m, \lambda)$. Denote also

$$\tilde{\sigma}_a(m, \lambda) = \sigma_a(m, \lambda) - \sigma_a^{(0)}(m, \lambda).$$

We have

$$\left.\frac{\partial I_k}{\partial t}\right|_{t=t_0} = \left.\frac{\partial I_k}{\partial \sigma_a(m, \lambda)}\right|_{\sigma_a(m,\lambda)=\sigma_a^{(0)}(m,\lambda)} = \left.\frac{\partial I_k}{\partial \tilde{\sigma}_a(m, \lambda)}\right|_{\tilde{\sigma}_a(m,\lambda)=0}.$$

Further,

$$\begin{aligned} \varphi_k^*(\boldsymbol{r}_n, \omega_n^*, \sigma_a, \lambda) &= c_1 \\ &\times \frac{\sigma_M^{(0)}[N(n), \lambda] g_M(\mu_n^*) + \{\sigma_a^{(0)}(N(n), \lambda) + \psi_{N(n)}^m \tilde{\sigma}_a[N(n), \lambda]\} g_a[\mu_n^*, N(n), \lambda]}{\sigma_M^{(0)}[N(n), \lambda] + \sigma_a^{(0)}[N(n), \lambda] + \psi_{N(n)}^m \cdot \tilde{\sigma}_a(m, \lambda)} \\ &\times \exp\left[-\sum_{i=1}^{i^*(n)} l_{n,i}^* \{\sigma_M^{(0)}[m^*(n, i), \lambda] + \sigma_a^{(0)}[m^*(n, i), \lambda] + \psi_{m^*(n,i)}^m \cdot \sigma_a(m, \lambda)\}\right]. \end{aligned}$$

Thus,

$$\begin{aligned} \frac{\partial \ln \varphi_k^*}{\partial \tilde{\sigma}_a(m, \lambda)} &= -L_{n,m}^{(1)} + \frac{\psi_{(Nn)}^m g_a(\mu_n^*, N(n), \lambda)}{\sigma_M^{(0)}[N(n), \lambda] g_M(\mu_n^*) + \sigma_a^{(0)}[N(n), \lambda] g_a[\mu_n^*, N(n), \lambda]} \\ &\quad - \frac{\psi_{N(n)}^m}{\sigma_M^{(0)}[N(n), \lambda] + \sigma_a^{(0)}[N(n), \lambda]}, \end{aligned} \tag{4.39}$$

because

$$\sum_{i=1}^{i^*(n)} l^*_{n,i}\psi^m_{m^*(n,i)} = L^*_{n,m}.$$

Now

$$\frac{\partial \ln Q_n(\lambda)}{\partial \tilde{\sigma}_a} = \sum_{p=1}^{n-1} \frac{\partial \ln k(\boldsymbol{x}_p, \boldsymbol{x}_{p+1}, \sigma_a, \lambda)}{\partial \tilde{\sigma}_a(m, \lambda)}$$

$$- \frac{\partial \ln \{\sigma_M^{(0)}[N(n), \lambda] + \sigma_a^{(0)}[N(n), \lambda] + \psi^m_{N(n)}\tilde{\sigma}_a[N(n), \lambda]\}}{\partial \tilde{\sigma}_a(m, \lambda)}$$

$$- \sum_{p=1}^{n} \frac{\partial}{\partial \tilde{\sigma}_a(m, \lambda)} \left\{ \sum_{i=1}^{i(p)} l_{p,i} \left[ \sigma_M^{(0)}[m(p, i), \lambda] + \sigma_a^{(0)}[m(p, i), \lambda] + \psi^m_{m(p,i)}\tilde{\sigma}_a(m, \lambda) \right] \right\}$$

$$+ \sum_{p=1}^{n-1} \frac{\partial \ln \left[\sigma_M^{(0)}[N(p), \lambda] g_M(\mu_p) + (\sigma_a^{(0)}[N(p), \lambda] + \psi^m_{p(N)}\tilde{\sigma}_a(m, \lambda)) g_a[\mu_p, N(p), \lambda]\right)\right]}{\partial \tilde{\sigma}_a(m, \lambda)}.$$

Finally,

$$\left.\frac{\partial \ln Q_n(\lambda)}{\partial \tilde{\sigma}_a(m, \lambda)}\right|_{\tilde{\sigma}_a(m,\lambda)=0} = \frac{\psi^m_{N(n)}}{\sigma_M^{(0)}[N(n), \lambda] + \sigma_a^{(0)}[N(n), \lambda]} - \sum_{p=1}^{n} L_{p,m}$$
$$+ \sum_{p=1}^{n-1} \frac{\psi^m_{N(p)} g_a[\mu_p, N(p), \lambda]}{\sigma_M^{(0)}[N(p), \lambda] g_M(\mu_p) + \sigma_a^{(0)}[N(p), \lambda] g_a[\mu_p, N(p), \lambda]}. \tag{4.40}$$

Combining (4.38, 39, 40), we obtain

$$\left.\frac{\partial I_k}{\partial \tilde{\sigma}_a(m, \lambda)}\right|_{\tilde{\sigma}_a(m,\lambda)=0} = E \sum_{n=1}^{N} Q_n(\boldsymbol{r}_n, \sigma_a^{(0)}, \lambda)\varphi^*_k(\boldsymbol{r}_n, \omega^*_n, \sigma_a^{(0)}, \lambda)\Psi_n(\lambda), \tag{4.41}$$

where

$$\Psi_n(\lambda) = -L^*_{n,m} - \sum_{p=1}^{n} L_{p,m}$$
$$+ \frac{\psi^m_{N(n)} g_a[\mu^*_n, N(n), \lambda]}{\sigma_M^{(0)}[N(n), \lambda] g_M(\mu^*_n) + \sigma_a^{(0)}[N(n), \lambda] g_a[\mu^*_n, N(n), \lambda]}$$
$$+ \sum_{p=1}^{n-1} \frac{\psi^m_{N(p)} g_a[\mu_p, N(p), \lambda]}{\sigma_M^{(0)}[N(p), \lambda] g_M(\mu_n) + \sigma_a^{(0)}[N(p), \lambda] g_a[\mu_p, N(p), \lambda]}.$$

We now attempt to justify the differentiation. For an arbitrary $\sigma_a \in (\sigma_a^{(0)} - \varepsilon, \sigma_a^{(0)} + \varepsilon)$, we have

$$\frac{\partial I_k(\sigma_a)}{\partial \sigma_a} = E \sum_{n=1}^{N} Q_n(\boldsymbol{r}_n, \sigma_a, \lambda)\varphi^*_k(\boldsymbol{r}_n, \omega^*_n, \sigma_a, \lambda)\Psi_n(\lambda).$$

For a bounded medium [i.e., $|\boldsymbol{r}_{n-1} - \boldsymbol{r}_n| < c < \infty$, and $\sigma(\boldsymbol{r}, \lambda) < c_2 < \infty$] it is easy to show that

$$Q_n(\boldsymbol{r}_n, \sigma_a, \lambda)\varphi_k^*(\boldsymbol{x}_n, \sigma_a, \lambda)\Psi_n(\lambda) \leqslant cq_\varepsilon^n n Q_n(\boldsymbol{r}_n, \sigma_a^{(0)}, \lambda)\varphi_k^*(\boldsymbol{x}_n, \sigma_a^{(0)}\lambda),$$

where $q_\varepsilon \to 1$ as $\varepsilon \to 0$.

If $\|K\| < 1$, and $\varepsilon$ is such that $\|K\| \cdot q_\varepsilon < 1$, then the derivative is dominated by a function that does not depend on a parameter and has a finite expectation. Therefore, the integral of the derivative with respect to probability measure converges uniformly if $|\sigma_a - \sigma_a^{(0)}| < \varepsilon$.

Notice that it is also possible to differentiate the intensity with respect to the indicatrix (provided that it is either piecewise constant or piecewise linear), the albedo, and the absorption coefficients.

## 4.7 Formulation of Inverse Atmospheric-Optics Problems

In a series of physical and technical problems, it is often desired to determine basic numerical characteristics of the process under study, where the precise mathematical statement of the problem is assumed known, and consequently a main description of the operator $L$, the coefficients of the equation, and a set of solutions, are given. The values of the coefficients of the equation that characterizes the operator $L$ are not known. The problem consists, then, of determining those coefficients, provided that some functionals of the problem are given; for example, the instrument readings. Such problems are usually called inverse problems. Even if a precise mathematical formulation is given, such problems are often ill posed in Cauchy's sense, i.e., a small variation of the initial data leads to unbounded increase of variation in the desired functions.

Moreover, there are often no solutions in terms of an appropriate class of functions. As a typical example, consider the Fredholm integral equation of the first kind:

$$\int_a^b k(x, t)\varphi(t)\,dt = u(x), \quad c \leqslant x \leqslant d. \tag{4.42}$$

This equation does not have a solution for an arbitrary function $u(x)$. For example, if $k(x, t) \in C^p(X)$ (the class of functions with continuous $p$th order derivatives with respect to $x$) for some $p$, then there is no function $\varphi(t) \in L_2$ that satisfies (4.42) if $u(x)$ is not sufficiently smooth. Sometimes it is useful to assume that the solution has derivatives of desired order. Approximate solutions obtained under this additional assumption are called regularized solutions, and the corresponding problems are called conditionally correct problems.

Consider some inverse atmospheric-optics problems that have been formulated by *Marchuk* [24]. Let us consider an equation in the operator form

$$Lf = \psi, \tag{4.43}$$

where $\psi, f \in F$.

Introduce a scalar product by $(g, h) = \int g(x)h(x)dx$ where the integration is performed over $\mathscr{D}$, the domain of definition of functions $h \in F$, $g \in F^*$, and $x$ is a set of variables of the problem (e.g., spatial and time coordinates, or velocity directions).

Consider also an operator $L^*$ adjoint to operator $L$. $L^*$ is defined by

$$(g, Lh) = (L^*g, h), \tag{4.44}$$

for arbitrary functions $g \in F^*$, $h \in F$. Introduce (for the present, formally) an equation

$$L^*f_p^* = p(x), \tag{4.45}$$

where $p(x)$ is, for the present, an arbitrary function, and $f_p^* \in F^*$.

Replacing $h$ and $g$ in (4.44) with the solutions of (4.43, 45) $f$ and $f_p^*$, we get $(f_p^*, Lf) = (f, L^*f_p^*)$ or $(f_p^*, \psi) = (f, p)$. This implies that $I_p(f)$ can be evaluated in two ways: either by solving (4.43) and using the formula $I_p(f) = (f, p)$ or by solving (4.45) and using

$$I_p(f) = I_\psi^*(f_p^*) = (f_p^*, \psi).$$

Assume that the operators $L$ and $L^*$ are uniquely defined by a set of parameters $\{\alpha_i\}$, $i = 1, \ldots, n$. Suppose that if these parameters are changed $\alpha_i' = \alpha_i + \delta\alpha_i$, then $L' = L + \delta L$, $f(x) \to f'(x)$, $I_p(f) \to I_p' = I_p + \delta I_p$. We now determine the connection between the variation $\delta L$ and the variation $\delta I_p$. The initial and changed parameters, operators, solutions, and functionals are called nonperturbed and perturbed, respectively. Consider a perturbed problem,

$$L'f' = (L + \delta L)f' = \psi, \tag{44.6}$$

and a nonperturbed adjoint problem,

$$L^*f_p^* = p. \tag{4.47}$$

Scalar multiplication of (4.46) by $f_p^*$ and of (4.47) by $f'$, and subtraction of one from the other yield

$$\delta I_p = -(f_p^*, \delta Lf'), \tag{4.48}$$

where

$$\delta I_p = (f', p) - (f_p^*, \psi) = I_p(f') - I_p(f).$$

It is here supposed that the function $\psi$ is not perturbed; otherwise, (4.48) takes the form

$$(f_p^*, \delta Lf' - \delta\psi) = -\delta I_p. \tag{4.49}$$

Formulation and solution of the inverse problems are based on (4.48) and (4.49). Assume that the solution of the nonperturbed problem (4.43) is known (and that, consequently, $I_{p_k}(f)$ are known), and that a set of functionals $I_{p_k}(f')$, $k = 1, \ldots, m$ of the solution of the perturbed problem are given. Then the right-hand sides of $(f_{p_k}^*, \delta Lf' - \delta\psi) = -\delta I_p$, $k = 1, 2, \ldots, m$ are known. If we consider the operator $L$ and the function $\psi$ to be linearly dependent on $\alpha_i$, then

$$\delta Lf' - \delta\psi = \sum_{i=1}^{n} \delta\alpha_i(A_i f' - \xi_i),$$

where $A_i$ are known operators, and $\xi_i$ are the given functions. Finally, we get

$$\sum_{i=1}^{n} a_{ik}\delta\alpha_i = -b_k,$$

$$a_{ik} = (\varphi_{pk}^*, A_i f' - \xi_i), \quad b_k = \delta I_{pk}, \quad k = 1, 2, \ldots, m.$$

To complete the definition of this system (whose coefficients depend on the unknown solution $f'$ of the perturbed problem), the method of successive approximations may be used or, if the perturbations are small, $f'$ may be merely replaced with $f$. We shall consider both of these approaches to determining the altitude dependence of the coefficient of aerosol scattering in Sect. 4.8. The technique just described can be easily generalized to a nonlinear operator $L$ by use of linearization (for example by Newton's method).

We now turn to a discussion of some inverse atmospheric-optics problems. Consider an equation of transfer in a spherical atmosphere in the integral form (see Sect. 2.5) $f = Kf + \psi$. For simplicity, assume the radiation to be monochromatic. As functionals of the problem, let us consider instrument readings of radiation intensity. That is, we know the integrals of the radiation intensity at some spatial positions over a solid angle that corresponds to the receiver angle of view,

$$I_p(f) = \int_{\mathscr{D}} \int_{\Delta\Omega} f\xi\delta(\boldsymbol{r} - \boldsymbol{r}_0)\, d\boldsymbol{r}\, d\boldsymbol{\Omega},$$

where $\xi$ is some given function of the instrument.

We shall try to finnd the coefficients of aerosol scattering in various layers of the atmosphere. Assume that the scattering coefficients are constant in each layer. That is, we want to obtain $\boldsymbol{\sigma} = (\sigma_1, \ldots, \sigma_n)$. In this case, $\psi$ is nonperturbed ($\psi$ is an external source, namely, the sun). The quantity $\delta Lf = \delta(f - Kf)$ can be expanded in Taylor's series:

$$\delta(f - Kf) \approx \sum_{i=1}^{n} \frac{\partial}{\partial \sigma_i}(f - K'f)\bigg|_{\boldsymbol{\sigma}=\boldsymbol{\sigma}_0} \Delta\sigma_i, \tag{4.50}$$

where $\Delta\boldsymbol{\sigma} = \boldsymbol{\sigma} - \boldsymbol{\sigma}_0$. Substituting (4.50) in (4.48) yields

$$\begin{aligned}(f_{p_k}^*, \delta Lf) &= \left(f_{p_k}^*, \sum_{i=1}^{n} \frac{\partial}{\partial \sigma_i}(f - K'f)\bigg|_{\boldsymbol{\sigma}=\boldsymbol{\sigma}_0} \Delta\sigma_i\right) \\ &= \sum_{i=1}^{n} \left(f_{p_k}^*, \frac{\partial}{\partial \sigma_i}(f - K'f)\bigg|_{\boldsymbol{\sigma}=\boldsymbol{\sigma}_0} \Delta\sigma_i\right), \quad k = 1, 2, \ldots, n,\end{aligned}$$

and finally

$$\sum_{i=1}^{n} \frac{\partial I_k}{\partial \sigma_i} \Delta\sigma_i = \delta I_{p_k} \quad \text{or} \quad A\Delta\boldsymbol{\sigma} = \boldsymbol{b}. \tag{4.51}$$

Thus, the system (4.51) is, generally speaking, unsolvable ($m > n$). Therefore, the problem is to find most suitable values of $\Delta\sigma_i$.

It is known that if the quantity $\|A\Delta\boldsymbol{\sigma} - \boldsymbol{b}\|^2 = \delta^2$ is minimized, we get a system of so-called normal equations: $A^*A\Delta\boldsymbol{\sigma} = A^*\boldsymbol{b}$. Equations (4.50) are called conditional equations. It is supposed that the measurement errors have a normal distribution with small standard deviation. Otherwise statistical weights must be introduced, whereby the equations are multiplied by quantities inversely proportional to the standard deviations of the measured quantities. Because the relative standard deviations of estimates of functionals usually fluctuate slowly, the functionals themselves can be taken as the statistical weights. They can be interpreted, then, as regularization coefficients:

$$A^*WA\Delta\sigma = A^*W\boldsymbol{b}. \tag{4.52}$$

This system can be solved by various methods, e.g., by Gauss's method, or the method of minimal discrepancies.

Since the system (4.52) is often ill posed, it is necessary to apply a regularization method due to *Tichonov* [25]. In this method, the expression

$$\|A^*WA\Delta\boldsymbol{\sigma} - A^*W\boldsymbol{b}\| + \alpha\Omega[\Delta\boldsymbol{\sigma}] = \delta,$$

is minimized, where $\Omega[\Delta\boldsymbol{\sigma}] = \Omega\Delta a \cdot \Delta\boldsymbol{\sigma}$, $\Omega$ is an approximation (in matrix form) for

$$\int_0^H \left| \sum_{k=1}^m q_k(x) \frac{d^k\sigma}{dx^k} \right| dx.$$

Here $H$ is the altitude in the atmosphere, $q_k(x) > 0$. It is usually assumed that for $k = 1$, $q_k(x) \equiv 1$, i.e., that the derivative of the solution $\Delta\sigma(x)$ is bounded. The regularization coefficient $\alpha$ can be obtained approximately from various arguments. For instance, in [26], $\alpha$ was taken to be

$$\alpha = \frac{n}{\Omega\Delta\sigma \cdot \Delta\sigma}, \tag{4.53}$$

where $n$ is the number of dimensions. However, because $\alpha$ depends on the unknown solution, the quantity $\Delta\boldsymbol{\sigma}$ in (4.53) is replaced by the right-hand side of the system. Then

$$\alpha = \frac{n}{A^* W\boldsymbol{b} \cdot \Omega A^* W\boldsymbol{b}},$$

It should be noted that the regularization method was applied by *Turchin* [27] to solve some inverse problems in a class of statistical functions.

## 4.8 Numerical Solution of the Inverse Problem of Determining the Altitude Dependence of the Coefficient of Aerosol Scattering

As in Sect. 4.7, we consider a formulation of inverse problems due to *Marchuk* [24]. Assume that there is a set of measured functionals $I_k(\sigma_1, \sigma_2, \ldots, \sigma_n) = (f, \varphi_k)$, $k = 1, \ldots, n_\theta$, where $f$ is the solution of the transfer equation $Lf = \psi$. We want to evaluate $(\sigma_1, \ldots, \sigma_n)$. If $(\sigma_1^{(0)}, \ldots, \sigma_n^{(0)})$ are initial estimates of these parameters, then from the perturbation theory we have

$$\sum_{i=1}^n a_{ik}\delta\sigma_i = \tilde{I}_k - I_k^{(0)},$$

provided that $L$ is linearly dependent on $\sigma_i$. Here, $I_k^{(0)} = I_k(\sigma_1^{(0)}, \ldots, \sigma_n^{(0)})$.

If the dependence is not linear, the problem can be solved by the method of successive approximations, in which small-perturbation formulas are used. In this case, the coefficients $a_{ik}$ are given by

$$a_{ik} = \frac{\partial I_k}{\partial \sigma_i}, \quad k = 1, 2, \ldots, n_\theta; \quad i = 1, 2, \ldots, n.$$

The Monte Carlo calculation of the derivatives is described in Sect. 4.6.

It is convenient to introduce weights that are inversely proportional to $I_n^{(0)}$. We then have the system to be solved,

$$\sum_{i=1}^{r} \frac{\partial I_k}{\partial \sigma_i} \cdot \frac{1}{I_k^{(0)}} \Delta\sigma_i = \ln \tilde{I}_k - \ln I_k^{(0)}, \quad k = 1, \ldots, n_\theta. \tag{4.54}$$

Notice that (4.54) can be regarded as the result of linearization of

$$\ln I_k(\sigma_1, \ldots, \sigma_n) = \ln \tilde{I}_k, \quad k = 1, 2, \ldots, n_\theta,$$

at the point $(\sigma_1^{(0)}, \sigma_2^{(0)}, \ldots, \sigma_n^{(0)})$. The method of least squares must be used to solve (4.54), because this system is overdetermined.

This method was used to solve the model inverse problem of finding the coefficients of aerosol scattering. An atmosphere of altitude 100 km was divided into 3 layers: 0–10 km, 10–40 km and 40–100 km, each having constant coefficients of aerosol scattering $\sigma_1$, $\sigma_2$, $\sigma_3$, respectively. As functionals $I_k$, integrals of the radiation intensity over the latitude angle $\theta$ lying in the plane $\varphi = 0°$, or $\varphi = \pi$ have been used. The index $k$ is defined as the index of the corresponding angular interval, as shown in Fig. 4.5.

The quantities $\boldsymbol{I}_k$ $(\sigma_1, \sigma_2, \sigma_3)$, $k = 1, \ldots, n_\theta$ (experimental results) were calculated from the optical mean values $\sigma_1 = 0.0149$, $\sigma_2 = 0.000623$, $\sigma_3 = 0.000357$ that were calculated for a smooth model $\sigma_a(h, \lambda_0)$. Notice that $\boldsymbol{I}_k$ also have random errors (as arise in measurements or experiments).

Indicatrices computed by use of a formula due to *Rosenberg* and *Gorĉakov* [51] were tabulated. The albedo was taken to be 0.845 (for snow) and 0.0545 (for grass, $\lambda = 0.55$ µm). Calculations were done for 3 different initial estimates for $\sigma$: $\sigma_1^{(0)} = 0.03$, $\sigma_2^{(0)} = 0.0003$, $\sigma_3^{(0)} = 0.00007$; for the line of sight toward the sun ($\psi = 0$) and away from the sun ($\psi = \pi$); and for two receiver positions $x_c = 0$ and $x_c = z_c$ at 400 km altitude.

Calculations show that the best results are obtained for the $\theta$ subdivision shown in Fig. 4.6. Examination of the derivatives $\partial I_k/\partial \sigma_i$ shows that for estimation of $\sigma_2$, $\sigma_3$, the functionals $I_6$, $I_7$, respectively, are the most informative, and functional $I_2$ (or $I_2$) for $\sigma_1$. Results of calculations are shown in Table 4.1, where

$$\delta\sigma_i^{(m)} = (\sigma_i^{(m)} - \sigma_i)/\sigma_i,$$

$\sigma_i^{(m)}$ = result of the $m$th iteration of the method of successive approximations; $(\partial I_{ki}/\partial \sigma_i) \cdot \Delta\sigma_i^{(m)}$ shows the variation of $k_i$th functional that is most informative for $\sigma_i$;

$$\Delta\sigma_i^{(m)} = \sigma_i^{(m)} - \sigma_i^{(m+1)};$$

$[D(\tilde{I}_{ki} - I_{ki}^{(m)})]^{1/2}$ is the standard deviation of $\tilde{I}_{ki} - I_{ki}^{(m)}$; and $P_a$ = albedo.

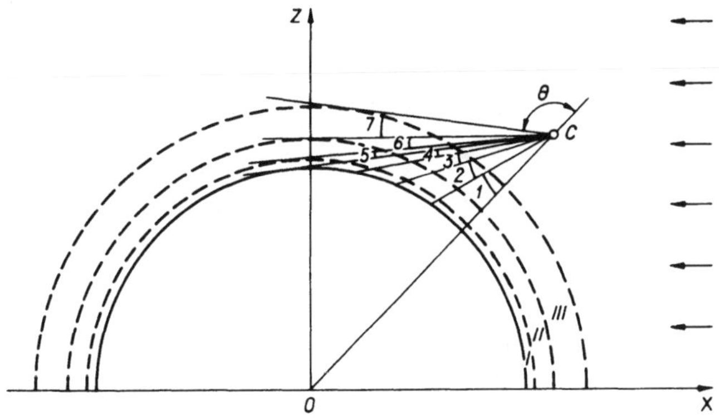


**Fig. 4.6.** Geometry of the model inverse problem

The cited results show that $\sigma_1$ is better determined for small $P_a$ than for large $P_a$. Numerical results show that, to obtain $\sigma_i$ to within 10%, $I_{ki}$ must be determined approximately to within 5%.

## 4.9 Simulation of Some Radiation Experiments by Taking into Account Absorption by Atmospheric Gases

In this section, an attempt will be made to obtain information about the radiation field of the atmosphere by a comparison of computational results with experimental data. Such information will be useful in formulation and solution of various inverse atmospheric-optics problems (determination of altitude dependence of the aerosol density, calculation of molecular components, etc.). On the basis of algorithms and programs described in this book, we consider a solution of some problems that arise in real observations of infrared radiation scattered in the atmosphere. Consider solar radiation with net flux of energy $E(\lambda)$ in a wavelength region $2 \leqslant \lambda \leqslant 3.2\ \mu m$ incident on the spherical-shell atmosphere in the direction $-\boldsymbol{x}$.

The atmosphere is divided into layers each having a constant coefficient of aerosol scattering. The phase function of the aerosol scattering $g(h_i, \mu_k)$ is tabulated for $h_i$ and $\mu_k$, the height and cosine of the scattering angle, respectively (linear interpolation is used between these points). Coordinates of the observation points were calculated by the formula

$$x_c = r_c \sin h_0, \quad z_c = r_c \cos h_0. \tag{4.55}$$

Here, $r_c$ is the distance from the center of the Earth to the observation point, and $h_0$ is the solar zenith angle. Calculations were carried out simultaneously

**Table 4.1.** Numerical solutions of model inverse problems

| Variant | Iteration | $i$ | $\delta\sigma_i^{(m)}$ | $K_i$ | $\frac{\partial I_{K_i}}{\partial \sigma_i} \cdot \Delta\sigma_i^{(m)}$ | $[\mathscr{D}(I_{K_i} - I_{K_i}^{(m)})]^{1/2}$ | $\frac{\frac{\partial I_{K_i}}{\partial \sigma_i} \cdot \Delta\sigma_i^{(m)}}{[\mathscr{D}(I_{K_i} - I_{K_i}^{(m)})]^{1/2}}$ | $\frac{\sqrt{\mathscr{D} I_{K_i}}}{I_{K_i}}$ | $P_a$ | $\varphi$ [rad.] | $x_c$ [km] | $z_c$ [km] | time [min] Computer BESM-6 |
|---|---|---|---|---|---|---|---|---|---|---|---|---|---|
| | | 1 | 0.68 | 2 | 0.000070392 | 0.000184 | 0.3826 | 0.0091 | | | | | |
| 1 | 1 | 2 | 0.12 | 6 | 0.000105 | 0.00001714 | 6.126 | 0.0185 | 0.854 | $\pi$ | 4787.82 | 4787.82 | 60 |
| | | 3 | 0.11 | 7 | 0.000117 | 0.0000124 | 9.435 | 0.072 | | | | | |
| | | 1 | 0.03 | 2 | 0.00127 | 0.0001776 | 7.151 | 0.024 | | | | | |
| 2 | 1 | 2 | 0.13 | 6 | 0.0000898 | 0.00003518 | 2.553 | 0.045 | 0.0545 | $\pi$ | 4787.82 | 4787.82 | 20 |
| | | 3 | 0.44 | 7 | 0.000138 | 0.00002414 | 5.717 | 0.08 | | | | | |
| | | 1 | 0.03 | 4 | 0.000548 | 0.0001824 | 3.004 | 0.02 | | | | | |
| 2 | 2 | 2 | 0.04 | 6 | 0.0000734 | 0.0000314 | 2.360 | 0.0397 | 0.0545 | $\pi$ | 4787.82 | 4787.82 | 15 |
| | | 3 | 0.22 | 7 | 0.0000245 | 0.0000156 | 1.571 | 0.137 | | | | | |
| | | 1 | 0.20 | 4 | 0.002098 | 0.0000676 | 31.036 | 0.01 | | | | | |
| 3 | 1 | 2 | 0.03 | 6 | 0.0000623 | 0.0001315 | 0.4737 | 0.023 | 0.0545 | 0 | 0 | 6771 | 20 |
| | | 3 | 0.27 | 7 | 0.0000387 | 0.0000045 | 8.6 | 0.043 | | | | | |

for several directions of view, and for different azimuthal angles $\varphi$. If $\varphi$ is fixed, the direction of view is defined by the latitude angle $\theta$; $\theta = \pi/2$ defines a horizontal direction (geometric horizon). Contributions from collision points that have not sent photons into the predetermined solid angle of observation were ignored. This technique reduces the computer time significantly. For comparison, the computed results were multiplied by scale factors

$$X(\lambda) = \frac{\pi \cdot E(\lambda)(R_0 + h)^2}{\Delta\theta},$$

where $R_0$ is the radius of the Earth, $h$ is the altitude of the atmosphere, and $\Delta\theta$ is the solid angle of view. Calculations were performed both for a clear and cloudy sky. In the latter case, reflection from clouds was sampled using Lambert's reflection law, and the photon weight after reflection was multiplied by

$$\frac{\mu a_g(\mu, \varphi)}{2\mu/2\pi} = \pi a_g(\mu, \varphi).$$

A differential albedo $a_g(\mu, \varphi)$ for clouds was obtained by the following approximation to the experimental data,

$$a_g(\mu, \varphi) = a_g(\mu, 0) - c(1 - \mu)\varphi, \quad \text{if} \quad \varphi \leqslant 70^\circ,$$

$$a_g(\mu, \varphi) = \text{const}, \qquad \text{if} \quad \varphi > 70^\circ,$$

so that

$$\int_0^{2\pi} d\varphi \int_0^1 \mu a_g(\mu, \varphi)\, d\mu = 1.$$

Here $\varphi$ is the azimuthal reflection angle, $\mu$ is the cosine of the angle between the radius vector of the reflection point and the direction of reflection, $c$ is a constant that depends on the solar zenith angle. It is easy to see by geometric arguments that

$$\varphi = \arccos \frac{cx - az}{[b^2(x^2 + z^2) + (az - cx)^2]^{1/2}} \quad \text{if} \quad cx - az \geqslant 0,$$

$$\varphi = \arccos \frac{az - cx}{[b^2(x^2 + z^2) + (az - cx)^2]^{1/2}} + \frac{\pi}{2} \quad \text{if} \quad cx - az < 0, \tag{4.56}$$

where $(a, b, c)$ is a unit vector of motion of the photon after reflection from clouds, $(x, 0, z)$ is the point at which the photon is reflected by the cloud.

Anisotropy of albedo was taken into account only for reflection of unscattered photons. In (4.56), there is no $y$ coordinate, because, owing to the axial symmetry of the system, the first photon runs were simulated in the plane $y = 0$.

The trajectories were simulated at a fictitious wavelength for which the appropriate physical characteristics of atmosphere were assumed to have small statistical errors at various altitudes.

On the basis of the importance-sampling technique, a photon that reaches the Earth was simulated with probability $\exp(-\tau_0)\cdot P_a[1 - \exp(-\tau_0) + \exp(-\tau_0)\cdot P_a]^{-1}$, where $\tau_0$ is the optical length of the path to the Earth's surface (it is assumed that the photon is directed downward). The initial coordinates were simulated in a similar manner.

For twilight conditions, the atmospheric model $\exp[-\tau_0(z)] = \text{const}$ was used, where $R_0 \leqslant z \leqslant R_0 + h$, $\tau_0(z)$ is the optical depth of the atmosphere at the height $z$ and in a direction perpendicular to the axis $z$. Calculations were made by use of the dependent-sampling method, simultaneously for fourteen wavelengths.

Absorption by atmospheric gases was taken into account by use of the transmission function that depends on the trajectory

$$P_1(\lambda) = \exp\left\{-\sum_{k=1}^{N} \beta_k(\lambda)\left[\int_{(L)} q_k'(\lambda, z)\,dl\right]^{m_k(\lambda)}\right\}, \tag{4.57}$$

where $k$ is the serial number of the atmospheric gas ($N = 5$, in the order: $CO$, $H_2O$, $N_2O$, $CO_2$, $CH_4$), $q_k'(\lambda, z)$ is the effective concentration of $k$th gas at the height $z$, $\beta_k(\lambda)$ and $m_k(\lambda)$ are piecewise-constant functions that correspond to a given meteorological condition, $(L)$ is the photon trajectory from the source to the receiver. Of particular interest are the intensity components $I_a$, due to photons reflected from the Earth (or clouds) without scattering; $H_1$, due to photons that have undergone a single scattering in the atmosphere before reaching the Earth (or clouds); $H$, due to photons that have undergone a single scattering in the atmosphere after being reflected from the Earth (or clouds)-so-called subillumination; and $I_1$ due to photons that have undergone no more than a single scattering.

Computed values of these components and experimental data for comparison with them are shown in Tables 4.2–7. The computed results in Tables 4.2, 3 agree satisfactorily with the experimental data ($I_e$). The poor agreement of the computed results with the experimental data in Table 4.3 was probably caused by the use of a crude atmospheric model. Calculations for a more precise model agree well with the experimental data.

The calculations (to within 10%) for horizontal directions of view, for 13 observation points and for 14 wavelengths, simultaneously, required 60–90 minutes of BESM-6 computer time. For other directions of view the accuracy is less (Table 4.3).

The dependence of the intensity in twilight conditions on the cloudiness ($I_{max}$ = value of the intensity for height 17 km and for a cloudy sky, $\sigma$ is the standard deviation in the present calculations) is shown in Table 4.5.

**Table 4.2.** Observations above 12 km; $\varphi = 20°$, $h_0 = 22–28°$

| Height [km] | $I/I_e$ [%] | $I_a/I$ [%] | $I_1/I$ [%] | $H_1/I$ [%] | $H/I$ [%] | $\sigma/I$ [%] | Height [km] | $I/I_e$ [%] | $I_a/I$ [%] | $I_1/I$ [%] | $H_1/I$ [%] | $H/I$ [%] | $\sigma/I$ [%] |
|---|---|---|---|---|---|---|---|---|---|---|---|---|---|
| Sight below the geometric horizon | | | | | | | Sight toward the geometric horizon (geometric-horizon sight) | | | | | | |
| 13 | 0.12 | 99.4 | 0.6 | 0 | 0 | 47.0 | 13 | 7.12 | 0 | 94.5 | 76.5 | 18.1 | 19.2 |
| 15 | 0.41 | 52.8 | 47.1 | 46.9 | 0 | 34.7 | 15 | 4.21 | 0 | 90.6 | 85.3 | 5.3 | 10.9 |
| 17 | 0.71 | 19.2 | 80.3 | 78.9 | 0.25 | 57.1 | 17 | 3.42 | 0 | 96.1 | 90.5 | 5.6 | 9.1 |
| 19 | 0.27 | 54.1 | 45.6 | 38.0 | 5.5 | 15.7 | 19 | 2.73 | 0 | 95.9 | 90.2 | 5.7 | 4.8 |
| 21 | 0.42 | 44.9 | 54.5 | 49.7 | 2.3 | 8.3 | 21 | 2.78 | 0 | 98.5 | 93.5 | 4.9 | 5.8 |
| 23 | 0.43 | 49.5 | 49.5 | 38.2 | 8.8 | 11.5 | 23 | 2.13 | 0 | 97.3 | 93.1 | 4.1 | 6.6 |
| 25 | 0.59 | 52.7 | 45.4 | 39.9 | 2.8 | 8.7 | 25 | 1.16 | 0 | 95.4 | 91.3 | 4.0 | 5.8 |
| 27 | 0.57 | 55.2 | 43.3 | 37.9 | 2.7 | 5.6 | 27 | 0.95 | 0 | 99.5 | 92.8 | 6.7 | 7.9 |
| 29 | 0.57 | 51.4 | 47.5 | 40.5 | 4.7 | 6.7 | 29 | 0.65 | 0 | 99.4 | 94.0 | 5.4 | 8.8 |
| 31 | 0.43 | 63.9 | 35.9 | 28.2 | 4.5 | 10.9 | 31 | 0.43 | 0 | 99.2 | 83.6 | 15.6 | 13.1 |

**Table 4.3.** Observations above 4 km; $\varphi = 20°$, $h_0 = 25°$

| Height [km] | $I/I_e$ [%] | $I_a/I$ [%] | $I_1/I$ [%] | $H_1/I$ [%] | $H/I$ [%] | $\sigma/I$ [%] |
|---|---|---|---|---|---|---|
| Sight below the geometric horizon | | | | | | |
| 5 | — | 0 | 0 | — | — | — |
| 7 | — | 51.8 | 44.4 | — | — | 36.6 |
| 9 | — | 57.2 | 7.9 | — | — | 34.7 |
| 11 | — | 22.2 | 64.7 | 50.0 | 8.8 | 36.7 |
| 13 | 6.20 | 7.6 | 77.5 | 73.4 | 1.4 | 26.4 |
| 15 | 4.30 | 3.4 | 21.2 | 19.8 | 0.4 | 50.7 |
| 17 | 3.14 | 8.7 | 86.2 | 70.2 | 13.6 | 15.2 |
| 19 | 2.85 | 7.6 | 74.8 | 68.6 | 3.4 | 12.2 |
| 21 | 2.73 | 7.9 | 79.5 | 67.8 | 9.0 | 10.5 |
| 23 | 2.07 | 9.6 | 69.3 | 58.9 | 7.4 | 13.4 |
| 25 | 1.19 | 9.2 | 76.5 | 67.2 | 6.4 | 11.1 |
| 27 | 0.91 | 10.0 | 79.4 | 68.4 | 8.2 | 7.2 |
| 29 | 0.71 | 7.0 | 75.0 | 64.1 | 7.4 | 15.4 |
| Geometric-horizon sight | | | | | | |
| 5 | — | 0 | 84.5 | 80.5 | 4.1 | 17.9 |
| 7 | — | 0 | 92.8 | 90.3 | 2.5 | 21.5 |
| 9 | — | 0 | 91.6 | 88.2 | 3.4 | 16.7 |
| 11 | — | 0 | 90.8 | 85.4 | 5.4 | 11.6 |
| 13 | 0.76 | 0 | 92.3 | 86.7 | 5.7 | 8.3 |
| 15 | 2.58 | 0 | 92.3 | 82.2 | 10.0 | 10.4 |
| 17 | 1.10 | 0 | 88.3 | 85.9 | 2.4 | 5.9 |
| 19 | 1.23 | 0 | 93.3 | 90.5 | 2.9 | 6.9 |
| 21 | 1.38 | 0 | 95.1 | 91.5 | 3.6 | 6.0 |
| 23 | 1.33 | 0 | 93.8 | 91.9 | 1.9 | 6.1 |
| 25 | 1.75 | 0 | 96.1 | 93.5 | 2.6 | 5.7 |
| 27 | 1.83 | 0 | 96.0 | 93.2 | 2.8 | 7.0 |
| 29 | 2.82 | 0 | 95.3 | 92.8 | 2.4 | 6.5 |

Table 4.4. Observations above 4 km; $\varphi = 150°$, $h_0 = 25°$

| Height [km] | $I_a/I$ [%] | $I_1/I$ [%] | $H_1/I$ [%] | $H/I$ [%] | $\sigma/I$ [%] |
|---|---|---|---|---|---|
| | | Sight below the geometric horizon | | | |
| 5 | 0 | 10.0 | 0 | 0 | 99.4 |
| 7 | 13.3 | 81.4 | 13.0 | 60.0 | 60.8 |
| 9 | 11.0 | 75.9 | 59.4 | 4.4 | 20.6 |
| 11 | 9.4 | 80.2 | 46.3 | 19.8 | 18.0 |
| 13 | 7.2 | 87.9 | 35.9 | 45.7 | 41.4 |
| 15 | 7.3 | 85.6 | 47.6 | 30.5 | 25.4 |
| 17 | 7.5 | 58.3 | 41.6 | 10.2 | 28.6 |
| 19 | 7.7 | 77.8 | 44.8 | 26.3 | 13.9 |
| 21 | 5.2 | 57.3 | 37.4 | 15.2 | 28.9 |
| 23 | 8.0 | 64.7 | 44.4 | 14.1 | 16.9 |
| 25 | 9.6 | 75.4 | 38.9 | 29.7 | 25.6 |
| 27 | 12.1 | 73.2 | 50.5 | 14.2 | 9.8 |
| 29 | 10.9 | 77.8 | 58.7 | 10.1 | 7.3 |
| | | Geometric-horizon sight | | | |
| 5 | — | 54.8 | 49.6 | 5.1 | 22.0 |
| 7 | — | 66.6 | 61.8 | 4.7 | 16.7 |
| 9 | — | 65.5 | 53.2 | 12.2 | 23.2 |
| 11 | — | 80.8 | 69.4 | 11.4 | 8.1 |
| 13 | — | 71.4 | 62.4 | 9.1 | 14.7 |
| 15 | — | 83.2 | 74.0 | 9.1 | 7.5 |
| 17 | — | 89.6 | 74.3 | 15.3 | 6.7 |
| 19 | — | 86.8 | 73.4 | 13.3 | 8.9 |
| 21 | — | 87.6 | 74.4 | 13.1 | 8.9 |
| 23 | — | 84.7 | 66.9 | 17.9 | 7.8 |
| 25 | — | 84.9 | 74.1 | 10.8 | 7.1 |
| 27 | — | 89.3 | 75.2 | 14.1 | 6.6 |
| 29 | — | 84.1 | 69.0 | 15.1 | 6.3 |

**Table 4.5.** Observations in twilight; $\varphi = 0$, $h_0 = -8°$

| Observation height [km] | Geometric horizon | | | Above the geometric horizon | | |
|---|---|---|---|---|---|---|
| | $I/I_{max}$ [%] | $I_1/I$ [%] | $\sigma/I$ [%] | $I/I_{max}$ [%] | $I_1/I$ [%] | $\sigma/I$ [%] |
| | Clouds at 10 km | | | | | |
| 10 | 69.7 | 74.3 | 21.8 | 2.75 | 96.9 | 14.6 |
| 11 | 90.1 | 80.8 | 15.7 | 2.68 | 96.9 | 15.9 |
| 13 | 92.0 | 96.0 | 5.8 | 2.68 | 99.7 | 15.2 |
| 15 | 92.2 | 99.3 | 6.1 | 2.56 | 99.7 | 18.5 |
| 17 | 100.0 | 99.4 | 6.6 | 2.64 | 99.9 | 18.9 |
| 19 | 96.3 | 98.9 | 6.0 | 2.50 | 99.8 | 21.2 |
| 21 | 92.5 | 98.8 | 5.9 | 2.58 | 99.8 | 21.8 |
| 23 | 88.7 | 99.5 | 5.6 | 2.84 | 99.3 | 17.6 |
| 25 | 80.1 | 98.5 | 5.6 | 2.41 | 98.7 | 15.8 |
| 27 | 73.8 | 97.3 | 5.3 | 2.43 | 92.5 | 16.0 |
| 29 | 72.9 | 97.0 | 5.7 | 2.74 | 93.1 | 15.1 |
| 31 | 66.0 | 97.8 | 7.3 | 2.77 | 99.8 | 15.3 |
| 33 | 55.2 | 99.1 | 7.4 | 2.56 | 99.9 | 16.4 |
| | Clear sky | | | | | |
| 0 | 0.05 | 66.1 | 18.7 | 1.86 | 95.2 | 9.3 |
| 3 | 8.3 | 85.9 | 8.5 | 3.65 | 99.1 | 8.4 |
| 5 | 41.7 | 78.4 | 12.4 | 4.04 | 99.7 | 9.6 |
| 7 | 89.4 | 88.2 | 9.6 | 4.11 | 93.5 | 13.1 |
| 9 | 102.7 | 95.8 | 6.8 | 3.15 | 97.9 | 14.5 |
| 11 | 123.8 | 97.4 | 6.1 | 2.98 | 97.5 | 12.0 |
| 13 | 184.0 | 81.8 | 13.3 | 4.69 | 67.3 | 27.5 |
| 15 | 169.1 | 94.8 | 16.1 | 3.52 | 99.1 | 12.2 |
| 17 | 182.4 | 94.8 | 5.8 | 3.24 | 98.7 | 13.5 |
| 19 | 182.2 | 94.5 | 6.7 | 3.01 | 98.3 | 15.5 |
| 25 | 143.9 | 98.8 | 7.8 | 2.72 | 99.4 | 12.5 |
| 31 | 117.4 | 99.1 | 7.5 | 2.51 | 98.9 | 17.0 |
| 35 | 84.8 | 98.6 | 9.3 | 1.98 | 96.3 | 13.8 |

**Table 4.6.** Observations above 10 km; $\varphi = 20°$

| Height [km] | $(\sigma_2 - \sigma_1)\cdot 10^{-3}$ [cm$^{-1}$] | $(\sigma_3 - \sigma_1)\cdot 10^{-3}$ [cm$^{-1}$] | $(\sigma_4 - \sigma_1)\cdot 10^{-3}$ [cm$^{-1}$] | $(\sigma_5 - \sigma_1)\cdot 10^{-3}$ [cm$^{-1}$] |
|---|---|---|---|---|
| 11 | 0 | −2.78 | −2.78 | 0 |
| 13 | 0 | −1.14 | −1.14 | 0 |
| 15 | 0 | −0.344 | −0.344 | 0 |
| 17 | 0 | −0.508 | −0.508 | 0 |
| 19 | 0 | −0.504 | −0.504 | 0 |
| 21 | 0 | −0.177 | −0.177 | 0 |
| 23 | 0 | −0.111 | −0.111 | 0 |
| 25 | 0 | −0.067 | −0.067 | 0 |
| 27 | 0 | −0.022 | −0.022 | 0 |
| 29 | 0 | −0.0024 | −0.0024 | 0 |
| 31 | −0.0168 | 0 | −0.00003 | −0.00003 |
| 33 | −0.0207 | 0 | 0.008 | 0.008 |
| 35 | −0.0149 | 0 | 0.0173 | 0.0173 |
| 37 | −0.0127 | 0 | 0.0265 | 0.0265 |
| 41 | −0.0083 | 0 | 0.0304 | 0.0304 |
| 54 | 0.0005 | 0 | 0.0424 | 0.0424 |
| 82 | 0.052 | 0 | 0.05 | 0.05 |

**Table 4.7.** Observations above 10 km; $\varphi = 20°$, $h_0 = 18$–$20°$

| Height [km] | $\frac{I(\sigma_2)}{I(\sigma_1)}$ [%] | $\frac{I(\sigma_3)}{I(\sigma_1)}$ [%] | $\frac{I(\sigma_4)}{I(\sigma_1)}$ [%] | $\frac{I(\sigma_5)}{I(\sigma_1)}$ [%] | Height [km] | $\frac{I(\sigma_2)}{I(\sigma_1)}$ [%] | $\frac{I(\sigma_3)}{I(\sigma_1)}$ [%] | $\frac{I(\sigma_4)}{I(\sigma_1)}$ [%] | $\frac{I(\sigma_5)}{I(\sigma_1)}$ [%] |
|---|---|---|---|---|---|---|---|---|---|
| | Geometric-horizon sight | | | | | Sight above the geometric horizon | | | |
| 11 | 99.7 | 27.2 | 28.3 | 101.0 | 11 | 100.0 | 35.8 | 68.9 | 125.9 |
| 13 | 99.5 | 37.7 | 41.2 | 102.6 | 13 | 103.4 | 39.4 | 149.9 | 208.1 |
| 15 | 99.3 | 33.2 | 43.9 | 109.2 | 15 | 96.2 | 44.6 | 148.4 | 202.4 |
| 17 | 98.6 | 32.1 | 50.5 | 116.4 | 17 | 97.0 | 48.3 | 160.6 | 212.9 |
| 19 | 97.8 | 35.9 | 66.3 | 128.7 | 19 | 100.5 | 45.2 | 188.3 | 241.4 |
| 21 | 95.6 | 39.3 | 80.4 | 139.4 | 21 | 97.0 | 59.3 | 256.7 | 296.8 |
| 23 | 94.3 | 48.4 | 111.9 | 162.2 | 23 | 100.5 | 70.5 | 392.3 | 422.2 |
| 25 | 91.7 | 60.7 | 138.2 | 176.9 | 25 | 99.2 | 72.7 | 325.9 | 352.0 |
| 27 | 83.8 | 83.8 | 233.5 | 249.9 | 27 | 98.9 | 92.9 | 558.2 | 565.3 |
| 29 | 75.7 | 96.9 | 282.3 | 285.8 | 29 | 63.0 | 100.2 | 506.8 | 501.0 |
| 31 | 46.2 | 99.6 | 343.9 | 345.4 | 31 | 103.5 | 100.2 | 860.2 | 860.8 |
| 33 | 57.1 | 99.4 | 436.9 | 438.9 | 33 | 155.4 | 102.0 | 1350.1 | 1314.2 |
| 35 | 70.5 | 99.5 | 556.7 | 558.5 | 35 | 132.3 | 100.3 | 1137.9 | 1113.9 |

The dependence of the radiance on various coefficients of aerosol scattering (see Table 4.6) is shown in Table 4.7.

## 4.10 Numerical Investigation of Radiation-Field Characteristics in a Spherical-Shell Atmosphere

In this section, we study the scattered and absorbed radiance for the Earth's atmosphere. The problem of constructing adequate (statistically justified) models of the Earth's atmosphere is one of the most fundamental problems in atmospheric optics. Such a model must be based on numerous and various qualitative and quantitative comparisons between experimental data and computational results.

Computational results can be obtained by solving the direct problems of atmospheric optics, that is, to establish connections, by solving the transfer equation, between the characteristics of the field of scattered radiation and parameters of the optics-meteorological atmospheric model.

Therefore, we have concentrated our attention on calculations of spectral, angular, and spatial distributions of intensity and polarization, as well as on their sensitivity towards changes of some parameters of the atmospheric model. In order to investigate more precisely the applicability of the single-scattering approximation, it is interesting to study the contribution due to multiply scattered radiation.

We present here estimates of multiple-scattering contributions to the total radiance for various illumination and observation conditions. It is also important to study the errors that occur in calculations of the intensity of scattered radiation, due to neglect of polarization. These errors can be investigated by comparison between the solutions of the transfer equation with and without taking polarization into account. Such a comparison has been carried out in detail by classical methods only for pure Rayleigh scattering in the atmosphere. For example, in [28] it is shown that the error that results from neglect of polarization increases with increase of optical depth in a Rayleigh-scattering atmosphere. However for realistic models of the atmosphere, there are no estimates of these errors, because the scattering matrixes for light scattered in the atmosphere and for reflections from various ground surfaces are not known sufficiently accurately. Furthermore, the lack of such studies can be explained by the difficulties that arise when the transfer equation that takes polarization into account is solved by classical methods of computational mathematics.

A simplified atmospheric model is considered with the following features:

1) light scattering is assumed to be isotropic,

2) $\varphi_0$, the albedo for single scattering (probability of photon's survival after a single scattering) is taken to be 1 and 0.5,

3) the extinction coefficient dependence on the height $h$ is taken to be $\sigma(h) = \sigma(0) \exp(-h/H_0)$, where $H_0 = 8$ km,

4) the total optical thickness of the atmosphere in the vertical direction is taken to be $\tau_0 = \sigma(0)H_0 = 0.1, 0.2, 0.3, 0.5, 1$,
5) $P_a$, the albedo for the Earth's surface is taken to be 0.2 and 0.8.
The results of the solution of this problem are of particular interest because:
i) They reflect the potential of the Monte Carlo method, the accuracy of which does not depend significantly on the characteristics of the radiation model.
ii) These results can be used as standards for comparisons with results obtained by other techniques of calculation of multiply scattered light in spherical-shell atmospheres.
iii) The physical dependences were obtained for a simple atmospheric model that does not include any unspecified event.

In the calculations, the coefficient $\sigma(h)$ was approximated by a step function. The atmosphere of altitude 100 km was divided into 36 layers (from 0 to 6 km with step $\Delta h = 1$ km; 6–15 km, $\Delta h = 1.5$ km; 15–35 km, $\Delta h = 2$ km; 35–100 km, $\Delta h = 5$ km) each having a constant $\sigma(h)$ so that the optimal thickness of atmosphere coincided with that specified.

The intensity was calculated for solar zenith angles of observation: $\psi = 30$, 60, 80, 82, 84, 86, 87, 89, 91, 92, 93, 94, 96°;
a) for $H = 0$ ($H$ = height of the observation point), the radiance of the sky was calculated for the zenith direction (in direction $\omega_0$) and at the horizon, for azimuth angles $\varphi = 0°$ ($\omega_1$) and $\varphi = 180°$ ($\omega_2$),
b) for $H = 250$ km, the radiance of the nadir ($\omega_0$) and horizon was calculated for the same azimuth angles (azimuth angle $\varphi$ is measured from the sun's vertical).
The radiance was calculated by use of the dependent-sampling method, simultaneously for all values of $\psi$, $\varphi$, $P_a$, $\tau_0$, $\varphi_0$. For these parameters, the intensity of singly and doubly scattered light was also calculated. The results are given in $S$ units of measurement, where $\pi S$ is the solar constant (i.e., the incident solar radiation at the top of the atmosphere). Therefore the Monte Carlo results were multiplied by the quantity $\pi^2 R_2^2$, where $R_2 = 6471$ km is the exterior radius of the atmosphere. All calculations for this problem were performed in 120 minutes of BESM-6 computer time. The use of the importance-sampling technique for simulating the first coordinates of the photon trajectory (see Sect. 4.4) significantly reduced the computation time. The relative errors of the estimates of intensity are less than 10%, except for some observation variants. Comparatively large errors were obtained for the direction of view $\omega_2$, $H = 0$, for the twilight region.

Let us comment on the computational results. Radiance as a function of $\psi$ and $\omega$ is shown in Table 4.8. The radiance of the sky decreases smoothly as the solar elevation $\psi$ increases to 90°: at $H = 0$, for three directions of view, and at $H = 250$ km for directions $\omega_0$ and $\omega_2$; for $\psi$ near 90°, the intensity rapidly decreases (becomes one-third normal), then it decreases smoothly as $\psi$ approaches 94°–95°, and finally the intensity again decreases rapidly as $\psi$ increases further.

**Table 4.8.** Dependence of radiance $I$ on observation position ($\psi$, $H$) and view direction $\omega$ ($P_a = 0.8$, $\tau_0 = 0.1$, $\varphi_0 = 1$)

| $\psi$ [°] | $H = 0$ | | | $H = 250$ km | | |
|---|---|---|---|---|---|---|
| | $\omega_0$ | $\omega_1$ | $\omega_2$ | $\omega_0$ | $\omega_1$ | $\omega_2$ |
| 30 | $5.91 \times 10^{-2}$ | $1.42 \times 10^{-2}$ | $1.54 \times 10^{-2}$ | $2.31 \times 10^{-1}$ | $4.11 \times 10^{-1}$ | $5.68 \times 10^{-1}$ |
| 60 | $4.41 \times 10^{-2}$ | $1.08 \times 10^{-2}$ | $1.00 \times 10^{-2}$ | $1.75 \times 10^{-1}$ | $5.56 \times 10^{-1}$ | $3.79 \times 10^{-1}$ |
| 80 | $2.79 \times 10^{-2}$ | $6.77 \times 10^{-3}$ | $5.91 \times 10^{-3}$ | $1.25 \times 10^{-1}$ | $2.00 \times 10^{-1}$ | $3.26 \times 10^{-2}$ |
| 82 | $2.57 \times 10^{-2}$ | $6.01 \times 10^{-3}$ | $5.07 \times 10^{-3}$ | $1.19 \times 10^{-1}$ | $3.20 \times 10^{-1}$ | $3.26 \times 10^{-2}$ |
| 84 | $1.89 \times 10^{-2}$ | $4.44 \times 10^{-3}$ | $3.52 \times 10^{-3}$ | $7.55 \times 10^{-2}$ | $3.52 \times 10^{-1}$ | $1.18 \times 10^{-2}$ |
| 86 | $1.81 \times 10^{-2}$ | $3.29 \times 10^{-3}$ | $2.61 \times 10^{-3}$ | $7.31 \times 10^{-2}$ | $3.53 \times 10^{-1}$ | $3.72 \times 10^{-3}$ |
| 87 | $1.54 \times 10^{-2}$ | $2.81 \times 10^{-3}$ | $1.83 \times 10^{-3}$ | $3.01 \times 10^{-2}$ | $3.53 \times 10^{-1}$ | $9.05 \times 10^{-4}$ |
| 88 | $1.28 \times 10^{-2}$ | $2.15 \times 10^{-3}$ | $1.47 \times 10^{-3}$ | $3.22 \times 10^{-2}$ | $3.49 \times 10^{-1}$ | $4.09 \times 10^{-4}$ |
| 89 | $1.02 \times 10^{-2}$ | $1.79 \times 10^{-3}$ | $8.46 \times 10^{-4}$ | $2.73 \times 10^{-2}$ | $3.39 \times 10^{-1}$ | $1.73 \times 10^{-4}$ |
| 91 | $4.39 \times 10^{-3}$ | $5.79 \times 10^{-4}$ | $8.91 \times 10^{-5}$ | $5.06 \times 10^{-3}$ | $3.14 \times 10^{-1}$ | $6.69 \times 10^{-5}$ |
| 92 | $3.48 \times 10^{-3}$ | $2.86 \times 10^{-4}$ | $1.66 \times 10^{-4}$ | $3.50 \times 10^{-3}$ | $3.02 \times 10^{-1}$ | $8.96 \times 10^{-6}$ |
| 93 | $1.41 \times 10^{-3}$ | $2.06 \times 10^{-4}$ | $3.72 \times 10^{-5}$ | $3.55 \times 10^{-3}$ | $2.97 \times 10^{-1}$ | $5.23 \times 10^{-7}$ |
| 94 | $5.13 \times 10^{-4}$ | $1.24 \times 10^{-4}$ | $2.38 \times 10^{-5}$ | $1.14 \times 10^{-3}$ | $2.81 \times 10^{-1}$ | $3.05 \times 10^{-7}$ |
| 96 | $4.58 \times 10^{-5}$ | $4.32 \times 10^{-5}$ | $3.09 \times 10^{-7}$ | $7.87 \times 10^{-4}$ | $2.56 \times 10^{-1}$ | $1.34 \times 10^{-9}$ |

The intensity at $H = 0$ decreases in directions $\omega_0$ and $\omega_1$, to less than $10^{-3}$ and in the direction $\omega_2$ to less than $10^{-5}$ times as $\psi$ varies from 30° to 96°. $I$ varies very little when observed above the atmosphere ($H = 250$ km) in the direction of the solar horizon $\omega_1$ as $\psi$ increases; in direction $\omega_2$ the increase of $I$ is maximal (becomes more than $10^8$ times normal).

The contribution due to single scattering when observed in the zenith direction, and in all three directions when observed at the top the atmosphere is almost invariant with solar elevation, whereas it decreases from 80% to 50% as $\psi$ varies from 30° to 96°, when observed from the Earth's surface in the horizon direction $\omega_1$ and $\omega_2$. These results are given in Table 4.9.

The single-scattering contribution decreases rapidly as the optical depth $\tau_0$ increases (Table 4.10). This decrease is maximal when the albedo of the single scattering $\varphi_0$ is taken to be 1. The reason is that the survival probability $\varphi_0$ affects the total intensity much more than the intensity of single-scattered radiation. The dependence of the single-scattering contribution on the albedo $P_a$ of the Earth's surface is shown in Table 4.11 for $\psi = 60°$, $\tau_0 = 0.1$, $\varphi_0 = 1$.

The radiance $I$ as a function of the parameter $\tau_0$ is shown in Table 4.12. For nadir and zenith, the quantity $I$ increases as $\tau_0$ decreases, whereas for $\psi = 30°$ and 60° it is approximately proportional to $\tau_0$. Further, $I$ increases slightly at the horizon when $H = 250$ km, whereas for large values of the albedo it even decreases as $\tau_0$ increases; when $H = 0$, the brightness near the horizon is 15 orders of magnitude less when $\tau_0 = 1$ than when $\tau_0 = 0.1$.

The estimates of the multiple-scattering contributions and dependences of the intensity and polarization on parameters of the atmospheric model are presented here. The following atmospheric model was used in our calculations.

**Table 4.9.** Single-scattering contribution to total intensity ($P_a = 0.8$, $\varphi_0 = \tau_0 = 0.1$)

| $\psi$ [°] | $H = 0$ | | | $H = 250$ km | | |
|---|---|---|---|---|---|---|
| | $\omega_0$ | $\omega_1$ | $\omega_2$ | $\omega_0$ | $\omega_1$ | $\omega_2$ |
| 30 | 77.2 | 80.1 | 77.8 | 78.2 | 78.0 | 78.1 |
| 60 | 78.3 | 79.6 | 78.9 | 83.0 | 78.2 | 78.6 |
| 80 | 76.0 | 71.6 | 75.6 | 79.9 | 80.4 | 94.3 |
| 82 | 78.4 | 73.4 | 76.7 | 79.4 | 80.7 | 98.3 |
| 84 | 82.2 | 76.1 | 72.9 | 80.5 | 78.9 | 98.0 |
| 86 | 76.9 | 72.8 | 71.7 | 78.1 | 79.5 | 98.1 |
| 87 | 79.5 | 71.2 | 70.8 | 87.7 | 79.9 | 98.1 |
| 88 | 78.1 | 71.9 | 63.8 | 83.3 | 79.2 | 97.9 |
| 89 | 80.7 | 68.6 | 44.6 | 92.3 | 79.7 | 97.8 |
| 91 | 93.7 | 49.3 | 0 | 96.1 | 78.7 | 94.7 |
| 92 | 66.7 | 66.2 | 0 | 86.4 | 79.2 | 93.4 |
| 93 | 81.3 | 48.3 | 0 | 91.3 | 78.7 | 89.7 |
| 94 | 90.1 | 50.3 | 0 | 94.1 | 80.5 | 26.2 |
| 96 | 86.5 | 55.6 | 0 | 78.8 | 79.4 | 0 |

i) $\sigma_a$, coefficients of aerosol scattering at $\lambda = 0.55\ \mu$m were determined from Fig. 4.7 (curve 2); the corresponding optical thickness of atmosphere is: $\tau_a^{(1)} = 0.178$. The standard values of the coefficients of molecular scattering used in the present calculations, correspond to a molecular optical thickness of atmosphere of $\tau_M = 0.09234$. In order to obtain the dependence of the polarization characteristics of the observed radiation on the turbidity of atmosphere, the calculations were done simultaneously for $\tau_a^{(2)} = 0.0545$, for the same coefficient of aerosol scattering $\sigma_a$. The corresponding values of the vertical transparency of atmosphere, taking into account molecular scattering, are: $T_1 = 0.763$, $T_2 = 0.864$.

ii) The matrix $R_a(\omega', \omega, r)$ for aerosol scattering was proposed by *Gorchakov* and *Rosenberg* [51]. It was calculated at $\lambda = 0.55\ \mu$m for the index of refraction $m = 1.5$, from Junge's distribution

$$\frac{dn}{d(\lg r)} = c \cdot r^{-3}, \quad 0.04\ \mu\text{m} \leqslant r \leqslant 10\ \mu\text{m}.$$

It was assumed that $R_a$ does not depend on altitude.

iii) Lambert's reflection law was used. In the case of surface reflection, the Stokes vector was calculated by use of the scattering matrix $R$ (see Sect. 4.4) such that $R_{11} = 2\mu$, $0 \leqslant \mu \leqslant 1$, whereas the other components were taken to be zero, i.e., the reflected light was assumed to be natural. For both mentioned values of $\tau_0$, the calculations were done simultaneously for two values of albedo: $P_a = 0.15, 0.055$. Monte Carlo calculations of the scattered light at $\lambda = 0.55\ \mu$m were done in the sun's vertical, with solar zenith angle 45°.

To estimate the angular distribution of the desired characteristics, the integrals of the components $I$, $Q$, $U$, $V$ were calculated at 6° intervals of the

**Table 4.10.** Dependence of contribution due to single-scattered radiation on total optical thickness $\tau_0$ and albedo of single scattering $\varphi_0$ [%] ($\psi = 60°$, $P_a = 0$)

| $\tau_0$ | $H = 0$ | | | | | | $H = 250$ km | | | | | |
|---|---|---|---|---|---|---|---|---|---|---|---|---|
| | $\omega_0$ | | $\omega_1$ | | $\omega_2$ | | $\omega_0$ | | $\omega_1$ | | $\omega_2$ | |
| | $\varphi_0 = 1$ | $\varphi_0 = 0.5$ | $\varphi_0 = 1$ | $\varphi_0 = 0.5$ | $\varphi_0 = 1$ | $\varphi_0 = 0.5$ | $\varphi_0 = 1$ | $\varphi_0 = 0.5$ | $\varphi_0 = 1$ | $\varphi_0 = 0.5$ | $\varphi_0 = 1$ | $\varphi_0 = 0.5$ |
| 0.1 | 84 | 92 | 84 | 92 | 85 | 93 | 88 | 95 | 86 | 93 | 82 | 91 |
| 0.2 | 74 | 87 | 75 | 87 | 73 | 87 | 81 | 91 | 75 | 88 | 75 | 88 |
| 0.3 | 66 | 83 | 67 | 83 | 64 | 82 | 72 | 88 | 68 | 84 | 71 | 85 |
| 0.5 | 50 | 76 | 54 | 76 | 48 | 74 | 55 | 82 | 59 | 80 | 64 | 82 |
| 1 | 17 | 60 | 27 | 61 | 15 | 52 | 35 | 73 | 50 | 75 | 57 | 80 |

**Table 4.11.** Single-scattering contribution to total intensity [%]

| $P_a$ | $H = 0$ | | | $H = 250$ km | | |
|---|---|---|---|---|---|---|
| | $\omega_0$ | $\omega_1$ | $\omega_2$ | $\omega_0$ | $\omega_1$ | $\omega_2$ |
| 0 | 84 | 84 | 85 | 89 | 86 | 82 |
| 0.2 | 82 | 83 | 83 | 87 | 84 | 81 |
| 0.8 | 78 | 79 | 79 | 83 | 78 | 78 |

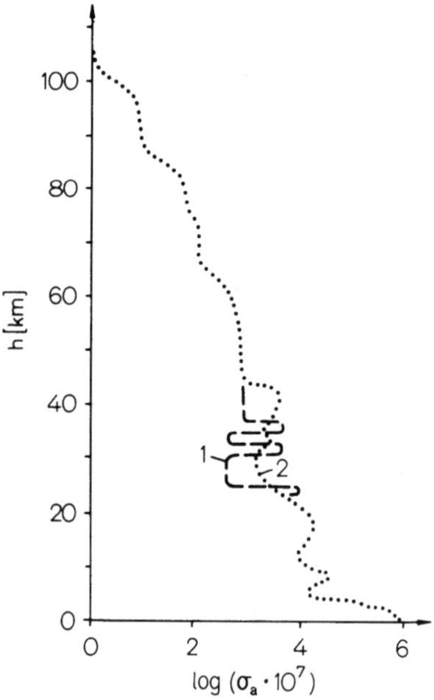


**Fig. 4.7.** Coefficient of aerosol scattering as function of altitude: (*1*) laminal model, (*2*) smooth model

**Table 4.12.** Dependence of $I$ on total optical thickness of atmosphere $\tau_0$, albedo $P_a$, and survival probability $\varphi_0$ ($\psi = 60°$)

| $\tau_0$ | $P_a = 0$ | | $P_a = 0.2$ | | $P_a = 0.8$ | |
|---|---|---|---|---|---|---|
| | $\varphi_0 = 1$ | $\varphi_0 = 0.5$ | $\varphi_0 = 1$ | $\varphi_0 = 0.5$ | $\varphi_0 = 1$ | $\varphi_0 = 0.5$ |
| | | | $H = 0, \omega = \omega_0$ | | | |
| 0.1 | $2.71 \times 10^{-2}$ | $1.23 \times 10^{-2}$ | $3.11 \times 10^{-2}$ | $1.41 \times 10^{-2}$ | $4.41 \times 10^{-2}$ | $1.94 \times 10^{-2}$ |
| 0.2 | $5.24 \times 10^{-2}$ | $2.24 \times 10^{-2}$ | $5.94 \times 10^{-2}$ | $2.51 \times 10^{-2}$ | $8.44 \times 10^{-2}$ | $3.35 \times 10^{-2}$ |
| 0.3 | $7.60 \times 10^{-2}$ | $3.03 \times 10^{-2}$ | $8.52 \times 10^{-2}$ | $3.33 \times 10^{-2}$ | $1.21 \times 10^{-1}$ | $4.33 \times 10^{-2}$ |
| 0.5 | $1.24 \times 10^{-1}$ | $4.07 \times 10^{-2}$ | $1.36 \times 10^{-1}$ | $4.37 \times 10^{-2}$ | $1.88 \times 10^{-1}$ | $5.41 \times 10^{-2}$ |
| 1 | $3.40 \times 10^{-1}$ | $4.90 \times 10^{-2}$ | $3.52 \times 10^{-1}$ | $5.08 \times 10^{-2}$ | $4.15 \times 10^{-1}$ | $5.73 \times 10^{-2}$ |
| | | | $\omega = \omega_1$ | | | |
| 0.1 | $5.68 \times 10^{-3}$ | $2.61 \times 10^{-3}$ | $6.91 \times 10^{-3}$ | $3.14 \times 10^{-3}$ | $1.08 \times 10^{-2}$ | $4.79 \times 10^{-3}$ |
| 0.2 | $1.26 \times 10^{-4}$ | $5.42 \times 10^{-5}$ | $1.55 \times 10^{-4}$ | $6.55 \times 10^{-5}$ | $2.53 \times 10^{-4}$ | $1.01 \times 10^{-4}$ |
| 0.3 | $2.71 \times 10^{-6}$ | $1.09 \times 10^{-6}$ | $3.36 \times 10^{-6}$ | $1.32 \times 10^{-6}$ | $5.67 \times 10^{-6}$ | $2.05 \times 10^{-6}$ |
| 0.5 | $1.27 \times 10^{-9}$ | $4.50 \times 10^{-10}$ | $1.59 \times 10^{-9}$ | $5.40 \times 10^{-10}$ | $2.81 \times 10^{-9}$ | $8.50 \times 10^{-10}$ |
| 1 | $6.83 \times 10^{-18}$ | $1.49 \times 10^{-18}$ | $8.42 \times 10^{-18}$ | $1.81 \times 10^{-18}$ | $1.45 \times 10^{-17}$ | $2.88 \times 10^{-18}$ |
| | | | $H = 250$ km, $\omega = \omega_0$ | | | |
| 0.1 | $2.55 \times 10^{-2}$ | $1.20 \times 10^{-2}$ | $6.10 \times 10^{-2}$ | $4.32 \times 10^{-2}$ | $1.75 \times 10^{-1}$ | $1.39 \times 10^{-1}$ |
| 0.2 | $4.89 \times 10^{-2}$ | $2.16 \times 10^{-2}$ | $8.18 \times 10^{-2}$ | $4.70 \times 10^{-2}$ | $1.93 \times 10^{-1}$ | $1.27 \times 10^{-1}$ |
| 0.3 | $7.19 \times 10^{-2}$ | $2.95 \times 10^{-2}$ | $1.02 \times 10^{-1}$ | $4.99 \times 10^{-2}$ | $2.11 \times 10^{-1}$ | $1.15 \times 10^{-1}$ |
| 0.5 | $1.22 \times 10^{-1}$ | $4.11 \times 10^{-2}$ | $1.48 \times 10^{-1}$ | $5.42 \times 10^{-2}$ | $2.52 \times 10^{-1}$ | $9.71 \times 10^{-2}$ |
| 1 | $2.28 \times 10^{-1}$ | $5.54 \times 10^{-2}$ | $2.53 \times 10^{-1}$ | $5.96 \times 10^{-2}$ | $3.47 \times 10^{-1}$ | $7.73 \times 10^{-2}$ |
| | | | $\omega = \omega_1$ | | | |
| 0.1 | $2.82 \times 10^{-1}$ | $1.30 \times 10^{-1}$ | $3.47 \times 10^{-1}$ | $1.58 \times 10^{-1}$ | $5.56 \times 10^{-1}$ | $2.45 \times 10^{-1}$ |
| 0.2 | $3.03 \times 10^{-1}$ | $1.30 \times 10^{-1}$ | $3.57 \times 10^{-1}$ | $1.51 \times 10^{-1}$ | $5.43 \times 10^{-1}$ | $2.15 \times 10^{-1}$ |
| 0.3 | $3.29 \times 10^{-1}$ | $1.33 \times 10^{-1}$ | $3.76 \times 10^{-1}$ | $1.48 \times 10^{-1}$ | $5.44 \times 10^{-1}$ | $1.97 \times 10^{-1}$ |
| 0.5 | $3.76 \times 10^{-1}$ | $1.39 \times 10^{-1}$ | $4.12 \times 10^{-1}$ | $1.48 \times 10^{-1}$ | $5.54 \times 10^{-1}$ | $1.78 \times 10^{-1}$ |
| 1 | $4.45 \times 10^{-1}$ | $1.48 \times 10^{-1}$ | $4.65 \times 10^{-1}$ | $1.50 \times 10^{-1}$ | $5.51 \times 10^{-1}$ | $1.59 \times 10^{-1}$ |

latitude angle $\theta$, except near the neutral points, where the interval was 3°. The angle $\theta$ was measured in the vertical plane through the sun, counterclockwise from the azimuth of the sun. The contribution $(I_1/I)$ of single-scattered radiation to the total intensity, and the dependence of the intensity on albedo $P_a$ are shown in Table 4.13 for various angles of view and for two different values of vertical transparency $T$.

To investigate the dependence of intensity on polarization, the calculations were done for the same indicatrix $R_{11}$ whereas the other components of the scattering matrix $R$ were taken to be zero. The corresponding results are given in Table 4.14. Intensity as a function of angle $\theta$ is shown in Fig. 4.8 for two values of $T$ and $P_a = 0.15$. The degree of polarization $P = Q/I$, as a function of angle $\theta$, is shown in Fig. 4.9 for two values of $T$ and different values of albedo. For comparison the curves of degree of polarization, reported by *Rosenberg* [29], are also shown in Fig. 4.9.

When $T = 0.864$, the calculated degree of polarization practically coincides with the experimental value; the displacement of the maximum of the experimental curve to the left from the expected position $\theta = 90°$ can be explained by the crude drawing or by fluctuations of atmospheric conditions during the experimental measurements. The reason for the satisfactory agreement for large values of the coefficient $T$ is evident, because, in this case, the polarization depends on molecular scattering, which is well determined.

The calculated values of the degree of polarization agree poorly with the experimental data, when $T = 0.763$, either because of a decrease of the contribu-

**Table 4.13.** Single-scattering contribution and influence of albedo on intensity [%]

| $\theta$ [°] | $I_1/I$ | | $I(P_a^2)/I(P_a^1)$ | | $\theta$ [°] | $I_1/I$ | | $I(P_a^2)/I(P_a^1)$ | |
|---|---|---|---|---|---|---|---|---|---|
| | $T=$ 0.763 | $T=$ 0.864 | $T=$ 0.763 | $T=$ 0.864 | | $T=$ 0.763 | $T=$ 0.864 | $T=$ 0.763 | $T=$ 0.864 |
| −42 | 61.4 | 72.5 | 94.9 | 93.7 | 36 | 80.6 | 83.0 | 97.3 | 95.6 |
| −36 | 71.8 | 78.9 | 96.0 | 94.8 | 42 | 78.9 | 81.3 | 96.3 | 94.5 |
| −30 | 77.2 | 83.4 | 97.2 | 96.0 | 48 | 69.3 | 75.7 | 95.0 | 93.1 |
| −24 | 78.2 | 84.4 | 97.7 | 96.6 | 54 | 68.9 | 75.1 | 94.5 | 92.5 |
| −18 | 84.0 | 87.7 | 98.5 | 97.7 | 60 | 65.6 | 70.5 | 92.6 | 89.4 |
| −12 | 90.6 | 91.5 | 98.7 | 97.9 | 66 | 58.0 | 66.8 | 91.0 | 88.4 |
| − 9 | 91.5 | 92.6 | 99.0 | 98.0 | 72 | 53.3 | 66.2 | 91.5 | 90.4 |
| − 6 | 92.9 | 93.9 | 99.3 | 98.5 | 78 | 60.5 | 68.3 | 90.1 | 89.7 |
| − 3 | 97.3 | 97.4 | 99.7 | 99.4 | 84 | 51.1 | 59.0 | 89.3 | 87.3 |
| 0 | 96.9 | 97.7 | 99.8 | 99.7 | 90 | 52.8 | 65.5 | 88.4 | 88.7 |
| 3 | 95.9 | 94.9 | 99.4 | 98.7 | 96 | 49.3 | 63.2 | 89.9 | 89.1 |
| 6 | 92.6 | 93.0 | 99.2 | 98.5 | 102 | 44.3 | 59.8 | 88.5 | 88.8 |
| 12 | 89.9 | 89.7 | 98.3 | 96.7 | 108 | 48.1 | 62.9 | 87.7 | 88.1 |
| 18 | 88.9 | 89.0 | 98.4 | 97.2 | 114 | 45.7 | 63.0 | 87.8 | 89.1 |
| 24 | 81.7 | 84.3 | 97.7 | 96.0 | 120 | 42.9 | 60.5 | 86.0 | 87.5 |
| 30 | 85.0 | 85.2 | 97.0 | 95.4 | 126 | 36.5 | 56.7 | 84.4 | 86.8 |
| | | | | | 132 | | | | |

**Table 4.14.** Relative error in intensity calculations due to neglect of polarization

| $\theta$ [°] | $T = 0.763$ $P_a = 0.15$ | $T = 0.864$ $P_a = 0.15$ | $T = 0.763$ $P_a = 0.055$ | $\theta$ [°] | $T = 0.763$ $P_a = 0.15$ | $T = 0.854$ $P_a = 0.15$ | $T = 0.763$ $P_a = 0.055$ |
|---|---|---|---|---|---|---|---|
| −42 | −0.22 | −0.3 | −0.21 | 36 | −0.36 | −0.51 | −0.37 |
| −36 | −0.3 | −0.8 | −0.3 | 42 | −0.32 | 0.4 | −0.31 |
| −30 | −0.44 | −0.6 | −0.43 | 48 | 0.2 | 0.16 | 0.24 |
| −24 | −0.34 | −0.7 | −0.33 | 54 | 0.2 | 0.23 | 0.21 |
| −18 | −0.17 | −0.5 | −0.19 | 60 | 1.0 | 2.23 | 1.1 |
| −12 | −0.3 | −0.6 | −0.26 | 66 | 1.56 | 1.53 | 1.59 |
| − 9 | −0.2 | −0.4 | −0.17 | 72 | 0.7 | 1.26 | 0.77 |
| − 6 | −0.1 | −0.2 | −0.1 | 78 | 1.1 | 1.57 | 1.2 |
| − 3 | −0.2 | −0.44 | −0.2 | 84 | 1.8 | 2.57 | 2.0 |
| 0 | −0.15 | −0.4 | −0.15 | 90 | 2.14 | 4.3 | 2.5 |
| 3 | −0.17 | −0.43 | −0.15 | 96 | 1.8 | 2.97 | 2.2 |
| 6 | −0.2 | −0.34 | −0.18 | 102 | 1.34 | 1.84 | 1.44 |
| 12 | −0.33 | −0.7 | −0.34 | 108 | 1.32 | 2.1 | 1.49 |
| 18 | −0.36 | −0.75 | −0.36 | 114 | 0.65 | 0.98 | 0.73 |
| 24 | −0.4 | −0.86 | −0.4 | 120 | 0.8 | 0.96 | 0.85 |
| 30 | −0.4 | −0.53 | −0.37 | 126 | 0.04 | 0.23 | 0.08 |
| | | | | 132 | | | |

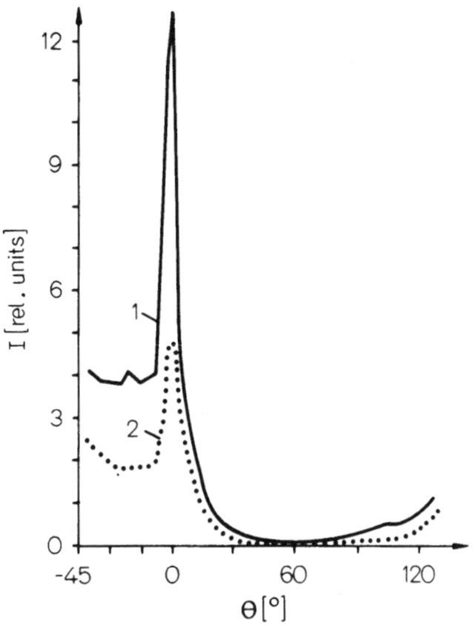


**Fig. 4.8.** Radiance as function of $\theta$ for $P_a = 0.15$, $H = 0$, $\psi = 45°$ (*1*) $T = 0.763$, (*2*) $T = 0.864$

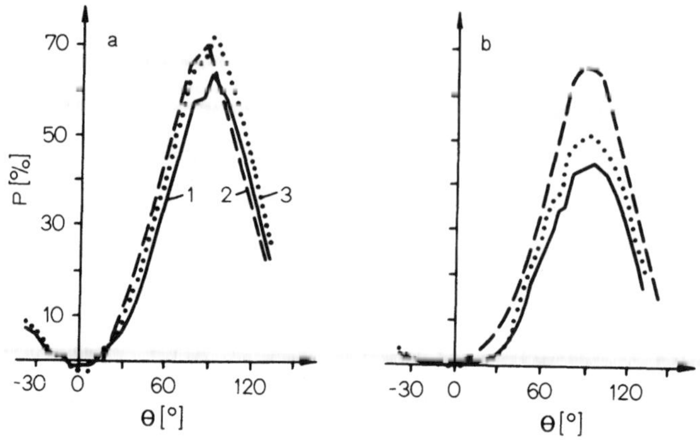


**Fig. 4.9a, b.** Degree of polarization as function of angle $\theta$ for $\psi = 45°$, $H = 0$ **(a)** $T = 0.804$; **(b)** $T = 0.763$; (*1*) experimental curve; (*2*) $P_a = 0.15$; (*3*) $P_a = 0.55$

tion from molecular scattering for the case of a small transparency, or because of the use of a crude model of aerosol scattering.

The point at which the degree of polarization becomes maximum according to the calculations, depends slightly on the optical thickness of the aerosol and on the albedo. The angle between the sun and the calculated neutral points is approximately 7°–13°, i.e., somewhat less than the experimental angle. Calculations also show (Fig. 4.9) that the positions of the neutral points vary slightly when the coefficient changes.

The calculations of intensity that take polarization into account are of particular interest. Note that the computational results obtained with and without taking polarization into account coincide for $\theta = 45°$, whereas the maximum difference (nearly 4%) is obtained in the region of maximum polarization for $\theta = 90°$. This difference depends on the turbidity of atmosphere, the albedo of the Earth's surface, and on the sun's position.

We now consider a problem of calculating some twilight effects. This problem as well as the atmospheric model was formulated by *Rosenberg* [30].

The scattering matrices are

$$\mathscr{D}_{ik}(\lambda, \theta) = \mathscr{D}^{\mathrm{M}}_{ik}(\lambda, \theta) + \mathscr{D}^{\mathrm{D}}_{ik}(\lambda, \theta) + \mathscr{D}^{\mathrm{TD}}_{ik}(\lambda, \theta)$$

$$\mathscr{D}_{ik} = \frac{\sigma}{4\pi} f \cdot \tilde{f}_{ik}, \quad \left(\tilde{f}_{ik} = \frac{\mathscr{D}_{ik}}{\mathscr{D}_{11}}\right), \quad i, k = 1, \ldots, 4, \tag{4.58}$$

where $\mathscr{D}_{ik}$, $\mathscr{D}^{\mathrm{D}}_{ik}$, $\mathscr{D}^{\mathrm{TD}}_{ik}$ are the scattering matrices of the molecular component of air, the haze, and foggy haze, respectively; $f_{ik}$ are the components of the reduced scattering matrices (see [30]), $f$ is the phase function (or scattering indicatrix).

The atmosphere is divided into three layers:

i) *The bottom layer* (0–7 km).

Measurements show the dependence $\sigma^{D} = \sigma^{D}(h, r, \lambda)$, cross section for scattering of atmospheric haze; $\sigma^{TD} = \sigma^{TD}(h, r)$, cross section for scattering of nebulous haze, where $r$ is the relative humidity. The height distribution density of condensation nuclei is given by $N(h) = N(0) \exp(-h/2.5)$.

If $r$ does not depend on height, we have

$$f^{D} = C(\theta)\cdot[\sigma^{D}(0, r_0, \lambda)]^{K(\theta)-1}. \tag{4.59}$$

Then, from (4.58), we get

$$\mathscr{D}_{ik}^{D} = \frac{1}{4\pi} C(\theta)\cdot[\sigma^{D}(0, r_0, \lambda)]^{K(\theta)} \tilde{f}_{ik}^{D} \, e^{-h/2.5}. \tag{4.60}$$

The empirical functions are assumed not to depend on the wavelength, and were taken from [30]. The problem was solved for $\lambda = 0.65\ \mu$m and values of $\lambda^{D}$ and $\sigma^{TD}$ given in Table 4.15. The matrix for scattering of haze was calculated from (4.60). For nebulous haze, $f^{TD}$ and $\tilde{f}^{TD}$ were assumed not to depend on $\sigma$, i.e.,

$$\mathscr{D}_{ik}^{TD} = \frac{1}{4\pi} \tilde{f}_{ik}^{TD}\cdot f^{TD}\cdot\sigma^{TD}(0, r_0)\, e^{-h/2.5}.$$

The values of $\tilde{f}_{ik}^{TD}$ and $f_{ik}^{TD}$, as well as the aureole parts of the indicatrices $f^{D}$ and $f^{TD}$ were taken from [30] The coefficients $\sigma^{D}$ and $\sigma^{TD}$ were tabulated for the layer according to

$$\sigma(h, \lambda) = \sigma(0, r_0, \lambda)\, e^{-h/2.5}, \quad \Delta h = 1 \text{ km}.$$

ii) *Stratospheric layer* (7–33 km).

The cross sections for scattering by haze $\sigma^{D}(h, \lambda)$ were calculated from the function $\sigma_a = 4 \exp(-|h - 19|/7.2)\sigma_M$. Standard coefficients of molecular scattering $\sigma_M(h)$ were used, and $\sigma^{TD}(h)$ were assumed to be zero. For scattering by haze, the scattering matrix of the bottom layer for $\sigma^{D} = 0.0588$ km$^{-1}$ was used.

**Table 4.15.** Values of $\sigma^{D}$ and $\sigma^{TD}$, km$^{-1}$

| Variant | $\sigma^{D}$ | $\sigma^{TD}$ |
|---|---|---|
| 1 | 0.0588 | 0.01 |
| 2 | 0.252 | 0.07 |
| 3 | 0.924 | 0.3 |

iii) *The high-altitude layer* (33–150 km).
For this, pure Rayleigh scattering was assumed, because the principal problem was to estimate the polarization. Absorption by ozone was taken into account. The corresponding absorption coefficients were taken from the tables of *Elterman* [31].

The intensity and polarization of singly and multiply scattered light in the vertical plane of the sun were calculated for $\psi = 91, 93, 95, 97, 99, 101, 103, 104°$ and for zenith angles $\theta = 0, \pm 15, \pm 30, \pm 45, \pm 60, \pm 75°$. The observation points were situated on the Earth's surface.

To determine the errors due to neglect of polarization for twilight conditions, the intensity was also calculated, for variant 1, without taking polarization into account. The error depends on observation conditions, i.e., on $(\psi, \theta)$, and is sometimes about 9%. The corresponding computed results are given in Table 4.16. As expected, the influence of the albedo $P_a$ of the Earth's surface on the intensity decreases as $\psi$ increases. For example, the ratio $I(P_a = 0.2)/I(P_a = 0.8)$ increases in direction of view $\theta = -75°$ from 0.95 to 0.98, and in direction $\theta = 0°$ from 0.93 to 0.994, as $\psi$ increases from 91° to 102°. (variant 1). The contribution $I_1/I$ of singly scattered radiation to the total intensity for variant 1 is given in Table 4.17. Approximately the same values of $I_1/I$ were obtained in variants 2 and 3.

It is interesting to note the dependence of the quantity $I_1/I$ on the albedo of the Earth's surface. The contribution from singly scattered radiation for $\psi = 91°$ decreases to approximately 3% as $P_a$ varies between 0.2 and 0.8.

Further, the influence of albedo on the quantity $I_1/I$ tends to zero as $\psi$ increases. The intensity of multiply scattered light at the zenith ($\theta = 0°$) as a function of $\psi$ for three variants of $\sigma^D$ and $\sigma^{TD}$ is shown in Fig. 4.10 (in relative units).

The intensity in this direction decreases, so that it is five orders of magnitude greater at 91° than at $\psi = 102°$. The relative statistical errors of these results

**Table 4.16.** Relative error in intensity calculations due to neglect of polarization, as a function of $\psi$ and $\theta$ ($\sigma^D = 0.0588\ \mathrm{km}^{-1}$, $\sigma^{TD} = 0.01\ \mathrm{km}^{-1}$). $\theta$ is measured in the vertical plane through the sun, from the zenith

| $\theta$ [°] \ $\psi$ [°] | 91 | 93 | 95 | 97 | 99 | 101 | 103 | 104 |
|---|---|---|---|---|---|---|---|---|
| −75 | 1.6 | −2 | 1.4 | −1 | −3.5 | −3.2 | −2 | 0 |
| −60 | 0.7 | 0.4 | 0.3 | 0.8 | −2 | −2.1 | −1.5 | 0 |
| −45 | 0.7 | 0.2 | 0.3 | 0.8 | −0.6 | −0.6 | 0 | 0 |
| −30 | 0.1 | 0.1 | 1.2 | 1.2 | 0.3 | 2 | 2.8 | 2.5 |
| −15 | 0.1 | 0.1 | 2.8 | 1.7 | 0.6 | 3.9 | 5.7 | 7.5 |
| 0 | −1.5 | 0.03 | 5 | 1 | 1 | 4.7 | 8.3 | 8.3 |
| 15 | −1.3 | 0.06 | 6 | −0.9 | 0.1 | 3.5 | 9.4 | 9.5 |
| 30 | 1.2 | 0.2 | 5 | −0.8 | −1.5 | 1.6 | 6.6 | 7.5 |
| 45 | 1.1 | 0.1 | 3.5 | −0.8 | −2.7 | 0.7 | 4.2 | 4.3 |
| 60 | 1 | 0.1 | 2.1 | 0.5 | −3.5 | 0.4 | 2 | 2 |
| 75 | 4 | 0.2 | 1.7 | 0.4 | −3.7 | 0.2 | 0.2 | 0.1 |

**Table 4.17.** Contribution $I_1/I$ from singly scattered radiation to total intensity [%] ($\sigma^D$ = 0.0588 km$^{-1}$, $\sigma^{TD}$ = 0.01 km$^{-1}$)

| $\theta$ [°] \ $\psi$ [°] | 91 | 93 | 95 | 97 | 99 | 101 | 103 | 104 |
|---|---|---|---|---|---|---|---|---|
| −75 | 52 | 34 | 10 | 0.3 | 0 | 0 | 0 | 0 |
| −60 | 62 | 51 | 43 | 19 | 1 | 0 | 0 | 0 |
| −45 | 63 | 59 | 50 | 35 | 10 | 0 | 0 | 0 |
| −30 | 65 | 62 | 56 | 41 | 17 | 0.2 | 0 | 0 |
| −15 | 62 | 57 | 56 | 47 | 22 | 2 | 0 | 0 |
| 0 | 58 | 61 | 56 | 51 | 26 | 5 | 0 | 0 |
| 15 | 62 | 61 | 56 | 54 | 29 | 9 | 0 | 0 |
| 30 | 60 | 60 | 55 | 55 | 32 | 16 | 0 | 0 |
| 45 | 60 | 57 | 54 | 56 | 36 | 22 | 0.7 | 0 |
| 60 | 59 | 56 | 49 | 54 | 41 | 29 | 19 | 0.5 |
| 75 | 47 | 49 | 43 | 39 | 58 | 42 | 48 | 48 |

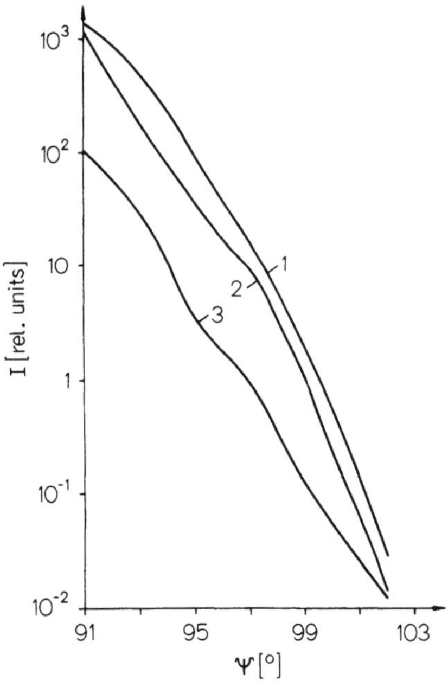


**Fig. 4.10.** Radiance of sky at zenith as function of $\psi$ (*1*) $\sigma^D = 0.0588$ km$^{-1}$, $\sigma^{TD} = 0.01$ km$^{-1}$; (*2*) $\sigma^D = 0.252$ km$^{-1}$, $\sigma^{TD} = 0.07$ km$^{-1}$; (*3*) $\sigma^D = 0.924$ km$^{-1}$; $\sigma^{TD} = 0.3$ km$^{-1}$. $\lambda = 0.65$ $\mu$m, $H = 0$, $P_a = 0.8$

are approximately 3–10%, except when deep-twilight conditions are considered, for which the error increases to approximately 40%, in some directions. All calculations for this problem were performed in approximately 180 min of BESM-6 computer time. The radiance, for two different values of $\psi$ and for two variants of $\sigma^D$ and $\sigma^{TD}$, as functions of angle, is shown in Fig. 4.11.

The degree of polarization of multiply scattered light at $\theta = 0$ as a function of $\psi$ is shown in Fig. 4.12, where a strong maximum appears near $\psi = 95°$. The degree of polarization as a function of direction of view $\theta$ is shown in Fig. 4.13, for various values of $\psi$ and two variants of $\sigma^D$ and $\sigma^{TD}$; the maximum is near $\theta = 0$.

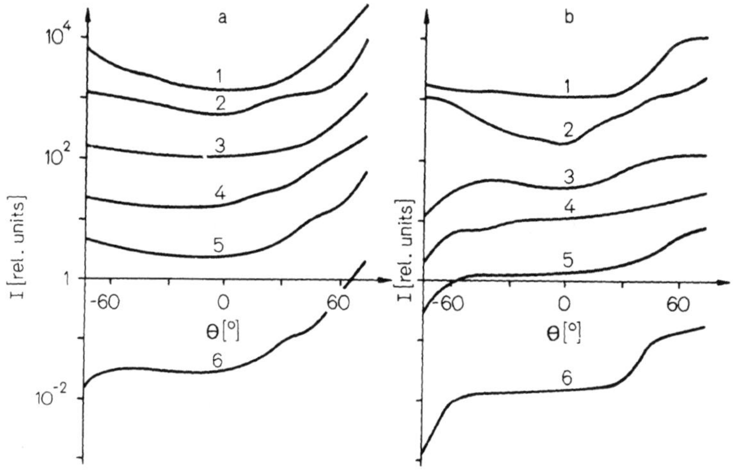


**Fig. 11a, b.** Radiance for different values of $\psi$ as function of angle $\theta$ (a) $\sigma^D = 0.0588\ km^{-1}$, $\sigma^{TD} = 0.01\ km^{-1}$; (b) $\sigma^D = 0.252\ km^{-1}$, $\sigma^{TD} = 0.07\ km^{-1}$; (*1*) $\psi = 91°$, (*2*) $\psi = 93°$, (*3*) $\psi = 95°$, (*4*) $\psi = 97°$, (*5*) $\psi = 99°$, (*6*) $\psi = 102°$; $\lambda = 0.65\ \mu m$, $H = 0$, $P_a = 0.8$

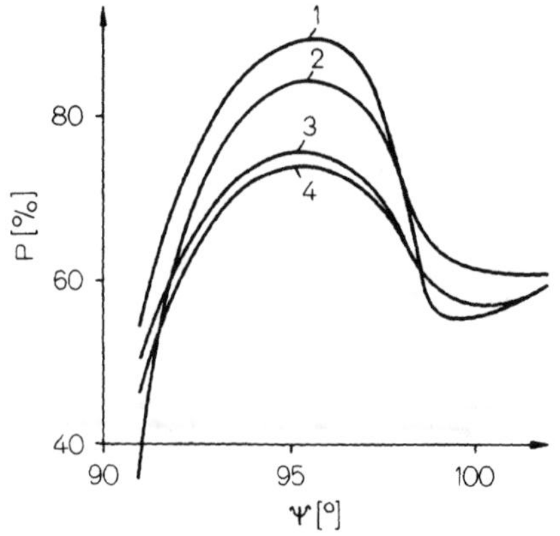


**Fig. 4.12.** Degree of polarization at zenith as function of $\psi$: (*1*) $\sigma^D = 0.0588\ km^{-1}$, $\sigma^{TD} = 0.01\ km^{-1}$, $P_a = 0.8$; (*2*) $\sigma^D = 0.0588\ km^{-1}$, $\sigma^{TD} = 0.01\ km^{-1}$, $P_a = 0.2$; (*3*) $\sigma^D = 0.252\ km^{-1}$, $\sigma^{TD} = 0.07\ km^{-1}$, $P_a = 0.8$; (*4*) $\sigma^D = 0.924\ km^{-1}$, $\sigma^{TD} = 0.3\ km^{-1}$, $P_a = 0.8$; $\lambda = 0.65\ \mu m$, $H = 0$

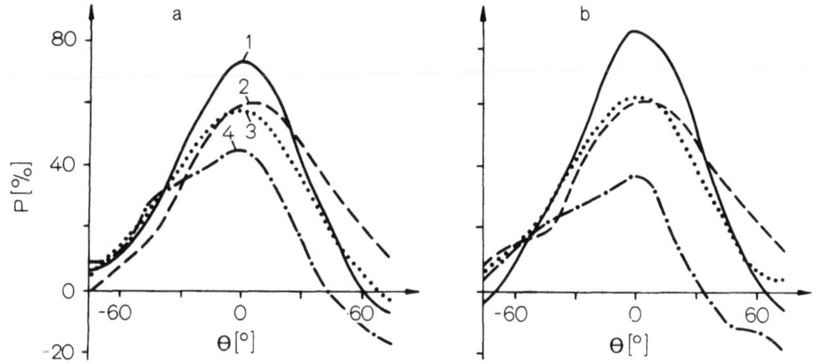


**Fig. 4.13a, b.** Degree of polarization in vertical plane through sun as function of angle $\theta$ **(a)** $\sigma^{D} = 0.0588\ \text{km}^{-1}$, $\sigma^{TD} = 0.01\ \text{km}^{-1}$; **(b)** $\sigma^{D} = 0.252\ \text{km}^{-1}$, $\sigma^{TD} = 0.07\ \text{km}^{-1}$; *(1)* $\psi = 95°$, *(2)* $\psi = 102°$, *(3)* $\psi = 99°$, *(4)* $\psi = 91°$, $H = 0$, $P_a = 0.8$, $\lambda = 0.65\ \mu\text{m}$

We now consider a problem of finding the angular and spectral distribution of the intensity and polarization of day skylight and twilight for various observation points at an altitude of $H = 250$ km. One of the principal purposes of these calculations is to investigate the role of singly scattered light.

The optical model of atmosphere used in our calculations was developed by the Department of Atmospheric Physics of the Leningrad State University. This model was described also by *Ivlev* and *Popov* [32, 33]. Two types of scattering are assumed, namely molecular and aerosol scattering. The extinction coefficient at altitude $h$ is taken to be

$$\sigma(\lambda, h) = \sigma_a(\lambda, h) + \sigma_M(\lambda, h) + \sigma_c^{(a)}(\lambda, h) + \sigma_c^{(0)}(\lambda, h),$$

where $\sigma_a$ and $\sigma_c$ are the coefficients of aerosol and molecular scattering, respectively, and $\sigma_c^{(a)}$, $\sigma_c^{(o)}$ are the coefficients of light absorption by aerosol and ozone, respectively.

In this notation, the albedo of single scattering (photon survival probability in one collision) is

$$\varphi_0(\lambda, h) = [\sigma_a(\lambda, h) + \sigma_M(\lambda, h)]/\sigma(\lambda, h).$$

The coefficients $\sigma_M(\lambda, h)$ and $\sigma_c^{(o)}(\lambda, h)$ for $0 \leqslant h \leqslant 50$ km were taken from the tables of *Elterman* [31]. For $50\ \text{km} \leqslant h \leqslant 100$ km they were calculated from

$$\sigma(\lambda, h) = \sigma(\lambda, 50)\frac{\rho(h)}{\rho(50)},$$

where $\rho$ is the density.

The coefficients $\sigma_a(\lambda, h)$ and $\sigma_c^{(a)}(\lambda, h)$, as well as $R_a(\lambda, \mu, h)$, the matrix for light scattering by the aerosol, were calculated from Mie theory.

The aerosol particles are known to be characterized by $m = \nu - i\kappa$, the index of refraction and $dN/dr = n(r, h)$, the size distribution of aerosol particles. In the model, more-complicated functions $n(r, h)$ were used, which were experimental functions tabulated for (i) the bottom layer of the atmosphere 0–5 km; (ii) troposphere 5–9 km; (iii) tropopause 9–17 km; (iv) lower stratosphere 17–24 km; (v) stratosphere 24–30 km; (vi) above 30 km.

The corresponding coefficients for these models $\sigma_a$, $\sigma^{(a)}$ and matrixes $R_a$ were calculated for the values of $\kappa$ and $\nu$ given in Table 4.18.

To determine the matrix $R_a$, the atmosphere was divided into three layers: 0–5 km, 5–17 km, and above 17 km, each having a constant matrix $R_a$ calculated according to fifth, third, and sixth model, respectively.

The problem was solved for 5 wavelengths: $\lambda = 0.4, 0.55, 0.6, 0.7, 0.8\ \mu$m when $P_a = 0.8$, and at $\lambda = 0.55$ when $P_a = 0$. For these values of $\lambda$, the total optical thickness of atmosphere measured in the vertical direction are shown in Table 4.19 for the corresponding coefficients $\sigma_a$, $\sigma_M$, $\sigma_c^{(a)}$, $\sigma_c^{(o)}$.

All calculations were performed for observation points at a height of 250 km. As shown in Fig. 3.2, the position of observation point is defined by the azimuth angle $\varphi$, the height $h(\theta)$, and $\psi$, where $\varphi$ is measured from sun's azimuth. The quantity $h(\theta)$ varies from 0 to 95 km, with $\Delta h = 4$ km (from 0 to 20 km), 10 km (20–50 km), and 15 km (above 50 km). The following variants were considered

$$\psi = 30°, 60°, 89°, \quad \varphi = 0, 90°, 180°;$$

$$\psi = 91°, 95°, 100°, \quad \varphi = 0, 45°, 90°.$$

**Table 4.18.** Values of $\nu$ and $\kappa$ for various values of $\lambda$

| $\lambda$ [$\mu$m] | Model 1, 2, 3, 6 | | Model 4 | | Model 5 | |
|---|---|---|---|---|---|---|
| | $\nu$ | $\kappa$ | $\nu$ | $\kappa$ | $\nu$ | $\kappa$ |
| 0.4 | 1.649 | 0.0082 | 1.540 | 0.0070 | 1.467 | 0.0055 |
| 0.55 | 1.647 | 0.0088 | 1.540 | 0.0037 | 1.467 | 0.0027 |
| 0.6 | 1.647 | 0.0093 | 1.540 | 0.0035 | 1.467 | 0.0025 |
| 0.7 | 1.648 | 0.0107 | 1.539 | 0.0033 | 1.466 | 0.0025 |
| 0.8 | 1.648 | 0.0120 | 1.539 | 0.0030 | 1.465 | 0.0025 |

**Table 4.19.** Optical thickness of atmosphere $\tau(\lambda)$

| $\lambda$ [$\mu$m] | $\tau_a$ | $\tau_M$ | $\tau_c^{(a)}$ | $\tau_c^{(o)}$ | $\tau$ |
|---|---|---|---|---|---|
| 0.4 | 0.203 | 0.364 | 0.0309 | 0 | 0.5979 |
| 0.55 | 0.19416 | 0.098 | 0.0176 | 0.031 | 0.3408 |
| 0.6 | 0.19425 | 0.069 | 0.0163 | 0.045 | 0.3245 |
| 0.7 | 0.1875 | 0.037 | 0.014 | 0.008 | 0.2465 |
| 0.8 | 0.183 | 0.021 | 0.0135 | 0.003 | 0.2205 |

For the cited parameters, the degree of polarization, and the intensity in $S$ units were calculated, where $\pi S$ is the net flux of solar radiation incident at the top of the atmosphere. These characteristics were obtained also for singly scattered light.

The relative errors of intensity calculations, due to neglect of polarization, were also calculated. For this, the intensity for all five values of wavelength was also calculated without taking polarization into account, using the same trajectories. The influence of polarization on the intensity depends on various factors, namely the optical thickness, the sun's position, the direction of view, and so on.

Relatively large errors (~2–6%) were obtained when $\lambda = 0.4\ \mu m$. These results are shown in Table 4.20. For the other values of $\lambda$, polarization has very little influence on intensity (less than 1%).

Calculations show that, the contribution $I_1/I$ of singly scattered radiation to the total intensity depends slightly on $\lambda$. Therefore, this contribution is shown in Table 4.21 for $\lambda = 0.55\ \mu m$ only.

For high sun positions ($\psi = 30°$, $60°$) this contribution is approximately 40–70%, dependent on the direction of view. The single-scattering contribution increases as $\psi$ increases, and reaches 99% for some directions of view; next, it decreases sharply and tends to zero.

The quantity $I_1/I$ depends also on the albedo of Earth's surface. For high sun positions, this dependence is strong, and for high values of $\psi$ (> 89°) the influence of albedo on the quantity $I_1/I$ is negligible. The intensity decreases as $h(\theta)$ increases, and is six orders of magnitude less at $h \geqslant 90$ km (Fig. 4.14). It is interesting to note how the point of maximum intensity varies with the sun's height and azimuthal angle $\varphi$. The intensity for $\psi = 30°$ reaches a maximum near the horizon.

**Table 4.20.** Relative errors in intensity calculations due to neglect of polarization [%] ($\lambda = 0.4\ \mu m$, $P_a = 0.8$)

| $h(\theta)$ [km] | $\psi = 30°$ | | | $\psi = 60°$ | | | $\psi = 89°$ | | |
|---|---|---|---|---|---|---|---|---|---|
| | $\varphi = 0$ | $\varphi = 90°$ | $\varphi = 180°$ | $\varphi = 0$ | $\varphi = 90°$ | $\varphi = 180°$ | $\varphi = 0$ | $\varphi = 90°$ | $\varphi = 180°$ |
| 0 | 3 | −2.6 | −3 | 1.3 | −0.9 | 6.3 | −2.6 | −2.7 | −0.2 |
| 4 | 4 | −2.7 | −3.7 | 3.6 | −3 | −2.3 | −3 | −3.2 | −0.2 |
| 8 | 2.3 | −3.6 | −4.6 | 0.3 | −4.2 | −2 | −2.6 | −3.6 | −0.2 |
| 12 | 3.3 | −5 | −3.7 | 0.1 | 0.7 | −4.2 | −2.6 | −4.4 | −0.2 |
| 16 | 3 | −5 | −4.4 | 0.2 | 1 | −3.7 | −2.2 | −1.3 | −0.2 |
| 20 | 3 | −5 | −4.2 | −0.5 | 3.2 | −4.6 | −2.3 | 0.2 | −0.4 |
| 30 | 3 | −6 | −2 | −0.4 | 2 | −4.6 | −2.1 | 1.6 | 0 |
| 40 | 2.3 | −5.4 | −1 | −0.4 | 3.8 | −4 | −2.4 | 5.8 | 0 |
| 50 | 2.4 | −5.5 | −1 | −0.5 | 5.6 | −3.7 | −2 | 0.1 | 0 |
| 65 | 2.8 | −6.5 | −1.4 | −0.5 | 0.6 | −3.6 | −2 | 2.1 | 0 |
| 80 | 2.5 | −4.5 | −1.8 | −0.3 | 1.2 | −2.1 | −1.2 | 0 | 0 |
| 95 | 3.2 | −4.5 | −2 | −0.3 | 2.4 | −3.2 | −0.4 | 0 | 0 |

**Table 4.21.** Contribution $I_1/I$ from single-scattered radiation to total intensity [%] ($\lambda = 0.55\ \mu$m)

| $h(\theta)$ [km] | $\psi = 30°$ | | | | $\psi = 89°$ | | | | $\psi = 95°$ | | | |
|---|---|---|---|---|---|---|---|---|---|---|---|---|
| | $\varphi = 0°$ | | $\varphi = 90°$ | $\varphi = 180°$ | $\varphi = 0°$ | | $\varphi = 90°$ | $\varphi = 180°$ | $\varphi = 0°$ | | $\varphi = 90°$ | $\varphi = 180°$ |
| | $P_a = 0$ | $P_a = 0.8$ | $P_a = 0.8$ | $P_a = 0.8$ | $P_a = 0$ | $P_a = 0.8$ | $P_a = 0.8$ | $P_a = 0.8$ | $P_a = 0$ | $P_a = 0.8$ | $P_a = 0.8$ | $P_a = 0.8$ |
| 0 | 71 | 27 | 29 | 39 | 69 | 66 | 65 | 91 | 84 | 82 | 71 | 56 |
| 4 | 80 | 30 | 33 | 39 | 86 | 81 | 86 | 92 | 88 | 86 | 76 | 58 |
| 8 | 76 | 32 | 31 | 39 | 88 | 83 | 85 | 92 | 88 | 86 | 77 | 42 |
| 12 | 77 | 32 | 32 | 42 | 87 | 83 | 80 | 93 | 86 | 84 | 79 | 59 |
| 16 | 75 | 30 | 30 | 41 | 91 | 87 | 93 | 93 | 92 | 91 | 84 | 75 |
| 20 | 77 | 31 | 30 | 40 | 92 | 88 | 97 | 94 | 95 | 93 | 78 | 78 |
| 30 | 84 | 33 | 30 | 40 | 94 | 91 | 99 | 95 | 98 | 97 | 87 | 87 |
| 40 | 88 | 38 | 37 | 49 | 95 | 91 | 92 | 96 | 98 | 97 | 88 | 94 |
| 50 | 88 | 37 | 33 | 43 | 96 | 94 | 91 | 96 | 98 | 98 | 88 | 99 |
| 65 | 89 | 39 | 37 | 47 | 96 | 94 | 98 | 97 | 98 | 98 | 89 | 99 |
| 80 | 88 | 39 | 31 | 38 | 97 | 95 | 99 | 98 | 99 | 99 | 90 | 99 |
| 95 | 86 | 35 | 24 | 27 | 97 | 96 | 98 | 95 | 99 | 99 | 92 | 97 |

**Table 4.22.** Albedo $P_a(\lambda)$ for snow and grass

| $\lambda\ \mu$m | 0.36 | 0.42 | 0.48 | 0.50 | 0.55 | 0.60 | 0.68 | 0.80 | 0.91 | 0.99 |
|---|---|---|---|---|---|---|---|---|---|---|
| Snow | 0.832 | 0.832 | 0.842 | 0.850 | 0.854 | 0.856 | 0.854 | 0.842 | 0.810 | 0.763 |
| Grass | 0.0263 | 0.0263 | 0.0278 | 0.0300 | 0.0545 | 0.0470 | 0.0437 | 0.3325 | 0.3775 | 0.3775 |

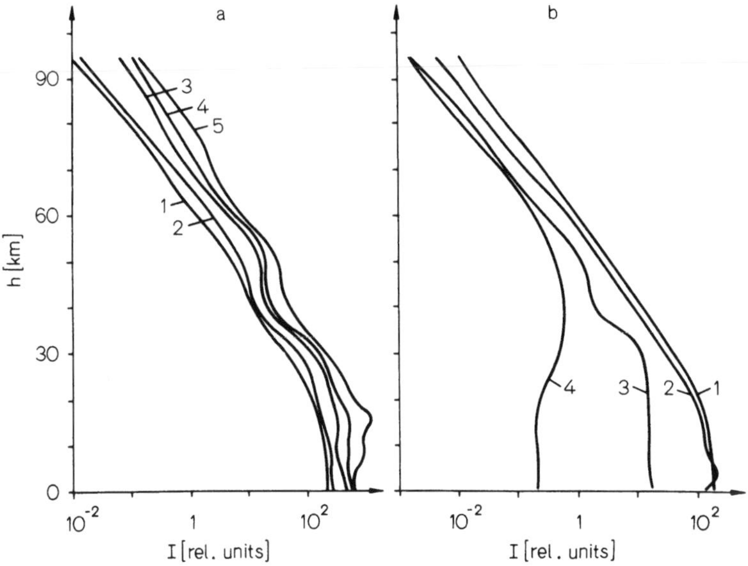


**Fig. 4.14a, b.** Intensity of multiply scattered radiation as function of altitude **(a)** $\varphi = 0°$; **(b)** $\varphi = 90°$; *(1)* $\psi = 30°$, *(2)* $\psi = 60°$, *(3)* $\psi = 89°$, *(4)* $\psi = 95°$, *(5)* $\psi = 100°$, $H = 250$ km, $\lambda = 0.55$ $\mu$m, $P_a = 0.8$

As $\psi$ increases, the altitude at which the intensity reaches a maximum also increases, as is particularly clearly seen at $\varphi = 90°$ (Fig. 4.14). The degree of polarization of multiply scattered radiation, for various values of the parameters $\lambda$, $\psi$, $\varphi$, $h(\theta)$, is shown in Figs. 4.15, 16.

There is a point $\psi = \psi_0$ at which the degree of polarization changes sign, when $\psi$ increases. The position of this point depends on the wavelength $\lambda$ and the direction of view (Fig. 4.15). The degree of polarization varies little when $h$ varies between 0 and 65 km. The degree of polarization at $\lambda = 0.55$ $\mu$m is shown as a function of altitude in Fig. 4.16 for $\varphi = 0$ and $\varphi = 180°$. Also shown is the influence of the albedo $P_a$ on the degree of polarization.

Calculations were done for (1) the illuminated region ($\psi = 30°$, $60°$, $89°$), and (2) the twilight region ($\psi = 91°$, $95°$, $100°$). All calculations were done in 120 minutes of BESM-6 computer time. The relative statistical errors of the computational results were approximately 10%. Both variants were used, with the same trajectories.

Consider now the problem of estimating the spectral radiance of the atmosphere near the horizon as observed at an altitude $10^5$ km. The problem was solved for seven models of vertical distribution of the aerosol cross section $\sigma_a(h, \lambda)$ proposed by the Department of Atmospheric Physics of the Leningrad State University [34]. The calculations were done for 10 values of wavelength $\lambda$ between 0.36 and 0.99 $\mu$m. The values of $\lambda$ and the corresponding values of albedo are given in Table 4.22.

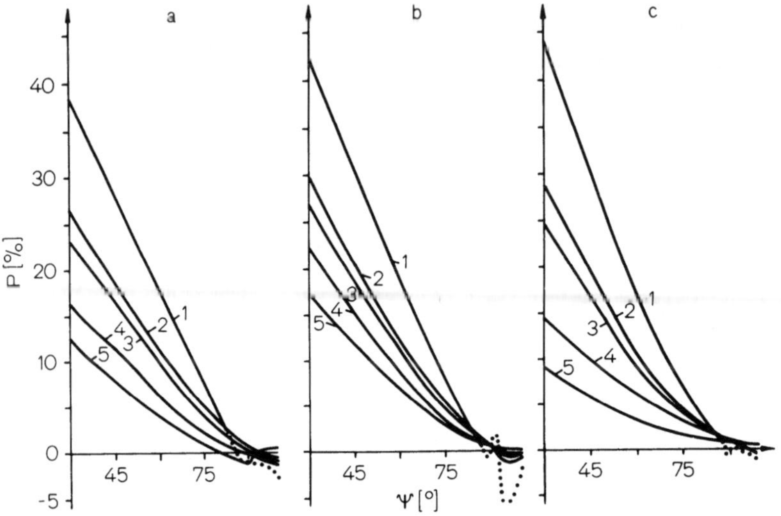


**Fig. 4.15a, b, c.** Degree of polarization of multiply scattered light as function of $\psi$ **(a)** $h(\theta) = 0$ (horizon); **(b)** $h(\theta) = 8$ km; **(c)** $h(\theta) = 20$ km; *(1)* $\lambda = 0.4\ \mu$m, *(2)* $\lambda = 0.55\ \mu$m, *(3)* $\lambda = 0.6\ \mu$m, *(4)* $\lambda = 0.7\ \mu$m, *(5)* $\lambda = 0.8\ \mu$m, $\varphi = 0°$, $P_a = 0.8$, $H = 250$ km

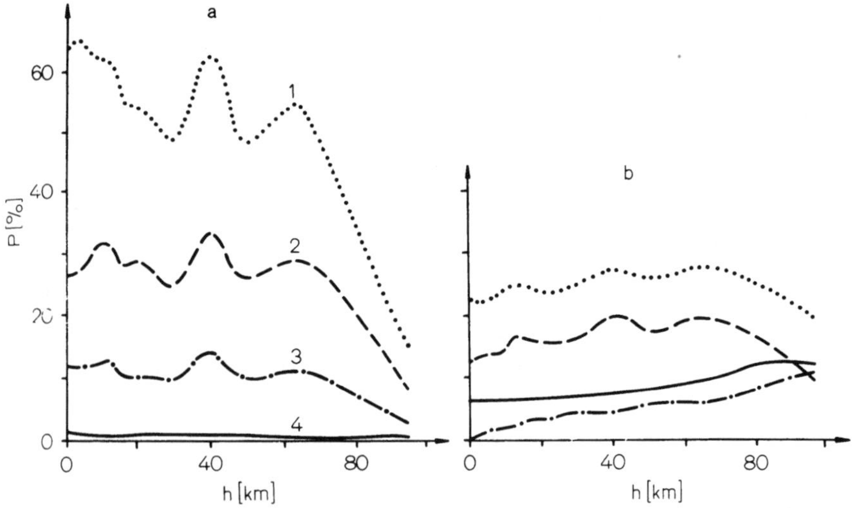


**Fig. 4.16a, b.** Degree of polarization as function of altitude at $\lambda = 0.55\ \mu$m **(a)** $\varphi = 0°$; **(b)** $\varphi = 180°$; *(1)* $\psi = 30°$, $P_a = 0$, *(2)* $\psi = 30°$, $P_a = 0.8$, *(3)* $\psi = 60°$, $P_a = 0.8$, *(4)* $\psi = 89°$, $P_a = 0.8$, $H = 250$ km

The curves of laminal and smooth models of $\sigma_a$ at $\lambda = \lambda_0 = 0.55\ \mu m$ are shown in Fig. 4.7. The coefficients of molecular scattering were defined by

$$\sigma_M(h, \lambda) = (\lambda_0/\lambda)^4 \cdot \sigma_M(h, \lambda_0);$$

the values of $\sigma_M(h, \lambda_0)$ were obtained from the standard model of [34] by dividing the atmosphere of altitude 100 km into 79 layers.

The indicatrix of aerosol scattering was assumed to be a linear function of altitude in each of the layers: 0–5; 5–22.5; 22.5–44; 44–72; 72–100 km, and was tabulated from the formula

$$f_a(\theta, h) = f(\theta)[\sigma_a(h)]^{K(\theta)}, \tag{4.61}$$

where $f(\theta)$ and $K(\theta)$ are empirical functions.

The variation of $\sigma_a$ within the four upper layers appears not to affect significantly the normalized indicatrices. Therefore, for these layers, we used constant aerosol indicatrices that correspond to mean optical values of $\sigma_a(h)$: $0.147 \times 10^{-2}$, $0.228 \times 10^{-3}$, $0.386 \times 10^{-4}$, $0.282 \times 10^{-5}\ \text{km}^{-1}$ calculated from the smooth model of $\sigma_a(h, \lambda_0)$. In the first layer, the indicatrix was interpolated with respect to altitude. Indicatrices for altitudes $h = 0$ and $h = 5$ km computed from (4.61) are $\sigma_a(0) = 0.06\ \text{km}^{-1}$, $\sigma_a(5) = 0.01\ \text{km}^{-1}$ also correspond to a smooth model. The aerosol indicatrix was assumed not to depend on wavelength. However, the total indicatrix depends on wavelength (see Sect. 4.1) Lambert's reflection law was assumed. The problem was solved for two observation variants: $H = 10^5$ km, $\psi = 0$ and $H = 10^5$ km, $\psi = 105°26'$ and the subdivisions of $h(\theta)$: $-8, -6, -4, -2, 0, 2, 4, 6, 8, 12, 16, 20, 24, 28$ km.

The computed results are the integrals of intensity over $\theta$ intervals (see the beginning of this chapter). Results were obtained for 7 models of $\sigma_a(h, \lambda)$ with the albedo of snow, and also for the smooth model with the albedo of grass, by simulating the same Markov chain for each observation variant.

The computational results are shown in Figs. 4.17–20. The feature of these results is a sharp maximum near the altitude of 12–20 km due to the presence of a maximum aerosol amount near 17 km.

## 4.11 Estimation of the Radiation Field in the Atmosphere

We assume collimated incident flux of solar radiation at the top ($z = H$) of a plane-parallel (inhomogeneous with altitude $z$) atmosphere. As usual, the optical characteristics are given: $\sigma_{as}(z, \lambda)$, the coefficient of aerosol scattering; $\sigma_{Ms}(z, \lambda)$, the coefficient of molecular scattering; $\sigma_{ac}(z, \lambda)$ and $\sigma_{Mc}(z, \lambda)$, the coefficients of aerosol and molecular absorbtion; $g_a(\mu_0, z, \lambda)$ and $g_M(\mu_0)$, the indicatrices of aerosol and molecular scattering, respectively. Here $\lambda$ is the wavelength, and $\mu_0$ is the cosine of the scattering angle. The indicatrix $g_M(\mu_0)$

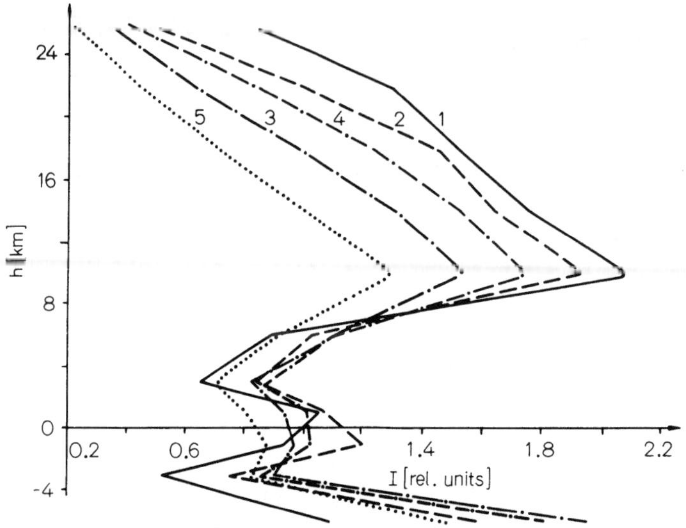


**Fig. 4.17.** Radiance as function of altitude for various wave-lengths $\lambda$ (*1*) $\lambda = 0.36\ \mu$m, (*2*) $\lambda = 0.48\ \mu$m, (*3*) $\lambda = 0.55\ \mu$m, (*4*) $\lambda = 0.68\ \mu$m, (*5*) $\lambda = 0.99\ \mu$m, (smooth model $\sigma_a$, albedo of snow, $\psi = 105°26'$, $H = 10^5$ km)

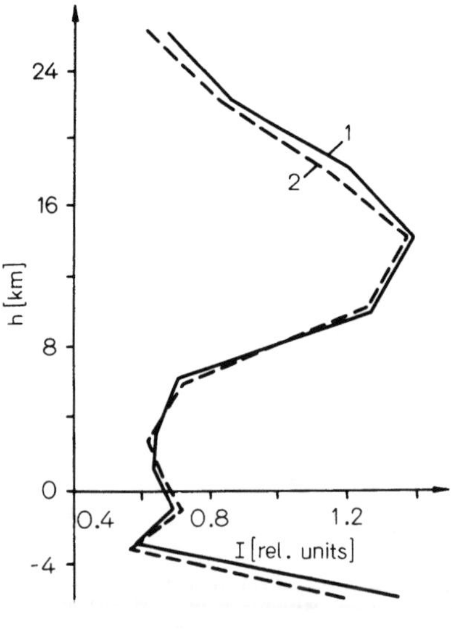


**Fig. 4.18.** Radiance as function of altitude for two models of $\sigma_a$ (*1*) smooth model, (*2*) laminal model $\psi = 0$, $H = 10^5$ km, $\lambda = 0.55\ \mu$m

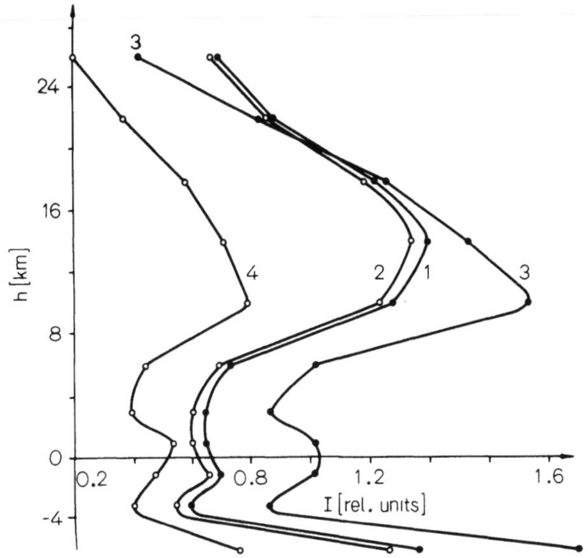


**Fig. 4.19.** Dependence of intensity on albedo of Earth's surface (*1*), (*2*): $\psi = 0$, (*3*), (*4*): $\psi = 105°26'$, (*1*), (*3*): albedo of snow; (*2*), (*4*): albedo of grass. $H = 10^5$ km, $\lambda = 0.55$ $\mu$m

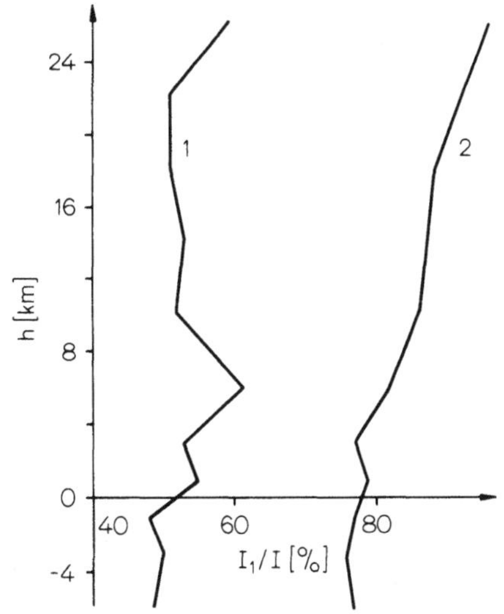


**Fig. 4.20.** Single-scattering contribution to total intensity (*1*) $\psi = 0$, (*2*) $\psi = 105°26'$; $H = 10^5$ km, $\lambda = 0.55$ $\mu$m, albedo of snow

does not depend on altitude $z$ or wavelength $\lambda$, and is defined by

$$g_M(\mu_0) = \frac{3}{8}(1 + \mu_0^2), \quad -1 \leqslant \mu_0 \leqslant 1. \tag{4.62}$$

The reflected light has equal flux per unit solid angle in every direction i.e., the Earth's surface is assumed to be a Lambert reflector (see the beginning of this chapter) having albedo $P_a: 0 \leqslant P_a(\lambda) \leqslant 1$. The problem is to estimate the solar radiation $E_k(\lambda)$ absorbed by the $k$th layer: $h_{k-1} \leqslant z \leqslant h_k$, $k = 1, \ldots, n_k$, $h_0 = 0$, $h_{n_k} = H$.

For convenience, step functions of altitude $z$

$$\sigma_{as}(z, \lambda) = \sigma_{as}^{(i)}(\lambda), \quad \sigma_{ac}(z, \lambda) = \sigma_{ac}^{(i)}(\lambda),$$

$$\sigma_{Ms}(z, \lambda) = \sigma_{Ms}^{(i)}(\lambda), \quad \sigma_{Mc}(z, \lambda) = \sigma_{Mc}^{(i)}(\lambda),$$

where $z_{i-1} \leqslant z \leqslant z_i$, $i = 1, \ldots, n_\sigma$, $z_0 = 0$, $z_{n_\sigma} = H$ are used to approximate the scattering and absorption coefficients. It is assumed also that

$$g_a(\mu_0, z, \lambda) = g_a^{(m)}(\mu_0, \lambda) \quad \text{if} \quad s_{m-1} \leqslant z \leqslant s_m, \quad m = 1, \ldots, n_g,$$
$$s_0 = 0, \quad s_{n_g} = H.$$

For convenience, the parameter $\lambda$ will be omitted.

The direction of propagation of light is defined by the polar angle $\theta$ and the azimuth angle $\varphi$; the $z$ axis is oriented upwards; $\varphi$ is measured clockwise, looking upward. Because the functions $\sigma_{as}$, $\sigma_{ac}$, $\sigma_{Ms}$, $\sigma_{Mc}$, $g_a$ do not depend on $x$ and $y$, the photon's position in space need be defined by only altitude $z$.

To estimate $E_k$, $k = 1, \ldots, n_k$, we may use

$$E_k = \Phi_\downarrow(h_k) + \Phi^\uparrow(h_{k-1}) - \Phi_\downarrow(h_{k-1}) - \Phi^\uparrow(h_k), \tag{4.63}$$

where

$$\begin{aligned} \Phi_\downarrow(h) &= \int_{-1}^{0}\int_{0}^{2\pi} \Phi(h, \mu, \varphi)|\mu|\, d\varphi\, d\mu, \\ \Phi^\uparrow(h) &= \int_{0}^{1}\int_{0}^{2\pi} \Phi(h, \mu, \varphi)\mu\, d\varphi\, d\mu. \end{aligned} \tag{4.64}$$

Here $\mu = \cos\theta$; $\Phi(h, \mu, \varphi)$ is the flux of light incident on the plane $z = h$ in the direction $(\mu, \varphi)$. To calculate the quantities $\Phi_\downarrow(h)$ and $\Phi^\uparrow(h)$, it is necessary to estimate the average number of photons that cross the plane $z = h$ with $\mu < 0$ and $\mu > 0$, respectively.

Another method to estimate $E_k$ (see Sect. 2.2) is based on the fact that the integral

$$\int_{h_{k-1}}^{h_k} dh \int_0^{2\pi} d\varphi \int_{-1}^{+1} \Phi(h, \mu, \varphi)\, d\mu,$$

equals $L_k$, the average length of the photon's trajectory within the layer $h_{k-1} \leqslant z \leqslant h_k$. Introduce the notation:

$$\sigma_{\mathrm{c}}^{(i)} = \sigma_{\mathrm{ac}}^{(i)} + \sigma_{\mathrm{Mc}}^{(i)} \quad \text{and} \quad \sigma_{\mathrm{s}}^{(i)} = \sigma_{\mathrm{as}}^{(i)} + \sigma_{\mathrm{Mc}}^{(i)}.$$

Let $z_i \leqslant h_{k-1} \leqslant z_{i+1}$ and $z_{i+n-1} \leqslant h_k \leqslant z_{i+n}$.

Then

$$\begin{aligned} E_k = \int_{h_{k-1}}^{h_k} dz \int_0^{2\pi} d\varphi \int_{-1}^{+1} \sigma_{\mathrm{c}}(z)\Phi(z, \mu, \varphi)\, d\mu &= \sigma_{\mathrm{c}}^{(i+1)} \int_{h_{k-1}}^{z_{i+1}} dz \int_0^{2\pi} d\varphi \int_{-1}^{+1} \Phi(z, \mu, \varphi)\, d\mu \\ &+ \sigma_{\mathrm{c}}^{(i+n)} \int_{z_{i+h-1}}^{h_k} dz \int_0^{2\pi} d\varphi \int_{-1}^{+1} \Phi(z, \mu, \varphi)\, d\mu \\ &+ \sum_{j=i+1}^{i+n-2} \sigma_{\mathrm{c}}^{(j+1)} \int_{z_j}^{z_j+1} dz \int_0^{2\pi} d\varphi \int_{-1}^{+1} \Phi(z, \mu, \varphi)\, d\mu = \sum_{j=i+1}^{i+n} \sigma_{\mathrm{c}}^{(j)} L_j, \end{aligned} \tag{4.65}$$

where $L_{i+1}$, $L_{i+n}$, $L_j$ are the average lengths of trajectories within the layers $h_{k-1} \leqslant z \leqslant z_{i+1}, z_{i+n-1} \leqslant z \leqslant h_k, z_j \leqslant z \leqslant z_{j+1}, (j = i + 2, \ldots, i + n - 1)$, respectively.

We now describe in detail the algorithm for solving the problem just formulated. Consider a photon that has undergone a collision in a plane $z = z'$, and moving in the direction $(\mu', \varphi')$. Assume that $z_{i-1} \leqslant z' \leqslant z_i$, and $s_{m-1} \leqslant z' \leqslant s_m$. It is known that the variance of the algorithm will be reduced if, instead of simulating the photon's absorption, the weight is transformed by:

$$Q = Q' \frac{\sigma_{\mathrm{s}}^{(i)}}{\sigma_{\mathrm{s}}^{(i)} + \sigma_{\mathrm{c}}^{(i)}}.$$

Furthermore, the trajectories are terminated if the photon escapes from the medium. In this case, the distribution density of $\mu_0$, the cosine of the scattering angle, is given by

$$g(v) = \frac{g_{\mathrm{a}}^{(m)}(v)\sigma_{\mathrm{as}}^{(i)} + g_{\mathrm{M}}(v)\sigma_{\mathrm{Ms}}^{(i)}}{\sigma_{\mathrm{as}}^{(i)} + \sigma_{\mathrm{Ms}}^{(i)}}. \tag{4.66}$$

The quantity $\mu_0$ is sampled from $g_{\mathrm{a}}^{(m)}(v)$ with probability $p = \sigma_{\mathrm{as}}^{(i)}/(\sigma_{\mathrm{as}}^{(i)} + \sigma_{\mathrm{Ms}}^{(i)})$, while it is sampled from the density $g_{\mathrm{M}}(v)$ with probability $1 - p$.

Assume that the phase function of aerosol scattering $g_a^{(m)}(\nu)$ is piecewise linear in each interval $(\nu_{j-1}, \nu_j)$, $j = 1, \ldots, n_\mu$, $\nu_0 = 1$, $\nu_{n_\mu} = 1$; furthermore,

$$-\sum_{j=1}^{n_\mu} \frac{(g_{j-1} + g_j)}{2} \Delta\nu_j = 1,$$

where

$$g_j = g_a^{(m)}(\nu_j), \quad \Delta\nu_j = \nu_j - \nu_{j-1} < 0.$$

In this case, $\mu_0$ is simulated as

$$\mu_0 = \nu_n - \frac{g_n \Delta\nu_n + [g_n^2(\Delta\nu_n)^2 - 2\Delta\nu_n(g_n - g_{n-1})\beta_n]^{1/2}}{g_n - g_{n-1}},$$

where

$$\beta_n = \alpha + \sum_{j=1}^{n} \frac{g_{j-1} - g_j}{2} \Delta\nu_j \leqslant 0,$$

if $\beta_{n-1} > 0$, and $\alpha$ is a random variable uniformly distributed between 0 and 1.

The $\mu_0$ simulation according to the density $g_M(\nu)$ is realized as

$$\mu_0 = \begin{cases} 8\alpha/3 - 1, & 0 \leqslant \alpha \leqslant 3/4, \\ \operatorname{sgn}(\alpha - 7/8)\max(\alpha_1, \alpha_2, \alpha_3), & 3/4 < \alpha \leqslant 1, \end{cases}$$

where $\alpha_1, \alpha_2, \alpha_3$ are independent. The coordinates of a new direction are given by

$$\varphi = 2\pi\alpha_4, \quad \mu = \mu'\mu_0 + [(1 - \mu'^2)(1 - \mu_0^2)]^{1/2} \cos\varphi.$$

To simulate the free-path length, a random variable $\alpha$ is chosen. Next a number $n$ is defined that satisfies the inequalities

$$0 < \sigma^{(i)} \frac{(z' - z_{i-1})}{\mu} - \sum_{j=1}^{n-1} \sigma^{(i-j)} l_{i-j} - \ln\alpha < \sigma^{(i-n)} \cdot l_{i-n}, \tag{4.67}$$

if $\mu < 0$; and

$$0 < -\sigma^{(i)} \frac{(z_i - z')}{\mu} - \sum_{j=1}^{n-1} \sigma^{(i+j)} l_{i+j} - \ln\alpha < \sigma^{(i+n)} l_{i+n}, \tag{4.68}$$

if $\mu > 0$.

Here

$$\sigma^{(j)} = \sigma_s^{(j)} + \sigma_c^{(j)}, \quad l_j = (z_j - z_{j-1})/|\mu|.$$

The free-path length $l$ is

$$l = \frac{z' - z_{i-1}}{\mu}\left[\frac{\sigma^{(i)}}{\sigma^{(i-n)}} - 1\right] + \sum_{j=1}^{n-1} l_{i-j}\left[1 - \frac{\sigma^{(i-j)}}{\sigma^{(i-n)}}\right] - \frac{\ln \alpha}{\sigma^{(i-n)}},$$

if $\mu < 0$, and

$$l = \frac{z_i - z'}{\mu}\left[1 - \frac{\sigma^{(i)}}{\sigma^{(i+n)}}\right] + \sum_{j=1}^{n-1} l_{i+j}\left[1 - \frac{\sigma^{(i+j)}}{\sigma^{(i+n)}}\right] - \frac{\ln \alpha}{\sigma^{(i+n)}},$$

if $\mu > 0$.

We assume that the photon escapes from the medium if $\mu > 0$, and reaches the Earth's surface if $\mu < 0$, provided that there is no $n$ that satisfies (4.67) or (4.68). In the latter case, the new direction of the photon's motion is $\varphi = 2\pi\alpha_1$, $\mu = \alpha_2^{1/2}$. The photon's weight is then multiplied by $P_a$, the reflection coefficient.

The quantities $E_k(\lambda_j)$, $j = 0, \ldots, n_j$, can be estimated by use of a single trajectory (see Sect. 4.6). The trajectory is constructed for a model for which the wavelength is $\lambda_0$.

Let $Q'(\lambda)$ be the weight of the photon that has undergone a scattering at $z'$ in the direction $(\mu', \varphi')$. If the free-path length $l$ is chosen, this weight is multiplied by

$$[\sigma^{(i)}(\lambda)/\sigma^{(i)}(\lambda_0)] \exp\{-[\tau(l) - \tau_0(l)]\}, \tag{4.69}$$

provided that a new collision occurs in the layer $z_{i-1} < z < z_i$.

Here $\tau$ and $\tau_0$ are the optical lengths of the free path for models that correspond to wavelengths $\lambda$ and $\lambda_0$, respectively. At a new collision point, the weight is multiplied by $\sigma_s^{(i)}(\lambda)/\sigma^{(i)}(\lambda)$, the survival probability; if $\mu_0$ is chosen, the weight is multiplied also by $g(\mu_0, \lambda)/g(\mu_0, \lambda_0)$, where $g(\nu, \lambda)$ is the total-scattering indicatrix given by (4.66). Thus the weight is transformed in one transition by

$$Q(\lambda) = Q'(\lambda)\frac{\sigma_s^{(i)}(\lambda)g(\mu_0, \lambda)}{\sigma^{(i)}(\lambda_0)g(\mu_0, \lambda_0)} \exp\{-[\tau(l) - \tau_0(l)]\}. \tag{4.70}$$

If it appears, after the free-path-length sampling, that the photon path intersects the plane $z = 0$ or $z = H$, then, instead of the factor (4.69), the weight $Q'(\lambda)$ must be multiplied by $\exp[-(\tau' - \tau_0')]$, where $\tau'$ and $\tau_0'$ are the optical lengths of trajectories from the collision point to the plane $z = 0$ or $z = H$ for models that correspond to wavelengths $\lambda$ and $\lambda_0$, respectively. When the photon goes into the layer $z_{i-1} \leqslant z \leqslant z_i$, the quantity $L_i(z)$ can be estimated, as simple ideas show, by storing the product of the length of the photon's path within this layer and the weight that the photon has at the end of the path.

Let $Q'(\lambda)$ be the photon's weight after scattering. Assume that the free-path length $l$ is chosen; if now the photon intersects the layer $z_{i-1} \leqslant z \leqslant z_i$ without a collision, then the value of

$$\frac{z_i - z_{i-1}}{|\mu|} Q'(\lambda) \exp[-(\tau - \tau_0)],$$

is stored. Here $\tau$ and $\tau_0$ are the optical lengths of the path between the scattering point and the plane $z = z_i$ if $\mu > 0$ (or the plane $z = z_{i-1}$ if $\mu < 0$) for models with $\lambda$ and $\lambda_0$, respectively. If the photon undergoes a collision within this layer, the quantity

$$\Delta l_i Q'(\lambda) \frac{\sigma^{(i)}(\lambda)}{\sigma^{(i)}(\lambda_0)} \exp\{-[\tau(l) - \tau_0(l)]\},$$

is stored, where $\Delta l_i$ is the length of the photon's path in this layer. When the photon intersects the plane $z = h_k$, the quantities $\Phi_\downarrow(h_k, \lambda)$ and $\Phi^\uparrow(h_k, \lambda)$ are estimated by storing the weight after the free-path length is simulated for $\mu < 0$ and $\mu > 0$, respectively.

In conclusion, let us consider a numerical example of estimating $E_k$, $k = 1, \ldots, n_k$ at $\lambda = 0.8$ $\mu$m. The optical model of the atmosphere developed at the Department of Atmospheric Physics of Leningrad State University was used (see Sect. 4.11).

Altitude profiles of the coefficients $\sigma_{as}$ and $\sigma_{ac}$ are shown in Fig. 4.21. Altitude distributions of the coefficients $\sigma_{Ms}$ and $\sigma_{Mc}$, compiled by *Elterman* [31], were used. The indicatrices of aerosol scattering are shown in Fig. 4.22. The indicatrix 1 was used for the layer $0 \leqslant z \leqslant 5$ km; 2 for the layer $5 \leqslant z \leqslant$

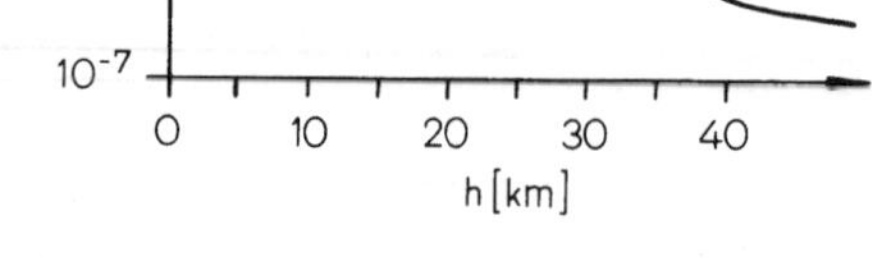

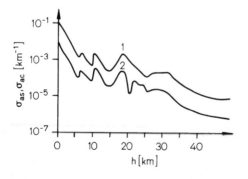

**Fig. 4.21.** Altitude profiles of coefficients (1) $\sigma_{as}$ and (2) $\sigma_{ac}$

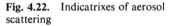

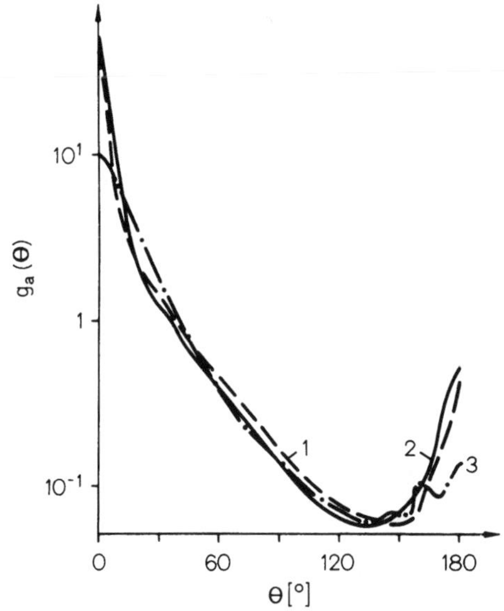


**Fig. 4.22.** Indicatrixes of aerosol scattering

17 km; and 3, for the layer $17 \leqslant z \leqslant 50$ km. In calculations it was assumed that $n_\sigma = 45$, $n_g = 3$, $n_k = 9$, $H = 50$ km. The cosine of angle between the $z$ axis and the direction of the incident light was taken to be 0.55. The results of estimations of $E_k$ ($k = 1, \ldots, 9$) for two values of albedo: $P_a = 0$, $P_a = 0.8$ are given in Table 4.23.

Here the following notation is used. The results obtained by the use of (4.63) are listed in column $E_k^{(1)}$, while $E_k^{(2)}$ corresponds to (4.65); $\beta_k$ is the relative standard deviation of the estimate $E_k^{(1)}$. The statistical error of the estimate $E_k^{(2)}$ is less than $\beta_k$.

**Table 4.23.** Altitude dependence of absorption of solar radiation in atmosphere

| $h_k$ [km] | $P_a = 0$ | | | $P_a = 0.8$ | | |
|---|---|---|---|---|---|---|
| | $E_k^{(1)}$ | $\beta_k$ | $E_k^{(2)}$ | $E_k^{(1)}$ | $\beta_k$ | $E_k^{(2)}$ |
| 0.50 | 0.0071 | 0.024 | 0.0070 | 0.0134 | 0.016 | 0.0135 |
| 1.00 | 0.0038 | 0.027 | 0.0039 | 0.0069 | 0.020 | 0.0070 |
| 1.95 | 0.0046 | 0.024 | 0.0046 | 0.0082 | 0.017 | 0.0082 |
| 3.00 | 0.0025 | 0.036 | 0.0026 | 0.0045 | 0.027 | 0.0045 |
| 4.25 | 0.0017 | 0.045 | 0.0017 | 0.0027 | 0.035 | 0.0028 |
| 5.55 | 0.0008 | 0.052 | 0.0007 | 0.0013 | 0.031 | 0.0012 |
| 7.20 | 0.0007 | 0.063 | 0.0007 | 0.0012 | 0.042 | 0.0012 |
| 11.75 | 0.0020 | 0.053 | 0.0020 | 0.0034 | 0.037 | 0.0033 |
| 50.00 | 0.0092 | 0.034 | 0.0089 | 0.0150 | 0.026 | 0.0147 |

## 4.12 Some Comparisons of Monte Carlo Techniques with Other Methods

In this section, the Monte Carlo technique will be compared numerically with some other methods.

The spatial-angular distribution of radiance obtained by the Monte Carlo method is compared here with those obtained by the method of characteristics with successive interpolation (see, for example, [35]). An atmospheric model defined by mean values of the optical parameters in the visible spectral region was used in present calculations at $\lambda = 0.66\ \mu$m (see [36]). The total cross section for aerosol and molecular scattering $\sigma_s(h)$ was used, as described in Sect. 4.1.

The atmosphere is divided into 100 layers of the same thickness of 1 km. The altitude profile of the coefficient $\sigma_s(h)$ is shown in Fig. 4.23. The corresponding optical thickness of atmosphere measured in vertical direction is equal to 0.38.

Absorption by atmospheric gases was not taken into account, i.e., the photon-survival probability was taken to be 1. The total aerosol and molecular scattering indicatrix was taken to be constant with altitude (see Fig. 25). Reflection from the ground was not taken into account.

Comparisons were made for two observation variants:

a) The angular distribution of the radiance that reaches the Earth's surface was calculated:

b) The radiance of the atmosphere at the altitude $H = 250$ km near the horizon was calculated for $h(\theta) = -36, -32, -28, \ldots, 40$ km ($\Delta h = 4$ km).

In both of these cases, the calculations were done for two positions of the sun ($\psi = 60°, 88°$) and two azimuth angles ($\varphi = 0, 180°$). The results obtained in a single-scattering approximation agree well (to within 5%, cf. Tables 4.24, 25). Statistical fluctuations of the Monte Carlo results at $H = 0$ are approximately 3% for $\psi = 88°$, and approximately 8% for $\psi = 60°$, whereas at $H = 250$ km they are 3–6% and 2–6%, respectively.

The estimates of the secondary-scattering contribution obtained by the characteristic method are somewhat larger than those obtained by the Monte Carlo method (see Tables 4.26, 27). Particular differences occur at $H = 250$ km near $\psi = 90°$ (see Table 4.27).

These differences also occur when other Monte Carlo algorithms are used, e.g., local methods and the method of adjoint-walk simulation, described in Sects. 4.2, 4.6, respectively. The Monte Carlo method is used when the scattered radiation is to be calculated, for few points of observation and directions of view, for example, in atmospheric-optics problems connected with estimating the spectral characteristics of the radiation field, or with obtaining the dependence of these characteristics on the parameters of the atmospheric model. To obtain the results to within $\delta = 3$–$10\%$ ($\delta$ is the relative statistical error) required approximately 60–180 minutes of BESM-6 computer time. For example, all cal-

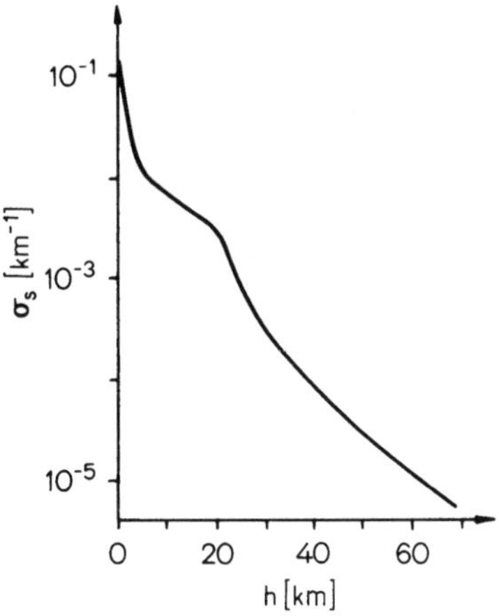


**Fig. 4.23.** Altitude profile of scattering coefficient

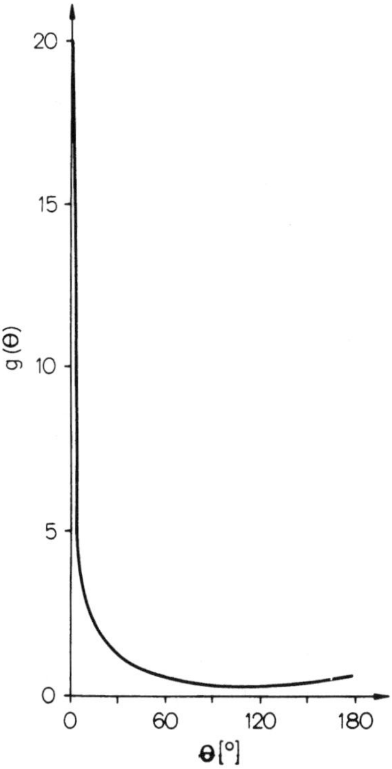


**Fig. 4.24.** Total scattering indicatrix

**Table 4.24.** Comparison of intensities of single-scattered light obtained by the Monte Carlo method (I) with those obtained by the characteristic method (II) ($H = 0$)

| $\theta°$ | $\psi = 60°$ | | | | $\psi = 88°$ | | | |
|---|---|---|---|---|---|---|---|---|
| | $\varphi = 0$ | | $\varphi = 180°$ | | $\varphi = 0$ | | $\varphi = 180°$ | |
| | I | II | I | II | I | II | I | II |
| 95 | 0.4409 | 0.3932 | $0.8965 \times 10^{-1}$ | $0.8636 \times 10^{-1}$ | $0.7466 \times 10^{-1}$ | $0.6705 \times 10^{-1}$ | $0.3270 \times 10^{-2}$ | $0.3016 \times 10^{-2}$ |
| 100 | 0.5247 | 0.4449 | $0.8252 \times 10^{-1}$ | $0.8016 \times 10^{-1}$ | $0.7925 \times 10^{-1}$ | $0.7736 \times 10^{-1}$ | $0.8950 \times 10^{-2}$ | $0.8342 \times 10^{-2}$ |
| 105 | 0.6763 | 0.5646 | $0.7388 \times 10^{-1}$ | $0.7112 \times 10^{-1}$ | $0.7965 \times 10^{-1}$ | $0.7438 \times 10^{-1}$ | $0.1065 \times 10^{-1}$ | $0.9983 \times 10^{-2}$ |
| 110 | 0.7890 | 0.6075 | $0.6520 \times 10^{-1}$ | $0.6186 \times 10^{-1}$ | $0.6102 \times 10^{-1}$ | $0.5812 \times 10^{-1}$ | $0.1030 \times 10^{-1}$ | $0.9806 \times 10^{-2}$ |
| 115 | 1.0560 | 0.8222 | $0.5730 \times 10^{-1}$ | $0.5371 \times 10^{-1}$ | $0.4629 \times 10^{-1}$ | $0.4585 \times 10^{-1}$ | $0.9611 \times 10^{-2}$ | $0.9513 \times 10^{-2}$ |
| 125 | 0.6816 | 0.6996 | $0.4698 \times 10^{-1}$ | $0.4491 \times 10^{-1}$ | $0.2924 \times 10^{-1}$ | $0.2784 \times 10^{-1}$ | $0.7893 \times 10^{-2}$ | $0.7485 \times 10^{-2}$ |
| 130 | 0.4059 | 0.4350 | $0.4261 \times 10^{-1}$ | $0.4279 \times 10^{-1}$ | $0.2332 \times 10^{-1}$ | $0.2190 \times 10^{-1}$ | $0.7167 \times 10^{-2}$ | $0.6717 \times 10^{-2}$ |
| 135 | 0.3074 | 0.3091 | $0.3919 \times 10^{-1}$ | $0.3599 \times 10^{-1}$ | $0.1888 \times 10^{-1}$ | $0.1886 \times 10^{-1}$ | $0.6681 \times 10^{-2}$ | $0.6676 \times 10^{-2}$ |
| 140 | 0.2139 | 0.2128 | $0.3655 \times 10^{-1}$ | $0.3399 \times 10^{-1}$ | $0.1538 \times 10^{-1}$ | $0.1525 \times 10^{-1}$ | $0.6278 \times 10^{-2}$ | $0.6232 \times 10^{-2}$ |
| 150 | 0.1385 | 0.1359 | $0.3546 \times 10^{-1}$ | $0.3213 \times 10^{-1}$ | $0.1023 \times 10^{-1}$ | $0.1051 \times 10^{-1}$ | $0.5772 \times 10^{-2}$ | $0.5697 \times 10^{-2}$ |
| 160 | $0.9586 \times 10^{-1}$ | $0.9248 \times 10^{-1}$ | $0.3604 \times 10^{-1}$ | $0.3474 \times 10^{-1}$ | $0.7931 \times 10^{-2}$ | $0.7656 \times 10^{-2}$ | $0.5364 \times 10^{-2}$ | $0.5186 \times 10^{-2}$ |
| 170 | $0.6803 \times 10^{-1}$ | $0.6657 \times 10^{-1}$ | $0.4191 \times 10^{-1}$ | $0.4036 \times 10^{-1}$ | $0.5990 \times 10^{-2}$ | $0.6091 \times 10^{-2}$ | $0.5139 \times 10^{-2}$ | $0.4896 \times 10^{-2}$ |
| 175 | $0.6072 \times 10^{-1}$ | $0.5862 \times 10^{-1}$ | $0.4665 \times 10^{-1}$ | $0.4538 \times 10^{-1}$ | $0.5691 \times 10^{-2}$ | $0.5645 \times 10^{-2}$ | $0.5254 \times 10^{-2}$ | $0.5039 \times 10^{-2}$ |
| 178 | $0.5544 \times 10^{-1}$ | $0.5364 \times 10^{-1}$ | $0.5082 \times 10^{-1}$ | $0.4832 \times 10^{-1}$ | $0.5497 \times 10^{-2}$ | $0.5423 \times 10^{-2}$ | $0.5356 \times 10^{-2}$ | $0.5126 \times 10^{-2}$ |
| 180 | $0.5264 \times 10^{-1}$ | $0.5024 \times 10^{-1}$ | $0.5264 \times 10^{-1}$ | $0.5024 \times 10^{-1}$ | $0.5426 \times 10^{-2}$ | $0.5275 \times 10^{-2}$ | $0.5246 \times 10^{-2}$ | $0.5274 \times 10^{-2}$ |

**Table 4.25.** Intensity of a single-scattered light ($H = 250$ km). I–Monte Carlo method; II–Characteristic method.

| $h(\theta)$ [km] | $\psi = 60°$ | | | | $\psi = 88°$ | | | |
|---|---|---|---|---|---|---|---|---|
| | $\varphi = 0$ | | $\varphi = 180°$ | | $\varphi = 0$ | | $\varphi = 180°$ | |
| | I | II | I | II | I | II | I | II |
| 40 | 0.0177 | 0.0189 | 0.0102 | 0.0101 | 0.0546 | 0.0539 | $0.579 \times 10^{-4}$ | $0.572 \times 10^{-4}$ |
| 36 | 0.0257 | 0.0279 | 0.0149 | 0.0150 | 0.0802 | 0.0791 | $0.683 \times 10^{-4}$ | $0.674 \times 10^{-4}$ |
| 32 | 0.0386 | 0.043 | 0.0230 | 0.0233 | 0.121 | 0.121 | $0.791 \times 10^{-4}$ | $0.782 \times 10^{-4}$ |
| 28 | 0.0654 | 0.072 | 0.0391 | 0.0391 | 0.204 | 0.201 | $0.904 \times 10^{-4}$ | $0.893 \times 10^{-4}$ |
| 24 | 0.126 | 0.137 | 0.0762 | 0.075 | 0.384 | 0.382 | $0.102 \times 10^{-3}$ | $0.101 \times 10^{-3}$ |
| 20 | 0.225 | 0.241 | 0.136 | 0.130 | 0.675 | 0.660 | $0.114 \times 10^{-3}$ | $0.112 \times 10^{-3}$ |
| 16 | 0.302 | 0.316 | 0.181 | 0.169 | 0.875 | 0.848 | $0.127 \times 10^{-3}$ | $0.125 \times 10^{-3}$ |
| 12 | 0.324 | 0.336 | 0.191 | 0.176 | 0.898 | 0.874 | $0.140 \times 10^{-3}$ | $0.138 \times 10^{-3}$ |
| 8 | 0.331 | 0.336 | 0.187 | 0.174 | 0.876 | 0.852 | $0.153 \times 10^{-3}$ | $0.151 \times 10^{-3}$ |
| 4 | 0.334 | 0.331 | 0.177 | 0.170 | 0.830 | 0.816 | $0.166 \times 10^{-3}$ | $0.164 \times 10^{-3}$ |
| 0 | 0.331 | 0.326 | 0.164 | 0.167 | 0.788 | 0.782 | $0.179 \times 10^{-3}$ | $0.177 \times 10^{-3}$ |
| −4 | 0.326 | 0.320 | 0.153 | 0.164 | 0.754 | 0.750 | $0.194 \times 10^{-3}$ | $0.191 \times 10^{-3}$ |
| −8 | 0.330 | 0.315 | 0.144 | 0.161 | 0.717 | 0.721 | $0.207 \times 10^{-3}$ | $0.206 \times 10^{-3}$ |
| −12 | 0.323 | 0.310 | 0.138 | 0.159 | 0.699 | 0.695 | $0.222 \times 10^{-3}$ | $0.220 \times 10^{-3}$ |
| −16 | 0.323 | 0.304 | 0.136 | 0.157 | 0.686 | 0.670 | $0.237 \times 10^{-3}$ | $0.235 \times 10^{-3}$ |
| −20 | 0.316 | 0.229 | 0.134 | 0.155 | 0.655 | 0.647 | $0.251 \times 10^{-3}$ | $0.250 \times 10^{-3}$ |
| −24 | 0.304 | 0.295 | 0.139 | 0.153 | 0.611 | 0.626 | $0.267 \times 10^{-3}$ | $0.265 \times 10^{-3}$ |
| −28 | 0.291 | 0.291 | 0.138 | 0.152 | 0.582 | 0.608 | $0.282 \times 10^{-3}$ | $0.278 \times 10^{-3}$ |
| −32 | 0.292 | 0.285 | 0.139 | 0.150 | 0.575 | 0.586 | $0.298 \times 10^{-3}$ | $0.296 \times 10^{-3}$ |
| −36 | 0.281 | 0.281 | 0.133 | 0.148 | 0.571 | 0.568 | $0.314 \times 10^{-3}$ | $0.312 \times 10^{-3}$ |

**Table 4.26.** Estimates of multiple-scattering contribution, [%] ($H = 0$). I–Monte Carlo method; II–Characteristic method

| $\theta$ [°] | $\psi = 60°$ | | | | $\psi = 88°$ | | | |
|---|---|---|---|---|---|---|---|---|
| | $\varphi = 0$ | | $\varphi = 180°$ | | $\varphi = 0$ | | $\varphi = 180°$ | |
| | I | II | I | II | I | II | I | II |
| 95 | 31 | 36 | 54 | 59 | 21 | 38 | 87 | 86 |
| 100 | 32 | 32 | 58 | 58 | 25 | 36 | 73 | 70 |
| 105 | 14 | 25 | 67 | 57.5 | 27 | 34.5 | 73 | 64.5 |
| 110 | 11 | 21 | 57 | 57 | 25 | 36 | 69 | 62 |
| 120 | 2 | 4 | 46 | 55.5 | 34 | 35.5 | 56 | 60 |
| 130 | 12 | 17 | 44 | 53 | 42 | 43 | 57 | 60.5 |
| 140 | 19 | 24 | 46 | 55 | 50 | 46 | 54 | 58 |
| 150 | 29 | 29 | 42 | 53 | 43 | 50 | 49 | 57.5 |
| 160 | 29 | 32 | 40 | 50 | 42 | 53 | 49 | 58 |
| 170 | 29 | 36 | 38 | 45.5 | 47 | 55.5 | 47 | 58.5 |
| 175 | 30 | 38 | 32 | 43 | 49 | 56.5 | 45 | 58 |
| 180 | 31 | 41 | 31 | 41 | 44 | 57 | 44 | 57 |

**Table 4.27.** Estimates of multiple-scattering contribution, [%] ($H = 250$ km). I: Monte Carlo method; II Characteristic method

| $h(\theta)$ [km] | $\psi = 60°$ | | | | $\psi = 88°$ | | | |
|---|---|---|---|---|---|---|---|---|
| | $\varphi = 0$ | | $\varphi = 180°$ | | $\varphi = 0$ | | $\varphi = 180°$ | |
| | I | II | I | II | I | II | I | II |
| 40 | 17 | 71.2 | 23 | 77 | 8.2 | 53 | 10 | 39 |
| 36 | 19 | 66.3 | 23 | 72 | 8.2 | 47 | 10.2 | 39 |
| 32 | 21 | 60 | 24 | 66 | 8.5 | 41 | 10.3 | 41 |
| 28 | 23 | 51 | 25 | 57 | 9.1 | 32 | 10.4 | 42 |
| 24 | 25 | 38 | 28 | 44 | 10.3 | 21 | 10.3 | 43 |
| 20 | 32 | 27 | 27 | 33 | 12.2 | 15 | 10.4 | 44 |
| 16 | 33 | 24 | 32 | 30 | 11 | 13 | 10.5 | 45 |
| 12 | 33 | 25 | 44 | 31 | 13 | 13.6 | 10.8 | 47 |
| 8 | 35 | 26 | 45 | 32 | 15 | 14.7 | 11.1 | 50 |
| 4 | 29 | 28 | 42 | 34 | 14 | 16 | 11.3 | 52 |
| 0 | 28 | 29 | 42 | 36 | 13.8 | 17 | 12 | 54 |
| −4 | 30 | 30 | 43 | 38 | 18 | 18 | 12.4 | 56 |
| −8 | 36 | 31 | 51 | 39 | 15.4 | 19 | 12.1 | 57.5 |
| −12 | 37 | 32 | 39 | 40 | 17.2 | 20 | 12.3 | 58.6 |
| −16 | 36 | 33 | 40 | 41 | 19 | 21 | 12 | 60 |
| −20 | 35 | 34 | 37 | 41 | 18 | 22 | 14 | 60.4 |
| −24 | 33 | 34.4 | 44 | 42 | 28 | 22 | 14.5 | 60.7 |
| −28 | 38 | 35 | 43 | 42 | 27 | 23 | 14.3 | 61 |
| −32 | 29 | 35 | 31 | 42 | 18 | 23 | 14.1 | 61 |
| −36 | 30 | 35.3 | 33 | 42 | 20 | 24 | 14 | 61.2 |

culations of the above problem for four observation points ($\psi = 60°$, $88°$; $H = 0$, 250 km), two values of azimuthal angle $\varphi$ and 20 directions of view were performed in 90 minutes of BESM-6 computer time.

The characteristic method calculates the field of radiance at all grid points. However the calculations were performed in 50–100 hours of BESM-6 computer time. Furthermore, in contrast to the Monte Carlo method, it is difficult to use the transmission function in the characteristic method or to take into account the light polarization.

Compare the Monte Carlo results for intensity and polarization of multiple scattered light with the tables compiled by *Coulson* et al. [37] for two observation variants: (1) $H = 0$ and (2) observation at the top of the atmosphere. The direction of view is defined by the zenith angle $\theta$; $\mu_0 = \cos\theta$ ($\mu > 0$ for upward observation, $\mu_0 < 0$ for downward observation). These calculations were done in order to investigate the efficiency and accuracy of the algorithm for calculating the polarization (see Sect. 4.4), and the modification of double local estimate (see Sect. 4.2). However, the calculated results can also be used to make some methodological conclusions. In particular, they show how sphericity influences the magnitude of the measured functionals.

Calculations were done for a Rayleigh atmosphere without absorption, with total optical thickness, measured in the vertical direction, $\tau = 0.1$, which corresponds to $\lambda = 0.546$ $\mu$m. In both observation variants, $\psi$ was taken to be

$53°8'$. The altitude of the atmosphere was taken to be 100 km. Downward observation was simulated by the Monte Carlo method for $H = 100$ km. We assumed unpolarized, collimated solar radiation at the top of the atmosphere with net flux $\pi S$. The radiation reflected from the Earth's surface was assumed to be unpolarized. The Earth's surface was assumed to be a Lambert reflector. The degree of polarization ($P$) and the intensity of radiation $I \times 10$ (in $S$ units) are given, for $H = 0$, in Tables 4.29, 28, respectively. The Monte Carlo results are listed in column MC, and the results reported by *Coulson* et al. [37] are given in column CDS. The relative errors of the Monte Carlo results are approximately 2–3%. They were obtained in approximately 40 minutes of BESM-6 computer time and agree well with the results of [37].

The differences of intensities of the order of 5–9% that occur for small values of $\mu_0$ (near the horizon) in Table 4.28 can probably be explained by the fact that our calculations were for a spherical atmosphere, whereas the Coulson et al. results were for a plane-parallel atmosphere. The differences (1–3%) in other cases are statistical fluctuations. The degree of polarization calculated by the Monte Carlo method agrees well with that computed by *Coulson* et al. The neutral points obtained by the Monte Carlo calculations are somewhat nearer to the sun (5° and 11°) than those reported in [37] (9° and 13°, respectively).

The contribution of single-scattered radiation obtained by the Monte Carlo method for the case $H = 0$, $\varphi = 0°$ varies from 82% at $\mu_0 = 0.02$ to 85% at $\mu_0 = 1$, and for the case $\varphi = 180°$, $H = 0$ from 76% to 85%.

The influence of sphericity on the intensity is particularly large for downward observation. Agreement is good for directions close to the nadir ($\mu_0 = -1$). The difference between MC and CDS results increases as the zenith angle decreases, and for $\mu_0 = -0.28$ they are already incompatible.

The contribution of the single-scattered radiation calculated by the Monte Carlo method is approximately 93% in the direction $\mu_0 = -0.02$ when $P_a = 0$; further, it decreases to 67% for $\mu_0 = -0.2$ and is approximately 85% for other directions. The behavior of the single-scattering contribution for $\varphi = 180°$ is roughly the same. The relative statistical errors of these results are 1–3% for $\varphi = 0°$ and 2–7% for $\varphi = 180°$.

Errors in intensity calculations due to neglect of polarization were estimated for a Rayleigh atmosphere with different optical thickness $\tau$: $0.1 \leqslant \tau \leqslant 1$. The computational results agree well with those obtained by *Germogenova* [28]. The largest errors were about 10% for $\tau = 1$.

We now consider a simplified radiation model of an atmosphere: (i) scattering is assumed to be isotropic; (ii) $\varphi_0$, the albedo for single scattering (i.e., the photon's survival probability) is taken to be 1; (iii) the extinction coefficient $\sigma(h)$ varies with altitude as

$$\sigma(h) = \sigma(0) \exp(-h/H_0),$$

where $H_0 = 8$ km; (iv) $\tau_0$, the total optical thickness of the atmosphere measured

**Table 4.28.** Comparison of intensity calculated by the Monte Carlo method (MC) with that reported by Coulson et al [37] (CDS) (Rayleigh atmosphere, $\tau = 0.1$, $H = 0$, $\psi = 53°8'$)

| $\mu_0$ | $\varphi = 0$ | | | | $\varphi = 180°$ | | | |
|---|---|---|---|---|---|---|---|---|
| | $P_a = 0$ | | $P_a = 0.25$ | | $P_a = 0$ | | $P_a = 0.25$ | |
| | MC | CDS | MC | CDS | MC | CDS | MC | CDS |
| 1.00 | 0.262 | 0.259 | 0.322 | 0.326 | 0.262 | 0.259 | 0.322 | 0.326 |
| 0.98 | 0.304 | 0.302 | 0.374 | 0.371 | 0.235 | 0.231 | 0.302 | 0.299 |
| 0.96 | 0.327 | 0.324 | 0.401 | 0.394 | 0.232 | 0.224 | 0.301 | 0.294 |
| 0.92 | 0.364 | 0.360 | 0.439 | 0.433 | 0.226 | 0.220 | 0.306 | 0.293 |
| 0.84 | 0.427 | 0.423 | 0.504 | 0.502 | 0.236 | 0.230 | 0.317 | 0.309 |
| 0.72 | 0.512 | 0.514 | 0.595 | 0.606 | 0.271 | 0.270 | 0.370 | 0.362 |
| 0.64 | 0.573 | 0.581 | 0.648 | 0.684 | 0.312 | 0.313 | 0.414 | 0.416 |
| 0.52 | 0.687 | 0.703 | 0.808 | 0.827 | 0.407 | 0.410 | 0.529 | 0.534 |
| 0.40 | 0.892 | 0.871 | 1.039 | 1.028 | 0.560 | 0.566 | 0.715 | 0.723 |
| 0.32 | 1.065 | 1.034 | 1.259 | 1.225 | 0.714 | 0.728 | 0.895 | 0.919 |
| 0.28 | 1.182 | 1.142 | 1.402 | 1.356 | 0.813 | 0.838 | 1.014 | 1.052 |
| 0.20 | 1.501 | 1.449 | 1.776 | 1.731 | 1.146 | 1.160 | 1.401 | 1.441 |
| 0.16 | 1.721 | 1.680 | 2.043 | 2.014 | 1.376 | 1.405 | 1.677 | 1.739 |
| 0.10 | 2.193 | 2.212 | 2.630 | 2.670 | 1.927 | 1.978 | 2.323 | 2.435 |
| 0.06 | 2.624 | 2.760 | 3.168 | 3.356 | 2.500 | 2.581 | 2.989 | 3.176 |
| 0.02 | 2.926 | 3.211 | 3.603 | 3.975 | 2.715 | 3.139 | 3.315 | 3.904 |
| −0.02 | $0.474 \times 10^{-5}$ | 0.343 | $0.541 \times 10^{-5}$ | 0.408 | $0.462 \times 10^{-5}$ | 0.351 | $0.523 \times 10^{-5}$ | 0.416 |
| −0.06 | $0.852 \times 10^{-5}$ | 0.268 | $0.958 \times 10^{-5}$ | 0.350 | $0.602 \times 10^{-5}$ | 0.287 | $0.714 \times 10^{-5}$ | 0.369 |
| −0.10 | $0.391 \times 10^{-4}$ | 0.203 | $0.431 \times 10^{-4}$ | 0.298 | $0.122 \times 10^{-4}$ | 0.227 | $0.182 \times 10^{-4}$ | 0.322 |
| −0.16 | $0.538 \times 10^{-3}$ | 0.143 | $0.628 \times 10^{-3}$ | 0.251 | $0.572 \times 10^{-3}$ | 0.171 | $0.738 \times 10^{-3}$ | 0.279 |
| −0.20 | $0.249 \times 10^{-2}$ | 0.177 | $0.880 \times 10^{-2}$ | 0.231 | $0.365 \times 10^{-2}$ | 0.147 | $0.996 \times 10^{-2}$ | 0.260 |
| −0.28 | 0.0428 | 0.0847 | 0.0923 | 0.205 | 0.144 | 0.115 | 0.241 | 0.235 |
| −0.32 | 0.0831 | 0.0735 | 0.245 | 0.196 | 0.114 | 0.104 | 0.235 | 0.227 |
| −0.40 | 0.0623 | 0.0571 | 0.219 | 0.183 | 0.0863 | 0.0877 | 0.220 | 0.213 |
| −0.52 | 0.0450 | 0.0413 | 0.193 | 0.170 | 0.0735 | 0.0707 | 0.214 | 0.200 |
| −0.64 | 0.0334 | 0.0316 | 0.178 | 0.163 | 0.0577 | 0.0585 | 0.199 | 0.190 |
| −0.72 | 0.0289 | 0.0272 | 0.174 | 0.159 | 0.0564 | 0.0517 | 0.201 | 0.184 |
| −0.84 | 0.0221 | 0.0231 | 0.165 | 0.157 | 0.0422 | 0.0425 | 0.187 | 0.176 |
| −0.92 | 0.0219 | 0.0222 | 0.164 | 0.156 | 0.0406 | 0.0362 | 0.178 | 0.170 |
| −0.96 | 0.0211 | 0.0225 | 0.163 | 0.157 | 0.0325 | 0.0325 | 0.173 | 0.167 |
| −0.98 | 0.0215 | 0.0232 | 0.165 | 0.158 | 0.0309 | 0.0304 | 0.173 | 0.165 |
| −1.00 | 0.0268 | 0.0261 | 0.166 | 0.161 | 0.0268 | 0.0261 | 0.166 | 0.161 |

**Table 4.29.** Comparison of degree of polarization, [%] obtained by the Monte Carlo method (MC) with those reported by Coulson et al. (31) (CDS) (Rayleigh atmosphere, $\tau = 0.1$, $H = 250$ km, $\psi = 53°8'$)

| $\mu_0$ | $\varphi = 0°$ | | | | $\varphi = 180°$ | | | |
|---|---|---|---|---|---|---|---|---|
| | $P_a = 0$ | | $P_a = 0.25$ | | $P_a = 0$ | | $P_a = 0.25$ | |
| | MC | CDS | MC | CDS | MC | CDS | MC | CDS |
| 0.02 | 15.9 | 15.8 | 12.3 | 12.5 | 19.4 | 18.4 | 16.1 | 14.5 |
| 0.06 | 13.4 | 13.3 | 10.6 | 10.8 | 21.7 | 21.2 | 18.5 | 17.1 |
| 0.10 | 11.7 | 11.0 | 9.49 | 9.08 | 24.3 | 24.2 | 20.5 | 19.6 |
| 0.16 | 8.71 | 8.00 | 6.96 | 6.63 | 28.0 | 29.1 | 23.6 | 23.5 |
| 0.20 | 5.99 | 6.18 | 4.87 | 5.5 | 29.4 | 32.6 | 24.8 | 26.2 |
| 0.28 | 3.78 | 3.05 | 2.80 | 2.55 | 40.4 | 40.3 | 32.5 | 32.1 |
| 0.32 | 2.97 | 1.73 | 2.25 | 1.45 | 44.1 | 44.5 | 35.2 | 35.2 |
| 0.40 | 0.74 | −0.40 | 0.61 | −0.34 | 53.2 | 53.3 | 41.6 | 41.7 |
| 0.52 | −0.94 | −2.27 | −0.89 | −1.93 | 67.0 | 67.5 | 51.5 | 51.8 |
| 0.64 | −1.02 | −2.32 | −0.79 | −1.96 | 81.0 | 81.2 | 61.0 | 61.2 |
| 0.72 | 0.16 | −0.99 | 0.51 | −0.83 | 88.1 | 88.4 | 65.8 | 66.0 |
| 0.84 | 5.43 | 4.32 | 4.82 | 3.65 | 89.3 | 91.8 | 68.1 | 68.3 |
| 0.92 | 12.8 | 12.4 | 10.7 | 10.3 | 81.4 | 83.8 | 61.6 | 63.0 |
| 0.96 | 20.0 | 19.9 | 16.1 | 16.4 | 71.4 | 73.6 | 55.1 | 56.1 |
| 0.98 | 26.5 | 26.2 | 21.6 | 21.4 | 64.1 | 65.3 | 50.1 | 50.3 |
| 1.00 | 44.1 | 44.5 | 36.7 | 35.3 | 44.1 | 44.5 | 36.7 | 35.3 |

in the vertical direction is taken to be $\sigma(0) \times H_0 = 0.3$; (v) $P_a$, the albedo of the Earth's surface is taken to be zero. The coefficient $\sigma(h)$ was approximated in our calculations with a step function, i.e., the atmosphere was divided into 36 layers, each having constant $\sigma(h)$. The Monte Carlo computational results (columns MC) are compared with exact analytical results proposed by *Minin* and *Smoktij* [38] (columns A) in Table 4.30 (calculations of the intensity of single-scattered radiation). The relative statistical errors of the Monte Carlo results are approximately 2–4%.

*Ivanov* et al. [38] have described an approximate method for estimating the degree of polarization in atmosphere. We compare now the degree of polarization calculated by their method with that by the Monte Carlo method. Calculations were done at $\lambda = 0.55$ µm, where Junge's distribution of particle size was used for the refractive index $m = 1.5$ and $0.04\ \mu m \leqslant r \leqslant 10\ \mu m$. The scattering matrix was assumed not to depend on altitude.

The total optical thickness of the atmosphere measured in the vertical direction was taken to be $\tau = \tau_a + \tau_M = 0.264$. The altitude profile of the

**Table 4.30.** Intensity of singly scattered radiation

| View direction | 86° | | 90° | | 94° | |
|---|---|---|---|---|---|---|
| | MC | A | MC | A | MC | A |
| Zenith | $1.59 \times 10^{-2}$ | $1.61 \times 10^{-2}$ | $4.49 \times 10^{-3}$ | $5.37 \times 10^{-3}$ | $3.82 \times 10^{-4}$ | $3.62 \times 10^{-4}$ |
| Nadir | $1.89 \times 10^{-2}$ | $1.88 \times 10^{-2}$ | $6.49 \times 10^{-3}$ | $6.85 \times 10^{-3}$ | $4.79 \times 10^{-4}$ | $4.89 \times 10^{-4}$ |

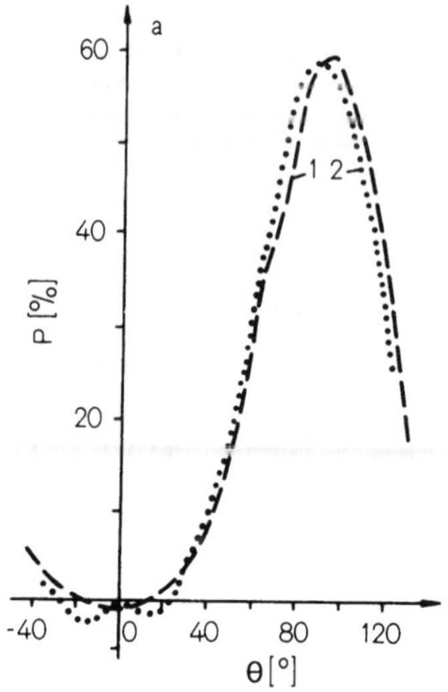

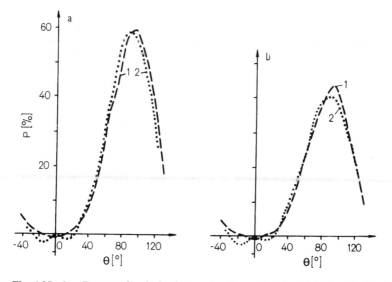

**Fig. 4.25a, b.** Degree of polarization as function of angle $\theta$ **(a)** $P_a = 0$; **(b)** $P_a = 0.25$; (*1*) Results of [38] (*2*) Monte Carlo results; $\tau = 0.264$, $\psi = 45°$, $H = 0$

coefficient $\sigma_a$ was assumed to be as shown in Fig. 4.7 (curve 1), and the coefficients of molecular scattering were taken from *Elterman*'s tables [31].

The observation point was situated on the Earth's surface, $\psi = 45°$. The degree of polarization ($P$) in the vertical plane through the sun is shown as a function of angle $\theta$ in Fig. 4.25 for $P_a = 0$ and $P_a = 0.25$. The angle $\theta$ is measured from a direction towards the sun (see Fig. 4.1) counterclockwise. The results agree well. The maxima $p_{max}^{(1)}$ and $p_{max}^{(2)}$ coincide to within 2–3%, and the deviation of position of $p_{max}^{(1)}$ from that of $p_{max}^{(2)}$ is approximately 3°. More noticeable deviation appears for neutral points. In the Monte Carlo calculations, the neutral points are at positions 14–16° from the sun when $P_a = 0$, and at positions 10–13° when $P_a = 0.25$; in calculations of *Pavlov* et al. [38] they are at positions 23–27° and 25–30°, respectively. The Monte Carlo results were obtained to within 3–4%, in 40 minutes of BESM-6 computer time. Notice that the approximate method for calculating the polarization characteristics of the scattered light is impractical for large optical depth of the aerosol, and does not take the sphericity into account.

# 5. Monte Carlo Algorithms for Solving Nonstationary Problems of the Theory of Narrow-Beam Propagation in the Atmosphere and Ocean

This chapter deals with calculational aspects of problems of utilizing high-power pulsed-laser lidary systems to obtain an information concerning the structures of the atmosphere and ocean. The problem can be formulated as follows. Consider a narrow light beam that propagates through an absorbing and scattering medium in a layer $h \leqslant z \leqslant H$. The light is emitted uniformly from a circular surface source of radius $R_s$ situated in the plane

$$a_s(x - x_s) + b_s(y - y_s) + c_s(z - z_s) = 0,$$

at a point $(x_s, y_s, z_s)$. The emission is also assumed to be isotropic within a circular conic solid angle $\Omega_s$ [with an angle $\gamma_s$ between its element and axis $\boldsymbol{n} = (a_s, b_s, c_s)$]. The time distribution density of the emitted photons is a function $p_t^{(0)}(\tau)$ such that

$$\int_0^\infty p_t^{(0)}(\tau)\, d\tau = 1.$$

Thus the total distribution density of the photons emitted by a source in the phase space $X$ of coordinates, directions, and time is given by the function $f_0(\boldsymbol{x}) = p_{\rho}^{(0)}(\boldsymbol{r})p_{\omega}^{(0)}(\omega)p_t^{(0)}(\tau)$, where

$$p_{\rho}^{(0)}(\boldsymbol{r}) = \begin{cases} (\pi R_s^2)^{-1} & \text{if} \quad \boldsymbol{r} \in S_s, \\ 0 & \text{otherwise,} \end{cases}$$

$$p_{\omega}^{(0)}(\omega) = \begin{cases} \dfrac{1}{2\pi(1 - \cos\gamma_s)} & \text{if} \quad \omega \in \Omega_s, \\ 0 & \text{otherwise.} \end{cases}$$

The receiver (detector) is a circle of radius $R_{\text{rec.}}$ ($R_{\text{det.}}$) with a center at the point $(x_r, y_r, z_r)$, The receiver is situated in the plane

$$a_r(x - x_r) + b_r(y - y_r) + c_r(z - z_r) = 0.$$

A unit vector $\boldsymbol{n}_s = (a_r, b_r, c_r)$ is parallel to the optical axis of the receiver. The light that reaches the receiver in a direction $\omega$ such that $|\omega \cdot \boldsymbol{n}_s| \geqslant \cos\gamma_r$, i.e., in the solid angle $\Omega_r$, is recorded. It is then desired to estimate the time distribution of intensity of the light recorded by the receiver. The total light intensity recorded by the receiver is equal to the integral

$$I(t) = \int_{S_r} ds \int_{\Omega_r} \Phi(r(s), \omega, t)\, d\omega, \tag{5.1}$$

of the intensity $\Phi[\boldsymbol{r}(s), \omega, t]$.

The integration is performed here over the receiver surface $S_r$ and over the solid angle $\Omega_r$. To obtain the mean intensity $\boldsymbol{I}(t)$, the integral (5.1) must obviously be divided by a constant $A = \pi R_r^2 \cdot 2\pi(1 - \cos \gamma_r)$.

The integrals

$$I_i = \int_{t_{i-1}}^{t_i} I(t)\, dt, \quad i = 1, 2, \ldots, n_t,$$

where $t_i$ are the points of the histogram, are evaluated by the Monte Carlo method. Dividing the quantity $I_i$ by $\Delta t_i = t_i - t_{i-1}$, we obtain an average intensity $\boldsymbol{I}_i$ for $i$th time-interval. To evaluate the quantity $I_i$, it is necessary (see Sect. 2.2) to estimate, by use of the Monte Carlo method, the number of photons that intersect the receiver surface in direction $\omega \in \Omega_r$, in $i$th time interval, and associated with a weight $Q = |\omega \cdot \boldsymbol{n}_s|^{-1}$. When $Q = 1$, this number of photons gives an estimate of illumination

$$E_i = \int_{t_{i-1}}^{t_i} E(t)\, dt.$$

The quantity $E(t)$ can be represented as

$$E(t) = \int_{S_r} ds \int_{\Omega_r} \Phi[r(s), \omega, t] \cdot |\omega n_s|\, d\omega. \tag{5.2}$$

Assume, for simplicity, that $\sigma(\boldsymbol{r}) = \sigma = \text{const}$, and $g(\mu, \boldsymbol{r}) = g(\mu)$. Special problems for an inhomogeneous medium will be formulated later.

## 5.1 Specific Features of the Calculations

For convenience, the origin of the coordinate system coincides with the detector, the $xy$-plane is parallel to the boundary of the layer (i.e., the $z$ axis is collinear with the inner normal to this boundary), and the optical axis $\boldsymbol{n}$ of the detector lies in the $xz$-plane so that the angle between $\boldsymbol{n}$ and $\boldsymbol{OX}$ is acute (Fig. 5.1). Let $\beta$ be an angle between the axis $OZ$ and the optical axis of the detector. Thus $x_r = y_r = z_r = 0$, $a_r = \sin \beta$, $b_r = 0$, $c_r = \cos \beta$.

The direction cosines $a_0$, $b_0$, $c_0$ characterize the direction of emission of the photon, and can be simulated by

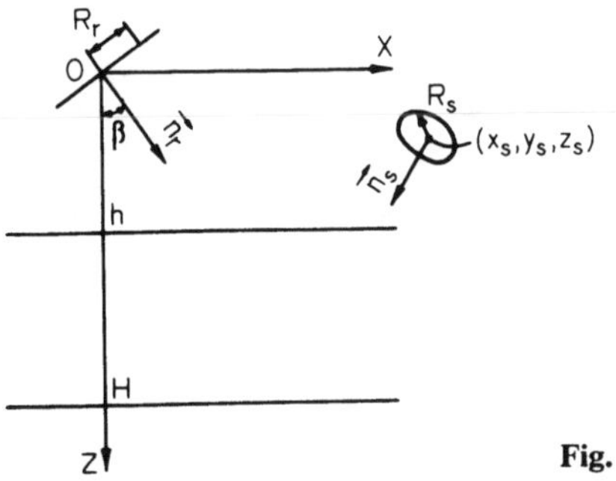


**Fig. 5.1.** Geometry of problem

1) $\mu = 1 - \alpha_1(1 - \cos \gamma_s)$,
2) $\xi_1 = 1 - 2\alpha_2, \eta_1 = 1 - \alpha_3$.

If $d = \xi_1^2 + \xi_2^2 \leqslant 1$, then

$$a_0 = a_s\mu - (b_s\xi_1 + a_sc_s\eta_1)\left[\frac{1 - \mu^2}{d(1 - c_r^2)}\right]^{1/2},$$

$$b_0 = b_s\mu + (a_s\xi_1 - b_sc_s\eta_1)\left[\frac{1 - \mu^2}{d(1 - c_r^2)}\right]^{1/2},$$

$$c_0 = c_r\mu + (1 - c_r^2)\eta_1\left[\frac{1 - \mu^2}{d(1 - c_r^2)}\right]^{1/2},$$

if $c_r \neq 1$ (see Sect. 2.2), and

$$a_0 = \xi_1\left[\frac{1 - \mu^2}{d}\right]^{1/2}, \quad b_0 = \eta_1\left[\frac{1 - \mu^2}{d}\right]^{1/2}, \quad c_0 = \mu,$$

if $c_r = 1$.

Here $\alpha_1, \alpha_2, \alpha_3, \alpha_4$ are independent random variables uniformly distributed between 0 and 1. If $d > 1$, new random numbers are selected and the procedure is repeated. When the coordinates $(x_0, y_0, z_0)$ of the point at which the photon leaves the source are to be determined, the area of this surface is usually assumed to be sufficiently small so that $x_0 = x_s, y_0 = y_s, z_0 = z_s$. However, a truncation error then appears when the quantity $I(t)$ is estimated. When the distance between the positions of the receiver and the source is small, this error may become significant for small values of $t$.

Assume that the optical axes of the receiver and source lie in the same plane. Then $b_s = 0$, and the coordinates $x_0, y_0, z_0$ can be sampled as

1. $\xi_2 = 1 - 2\alpha_4, \eta_2 = 1 - 2\alpha_5$
2. $x_0 = x_s + R_s\xi_2a_s, y = y_s + R_s\eta_2, z_0 = z_s + R_s\xi_2c_s$

if $\xi_2^2 + \eta_2^2 \leqslant 1$. Otherwise, the procedure is repeated. Assume that the time distribution density of the emitted photons is a $\delta$ function: $p_t^{(0)}(\tau) = \delta(\tau)$. Then $t_0 = 0$, i.e., the photons instantly leave the source. Later, we will consider an arbitrary time distribution $p_t(\tau)$.

Let us suppose that the source is placed outside of the layer, i.e., $z_s < h$. The coordinates of the point of intersection of the photon with the boundary of layer are

$$x = x_0 + \frac{a_0(h - z_s)}{c_0}, \quad y = y_0 + \frac{b_0(h - z_s)}{c_0}, \quad z = h,$$

provided that the photon has not undergone scattering before hitting this layer.

The light may also undergo either refraction or specular reflection from the layer's boundary. The direction cosines $a$, $b$, $c$, that characterize the photon's motion after refraction are calculated from

$$\frac{\sin\beta}{\sin\gamma} = n, \quad \sin\beta = (1 - c_0^2)^{1/2}, \quad \sin\gamma = (1 - c^2)^{1/2}, \quad a^2 + b^2 + c^2 = 1,$$

$$aa_0 + bb_0 + cc_0 = \cos(\beta - \gamma),$$

where $n$ is the refractive index. Thus, from

$$1 - c^2 = \frac{1}{n}(1 - c_0^2),$$

$$aa_0 + bb_0 = [(1 - c_0^2)(1 - c^2)]^{1/2},$$

$$a^2 + b^2 + c^2 = 1,$$

we get

$$a = \frac{a_0}{n}, \quad b = \frac{b_0}{n}, \quad c = \frac{1}{n}[n^2 - 1 + c_0^2]^{1/2}. \tag{5.3}$$

Figure 5.2 assumes that medium II is denser than medium I (for example water and air, where $n = 1.33$).

Otherwise the direction cosines must be calculated as

$$a = na_0, \quad b = nb_0, \quad c = n\left[\frac{1}{n^2} - 1 + c_0^2\right]^{1/2},$$

provided that $a_0 \leqslant 1/n$, $b_0 \leqslant 1/n$, $|c_0| \geqslant (n^2 - 1)^{1/2}/n$. If the latter condition is not satisfied, the photon will not escape from the layer (the intensity of the refracted light vanishes). The intensities of refracted light $I_1$ and of reflected

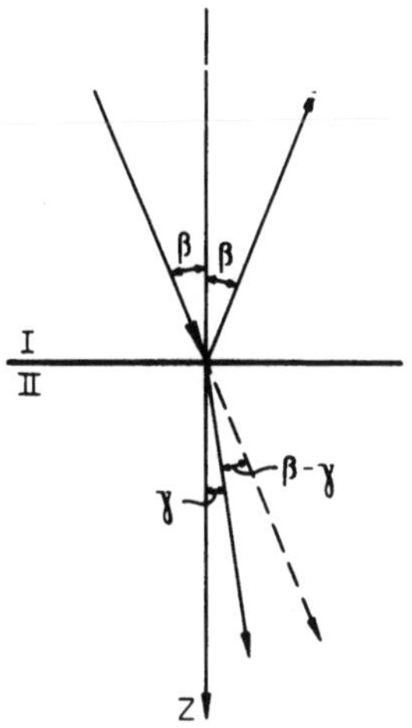


**Fig. 5.2.** Geometry of refraction and reflection

light $I_2$ are given by the Fresnel formula. The reflection coefficient $R[\beta, \gamma(\beta, n)]$ is given by

$$R[\beta, \gamma(\beta, n)] = \frac{1}{2}\left[\frac{\tan^2(\beta - \gamma)}{\tan^2(\beta + \gamma)} + \frac{\sin^2(\beta - \gamma)}{\sin^2(\beta + \gamma)}\right].$$

By (5.3) we get

$$R(c_0, n) = \frac{(c_0 - B)^2(c_0^2B^2 + A^2)}{(c_0 + B)^2(c_0B + A)^2}, \tag{5.4}$$

where

$$A = 1 - c_0^2, \quad B = (n^2 - 1 + c_0^2)^{1/2}.$$

Thus

$$I_1 = I_0[1 - R(c_0, n)], \quad I_2 = I_0R(c_0, n),$$

where $I_0$ is the radiance of light in the direction $(a_0, b_0, c_0)$.

Therefore, when the trajectories are simulated, the reflection must be sampled with probability $R$ while the refraction must be sampled with probability $1 - R$. When the light propagated from a medium II to medium I is considered, where medium II is denser than medium I, the reflection coefficient $R_1$ can be estimated as [39].

$$R_1(c_0, n) = \frac{n^2 - 1 + R(c_0, n)}{n^2}. \tag{5.5}$$

## 5.2 Features of the Use of Local Estimates

Local estimates of the distributions of $I(t)$ and $E(t)$ are especially well adapted to situations that involve detectors whose acceptance apertures are small. General properties of local estimates are described in Sects. 3.8–10. However, some modifications of local estimates will be needed in order to solve problems treated in this chapter. As in Sect. 3.8,

$$I_i = \int_{t_{i-1}}^{t_i} dt^* \int_{S_r} ds \int_{\Omega_r} \Phi[\boldsymbol{r}^*(s), \omega^*, t^*]\, d\omega^* = M\xi_1,$$

$$\xi_1 = \sum_{j=0}^{N} Q_j \varphi_1(\boldsymbol{x}_j, \boldsymbol{\rho}^*),$$

where

$$\varphi_1(\boldsymbol{x}_j, \boldsymbol{\rho}^*) = \frac{\exp[-\tau(\boldsymbol{r}_j, \boldsymbol{\rho}^*)]g(\mu^*)}{2\pi|\boldsymbol{r}_j - \boldsymbol{\rho}^*|^2 p(\boldsymbol{\rho}^*)} \Delta(\boldsymbol{l}^*)\Delta_i(t^*). \tag{5.6}$$

Here,

$$\boldsymbol{l}^* = (\boldsymbol{r}_j - \boldsymbol{\rho}^*)/|\boldsymbol{r}_j - \boldsymbol{\rho}^*|, \quad \boldsymbol{\mu}^* = (\omega \cdot \boldsymbol{l}^*);$$

$\Delta(\boldsymbol{l}^*)$ and $\Delta_i(t^*)$ are the characteristic functions of the domain $\Omega_r$ and of the $i$th time interval, respectively; $p(\boldsymbol{r}^*)$ is the distribution density of the random point $\boldsymbol{\rho}^*$ on the receiver surface. The presence of the function $\Delta_i(t^*)$ in (5.6) is due to the factor $\delta[t^* - (t_j + |\boldsymbol{r}_j - \boldsymbol{\rho}^*|/v)]$ in the kernel function $k(\boldsymbol{x}_j, \boldsymbol{\rho}')$, where $t_j$ is the time at which a collision occurs at point $\boldsymbol{r}_j$, and $v$ is the velocity of light in the medium. The integration over the acceptance aperture of the detector is performed by the Monte Carlo method, i.e., a random point $\boldsymbol{\rho}^*$ is sampled from the density $p(\boldsymbol{r}^*)$, next, the quantity (5.6) is evaluated. As a result, the estimation $\xi_1$ is unbiased, because of the function $p(\boldsymbol{r}^*)$ in (5.6).

Let us consider a detector that is positioned inside the scattering layer. As mentioned in Sect. 3.8, the local estimate has infinite variance, due to the divisor $|\boldsymbol{r}_j - \boldsymbol{\rho}^*|$ in (5.6). However, it can be made finite by appropriate choice of the density function $p(\boldsymbol{r}^*)$. Let $\boldsymbol{r}_j = (x_j, y_j, z_j)$ be the radius vector of the $j$th collision point, and let $\boldsymbol{r}^*$ be the radius vector of a point $(x^*, y^*, z^*)$ that lie in the receiver plane.

Let $(x_c, y_c, z_c)$ be the projection of the point $(x_j, y_j, z_j)$ on the receiver plane

$$x \sin\beta + z\cos\beta = 0,$$

and let $\boldsymbol{r}_r$ and $\boldsymbol{r}^*$ be the radius vectors (with origin at $(x_c, y_c, z_c)$) of the points $(x_j, y_j, z_j)$ and $(x^*, y^*, z^*)$, respectively (see Fig. 5.3).

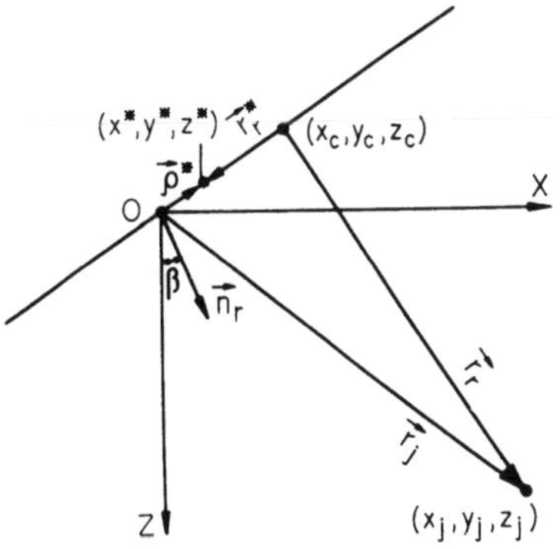


**Fig. 5.3.** Illustration of choice of $\rho^*$

Because

$$x_c = x_j \cos^2 \beta - z_j \cos \beta \sin \beta, \quad y_c = y_j,$$
$$z_c = z_j \sin^2 \beta - x_j \cos \beta \sin \beta,$$

we have

$$\boldsymbol{r}_r = (x_j \sin^2 \beta + z_j \cos \beta \sin \beta, 0, z_j \cos^2 \beta + x_j \cos \beta \sin \beta),$$
$$\boldsymbol{r}_r^* = (x^* - x_j \cos^2 \beta + z_j \cos \beta \sin \beta, y^* - y_j, z^* - z_j \sin^2 \beta + x_j \cos \beta \sin \beta),$$
$$|\boldsymbol{r}_r| = x_j \sin \beta + z_j \cos \beta.$$

Consider points $(x_r^*, y_r^*, z_r^*)$ that are distributed in a domain $G$ of the receiver plane with density

$$p(\boldsymbol{r}_r^*) = \frac{C}{|\boldsymbol{r}_r^* - \boldsymbol{r}_r|^2},$$

so that

$$\int_G p[\boldsymbol{r}_r^*(s)]\, ds = 1,$$

where $C$ is the normalization factor, and the domain $G$ is defined by

$$(x - x_c)^2 + (y - y_c)^2 + (z - z_c)^2 \leqslant R_r + \sqrt{x_c^2 + y_c^2 + z_c^2},$$
$$x \sin \beta + z \cos \beta = 0.$$

Writing this density in a polar coordinate system $(\rho, \varphi)$ in the receiver plane with origin at the point $N(x_c, y_c, z_c)$, the angle $\varphi$ measured clockwise when looking

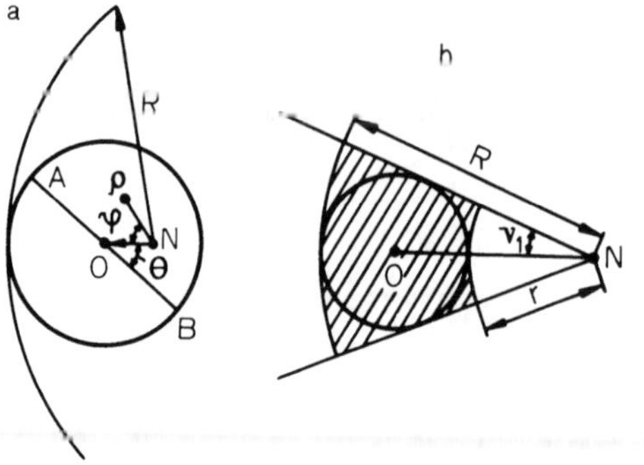


**Fig. 5.4a, b.** (a) Definition of $p(\boldsymbol{r}^*)$ and (b) choice of $\rho_r^*$

from the point $N$ to the center of the detector (Fig. 5.4), then the distance between the points $N$ and $O$ is given by

$$s = [(x_j \cos\beta - z_j \sin\beta)^2 + y_j^2]^{1/2}.$$

Let $R = R_r + s$.
If $s < R_r(a)$ then, clearly

$$p(\rho, \varphi) = \frac{C\rho}{\rho^2 + |\boldsymbol{r}_r|^2},$$

where

$$C = \frac{1}{\pi}\left[\ln \frac{R^2 + |\boldsymbol{r}_r|^2}{|\boldsymbol{r}_r|^2}\right]^{-1}, \tag{5.7}$$

and $0 \leqslant \rho \leqslant R$, $0 \leqslant \varphi \leqslant 2\pi$.

From this, we obtain the simulating formulas for $\rho$ and $\varphi$:

$$\rho = |\boldsymbol{r}_r| \cdot \left[\left(\frac{R^2 + |\boldsymbol{r}_r|^2}{|\boldsymbol{r}_r|^2}\right)^{\alpha_1} - 1\right]^{1/2}, \quad \varphi = 2\pi\alpha_2. \tag{5.8}$$

Now

$$\begin{aligned} x_r^* &= x_c + \rho\cos(\theta - \varphi)\cos\beta \\ &= x_j \cos^2\beta - z_j \cos\beta \sin\beta + \rho\cos(\theta - \varphi)\cos\beta, \\ y_r^* &= y_c + \rho\sin(\theta - \varphi) = y_j + \rho\sin(\theta - \varphi), \\ z_r^* &= z_c + \rho\cos(\theta - \varphi)\sin\beta \\ &= z_j \sin^2\beta - x_j \cos\beta\sin\beta + \rho\cos(\theta - \varphi)\sin\beta. \end{aligned} \tag{5.9}$$

Here $\theta$ is the angle between the direction $NO$ and the line $AB$ of intersection of the receiver plane with the $xz$-plane. This angle is measured clockwise when looking from the point $N$ to the point $O$, and $\sin\theta = y_j/s$.

Let $\Delta\rho(\boldsymbol{r}_r^*)$ be the characteristic function of the receiver surface, that is, we have $\Delta\rho(\boldsymbol{r}_r^*) = 1$ and $x^* = x_r^*$, $y^* = y_r^*$, $z^* = z_r^*$, if $x_r^{*2} + y_r^{*2} + z_r^{*2} \leqslant R_r^2$; otherwise $\Delta\rho(\boldsymbol{r}_r^*) = 0$, which corresponds to zero contribution to the estimate of $I(t)$.

If $s > R_r$ (i.e., the point $N$ is outside of the receiver surface) the probability of zero contribution may be large. Therefore, the point $(x_r^*, y_r^*, z_r^*)$ must be chosen from the hatched domain (see Fig. 5.4b) defined by the inequalities $r \leqslant \rho \leqslant R$, $-\nu_1 \leqslant \varphi \leqslant \nu_1$. Here $r = s - R_r$, $\nu_1 = \arcsin(R_r/s)$. Furthermore, the coefficient $C$ in (5.7) is given by

$$C = \left[\nu_1 \ln\left(\frac{R^2 + |\boldsymbol{r}_r|^2}{r^2 + |\boldsymbol{r}_r|^2}\right)\right]^{-1},$$

and the quantities $\rho$ and $\varphi$ are simulated as

$$\rho = \left[(r^2 + |\boldsymbol{r}_r|^2)\left(\frac{R^2 + |\boldsymbol{r}_r|^2}{r^2 + |\boldsymbol{r}_r|^2}\right)^{\alpha_1} - |\boldsymbol{r}_r|^2\right]^{1/2}, \quad \varphi = (1 - 2\alpha_2)\nu_1.$$

Because $|\boldsymbol{r}_j - \boldsymbol{\rho}^*| = |\boldsymbol{r}_r - \boldsymbol{r}_r^*|$ (see Fig. 5.2), the expression (5.6) takes the form

$$\varphi_1(\boldsymbol{x}_j, \boldsymbol{\rho}^*) = \frac{\exp[-\tau(\boldsymbol{r}_j, \boldsymbol{\rho}^*)]g(\mu^*)}{2\pi c}\Delta(\boldsymbol{l}^*)\Delta_i(t^*)\Delta\rho(\boldsymbol{r}_r^*).$$

In realistic hydro- and atmospheric-optics problems,

$$g(\mu^*) = g\left(\omega_j \cdot \frac{\boldsymbol{\rho}^* - \boldsymbol{r}_j}{|\boldsymbol{\rho}^* - \boldsymbol{r}_j|}\right) \leqslant g_0 < +\infty.$$

Therefore,

$$\varphi_1(\boldsymbol{x}_j, \boldsymbol{\rho}^*) \leqslant \frac{g_0}{2}\ln\left(\frac{R^2 + |\boldsymbol{r}_r|^2}{r^2 + |\boldsymbol{r}_r|^2}\right),$$

where $r = 0$ if case $a$ of Fig. 5.4 is considered. Thus, utilization of the function $p(\boldsymbol{r}^*)$ leads to evaluation of the function $\varphi_1(\boldsymbol{x}_j, \boldsymbol{\rho}^*)$ at each collision. Note that this function diverges as $-\ln(|\boldsymbol{r}_r|)$. As in [15], the variance of this estimate may be proved to be finite (see also Sect. 3.9).

Let us note one feature of the local estimate that leads to a decrease in efficiency if $\gamma_r$ is close to zero. Consider a point $\boldsymbol{r}_j$ that lies in the receiver field of view, e.g., the hatched domain $G$ in Fig. 5.5, defined by

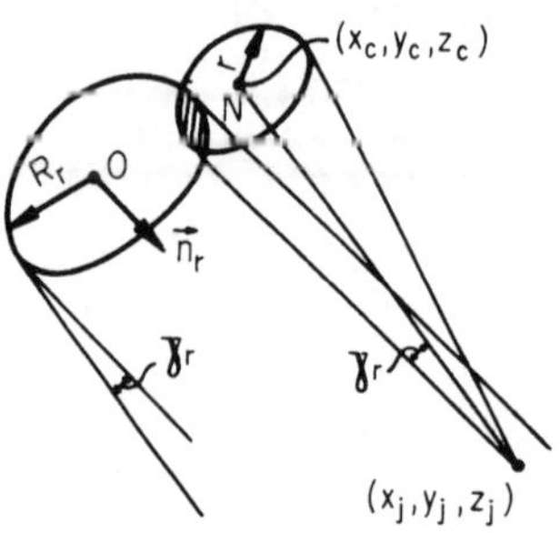


**Fig. 5.5.** Observation of scattered light

$$(x - x_c)^2 + (y - y_c)^2 + (z - z_c)^2 < r^2, \qquad x^2 + y^2 + z^2 \leqslant R_r^2,$$

$\sin\beta(x - x_c) + \cos\beta\,(z - z_c) = 0$, where $r = |\boldsymbol{r}_j - \boldsymbol{r}_c| \tan\gamma_r$. The area of this domain satisfies the inequality $0 < S \leqslant \pi R_r^2$. Furthermore, $S = \pi R_r^2$ if and only if $R_r + s \leqslant r$ (Fig. 5.6a). If $r + s \leqslant R_r$ then $S = \pi r^2$ (Fig. 5.6b). If $|R_r - r| \leqslant s$ (Fig. 5.6c), then

$$S = R_r^2 \arcsin\left(\frac{\delta}{R_r}\right) + r^2 \arcsin\left(\frac{\delta}{r}\right) - s\delta,$$

where

$$\delta = \frac{2}{3}[d(d - R_r)(d - r)(d - s)]^{1/2}, \quad d = (s + r + R_r)/2.$$

Let $\boldsymbol{\rho}^* = (x^*, y^*, z^*)$ be a point uniformly distributed on the receiver surface. This point defines uniquely the direction $\boldsymbol{l}^*$ in (5.6). The inequality $\boldsymbol{l}^* \cdot \boldsymbol{n}_s \geqslant \cos\gamma_r$ holds if and only if $(x^*, y^*, z^*) \in G$. It follows from this that, with probability $1 - S/(\pi R_r^2)$, the contribution from the collision point $\boldsymbol{r}_j$ to the estimate of $I(t)$ is not taken into account. It is more effective to sample the point $\boldsymbol{\rho}^*$ from the distribution density

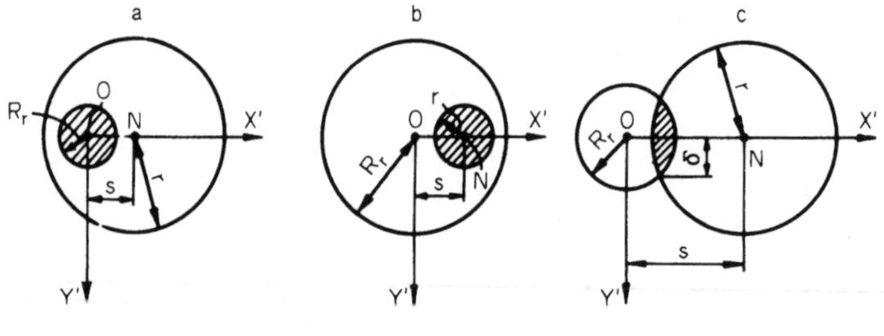


**Fig. 5.6a, b, c.** Geometry of $\rho^*$ choice

$$p(r^*) = \begin{cases} 1/\pi R_r^2, & \text{if } R_r + s \leqslant r, \\ 1/\pi r^2, & \text{if } r + s \leqslant R_r, \\ [R_r^2 \arcsin(\delta/R_r) + r^2 \arcsin(\delta/r) - s\delta]^{-1}, & \text{if } |R_r - r| \leqslant s; \end{cases} \tag{5.10}$$

i.e., the point $\rho^*$ is uniformly distributed in the domain $G$.

Let $x'$, $y'$, $z'$ be the coordinates of a point $\rho^*$ in a coordinate system with origin at the detector position $O$. The plane $X'O'Y'$ coincides with the receiver plane, and the axis $OX'$ passes through the point $(x_c, y_c, z_c)$ (the point $N$ in Fig. 5.6).

The coordinates $x'$, $y'$, $z'$ are sampled as

1) $\xi = 1 - 2\alpha_1, \eta = 1 - 2\alpha_2$
2) $x'' = h_1 + \xi h_2, y'' = \eta h_3, z'' = 0$, where

$$h_1 = \begin{cases} 0(\mathrm{a}) \\ s(\mathrm{b}) \\ (R_r + s - r)/2(\mathrm{c}) \end{cases} \qquad h_2 = \begin{cases} R_r(\mathrm{a}) \\ r(\mathrm{b}) \\ (R_r + r - s)/2(\mathrm{c}) \end{cases} \qquad h_3 = \begin{cases} R_r(\mathrm{a}) \\ r(\mathrm{b}) \\ \delta(\mathrm{c}). \end{cases}$$

If $(x'')^2 + (y'')^2 \leqslant R^2$ and $(x'' - s)^2 + (y'')^2 \leqslant r^2$, $x' = x''$, $y' = y''$, $z' = z''$. Otherwise the procedure is repeated. The coordinates $x^*$, $y^*$, $z^*$ are calculated as $x^* = (x' \cos\theta + y' \sin\theta)\cos\beta$, $y^* = -x'\sin\theta + y'\cos\theta$, $z^* = (x'\cos\theta + y'\sin\theta)\sin\beta$, where the angle $\theta$ is defined as in (5.9). Because $S \leqslant \pi r^2 \leqslant \pi|\boldsymbol{r}_j - \boldsymbol{\rho}^*|^2$, if $\gamma_r < \pi/4$, the contribution to the estimate $I_i$ from the collision at point $\boldsymbol{r}_j$ that is given by the expression

$$\frac{\exp[-\tau(\boldsymbol{r}_j, \boldsymbol{\rho}^*)]g(\mu^*)S}{2\pi|\boldsymbol{r}_j - \boldsymbol{\rho}^*|^2}\Delta_i(t^*),$$

is bounded, and the local estimate has therefore a finite variance.

The local estimate for estimating $I(t)$ for the case of a small angular aperture of the receiver is inefficient, owing to the function $\Delta(l^*)$ in (5.6). The double local estimate is preferable in such a case, (see Sect. 3.8). In practice, the double local estimate is obtained by choosing a random point $\boldsymbol{\rho}^*$ on the receiver surface, according to the density $p(\boldsymbol{r}^*)$, a random direction $\boldsymbol{\omega}^*$ and a random length $\lambda^*$ of an auxiliary run starting at $\boldsymbol{\rho}^*$, sampled from the distribution densities $p_\omega(\boldsymbol{\omega}^*)$ and $p_\lambda(\lambda)^*$, respectively. Then, the point $\boldsymbol{\rho}''$ is given by $\boldsymbol{\rho}'' = \boldsymbol{\rho}^* + \boldsymbol{\omega}^*\lambda^*$, and $I_i = M\xi_2$, where

$$\xi_2 = \sum_{j=0}^{N} Q_j \varphi_2(\boldsymbol{r}_j, \boldsymbol{\omega}_j, \lambda^*), \tag{5.11}$$

$$\varphi_2(\boldsymbol{r}_j, \omega_j, \lambda_j^*) = \frac{q \exp\{-[\tau(\boldsymbol{r}_j, \boldsymbol{\rho}'') + \tau(\boldsymbol{\rho}'', \boldsymbol{\rho}^*)]\}}{[2\pi|\boldsymbol{r}_j - \boldsymbol{\rho}''|]^2 p(\boldsymbol{\rho}^*) p_\omega(\omega^*) p_\lambda(\lambda^*)}$$
$$\times g\left[\left(\omega_j, \frac{\boldsymbol{\rho}'' - \boldsymbol{r}_j}{|\boldsymbol{\rho}'' - \boldsymbol{r}_j|}\right)\right] \cdot g\left[\left(\omega^*, \frac{\boldsymbol{r}_j - \boldsymbol{\rho}''}{|\boldsymbol{r}_j - \boldsymbol{\rho}''|}\right)\right] \Delta_i(t^*), \quad (5.12)$$
$$Q_j = q^j, \quad t_j + \frac{|\boldsymbol{r}_j - \boldsymbol{\rho}''| + |\boldsymbol{\rho}'' - \boldsymbol{\rho}^*|}{v} = t^*,$$

where, for the first run in the direction $\omega_0$, it is assumed that

$$g\left[\left(\omega_0, \frac{\boldsymbol{\rho}'' - \boldsymbol{r}_0}{|\boldsymbol{\rho}'' - \boldsymbol{r}_0|}\right)\right] = \begin{cases} [2\pi(1 - \cos\gamma_s)]^{-1}, & \text{if } \boldsymbol{n}_s \cdot \dfrac{\boldsymbol{\rho}'' - \boldsymbol{r}_0}{|\boldsymbol{\rho}'' - \boldsymbol{r}_0|} \geqslant \cos\gamma_s, \\ 0, & \text{otherwise.} \end{cases} \quad (5.13)$$

The factor $[p(\boldsymbol{\rho}^*) \cdot p_\omega(\omega^*) p_\lambda(\lambda^*)]^{-1}$ in (5.12) assures that the estimate $\xi_2$ is unbiased. If we adopt the simple formulas

$$p(\boldsymbol{r}^*) = (\pi R_r^2)^{-1}, \quad p_\omega(\omega^*) = [2\pi(1 - \cos\gamma_r)]^{-1}, \quad p_\lambda(l^*) = \sigma \exp\{-\sigma l^*\},$$

then

$$\varphi_2(\boldsymbol{r}_j, \omega_j, \lambda_j^*) = \frac{q \exp[-\tau(\boldsymbol{r}_j, \boldsymbol{\rho}'')]}{2\sigma|\boldsymbol{r}_j - \boldsymbol{\rho}''|^2} g\left[\left(\omega_j, \frac{\boldsymbol{\rho}'' - \boldsymbol{r}_j}{|\boldsymbol{r}_j - \boldsymbol{\rho}''|}\right)\right]$$
$$\times g\left[\left(\omega^*, \frac{\boldsymbol{r}_j - \boldsymbol{\rho}''}{|\boldsymbol{r}_j - \boldsymbol{\rho}''|}\right)\right] R_r^2(1 - \cos\gamma_r)\Delta_i(t^*).$$

The mathematical expectation of the quantity $\varphi_2(\boldsymbol{r}_0, \omega_0, \lambda_0^*)$ is equal to the intensity of single-scattered light. However, this method of estimation of $I_i^{(1)}$ does not take account of the geometry of the problem. Therefore, use of the estimate $\xi_2$ is inefficient for calculating the intensity of singly scattered light.

In Sect. 5.4, we shall consider an effective method for estimating $I_i^{(1)}$. Notice that the mathematical expectation of the random variable $\xi_2$ is equal to the illumination $E_i$, provided that, in (5.11), a weighted summation is performed, where the weight is given by $(\boldsymbol{n}_s \cdot (\boldsymbol{\rho}'' - \boldsymbol{\rho}^*)/|\boldsymbol{\rho}'' - \boldsymbol{\rho}^*|)$.

Assume that the detector is positioned outside the scattering layer, i.e., $h > 0$. If the light has undergone a refraction when passing through the layer, the contribution from the collision point $\boldsymbol{r}$ to the estimate of $I_i$ is calculated by choosing a random point $\boldsymbol{\rho}^*$ on the receiver surface according to the density $p(\boldsymbol{r}^*)$. An auxiliary quantity $\delta$ is obtained from the system

$$\frac{\sin\gamma}{\sin\beta} = n, \quad \frac{s - \delta}{h - z^*} = \tan\gamma, \quad \frac{\delta}{z_j - h} = \tan\beta, \quad (5.14)$$

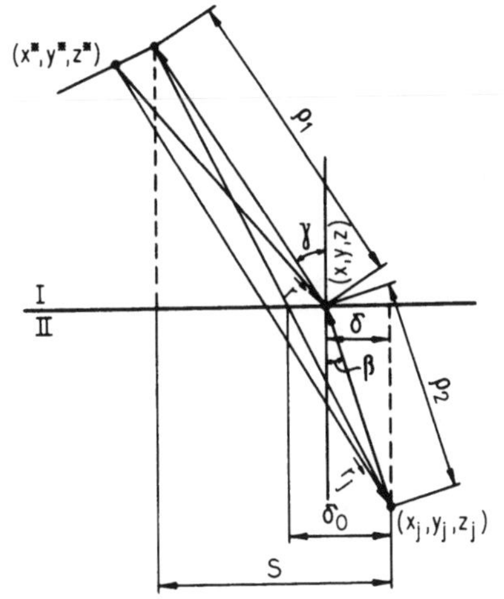


**Fig. 5.7.** Geometry of estimate that takes refraction into account

where $s = [(x_j - x^*)^2 + (y_j - y^*)^2]^{1/2}$ (see Fig. 5.7). It can easily be shown that $\delta$ satisfies

$$(1 - n^2)\delta^4 + 2s(n^2 - 1)\delta^3 + [s^2 + (z_j - h)^2 - n^2s^2 - n^2(h - z^*)^2]\delta^2 - 2s(z_j - h)^2\delta + s^2(z_j - h)^2 = 0.$$

This equation may be solved by an approximate method, for instance, the Newton method. The first approximate value can be taken as $\delta_0 = s(z_j - h)/(z_j - z^*)$. Next, the coordinates $(x, y, z)$ of the intersection of the path of the photon with the plane $z = h$

$$x = x_j + \frac{\delta}{s}(x^* - x_j), \quad y = y_j + \frac{\delta}{s}(y^* - y_j), \quad z = h,$$

and

$$\rho_1 = [(h - z^*)^2 + (s - \delta)^2]^{1/2}, \quad \rho_2 = [(z_j - h)^2 + \delta^2]^{1/2},$$

$$\mu = \omega_j \cdot \frac{\boldsymbol{r} - \boldsymbol{r}_j}{|\boldsymbol{r} - \boldsymbol{r}_j|} = \frac{1}{\rho_2}[a_j(x - x_j) + b_j(y - y_j) + c_j(z - z_j)],$$

are calculated.

The contribution from the photon's collision point $\boldsymbol{r}_j$ to the estimate of $I_1$ is

$$\varphi_1(\boldsymbol{r}_j, \omega_j, \boldsymbol{\rho}^*) = \frac{q_{\mathrm{II}} \exp[-(\sigma_{\mathrm{I}}\rho_1 + \sigma_{\mathrm{II}}\rho_2)]g(\mu)(1 - R)}{2\pi(\rho_1 + \rho_2)^2 n^2 p(\boldsymbol{\rho}^*)} \Delta_i(t^*)\Delta(\boldsymbol{l})^*, \tag{5.15}$$

provided that $\delta/\rho_2 < n^{-1}$. Otherwise, this contribution is taken to be zero (i.e., the intensity of the refracted light is zero).

The quantities $\sigma_{\text{I}}$ and $\sigma_{\text{II}}$ in (5.15) are the extinction coefficients for media I and II, respectively, and $q_{\text{II}}$ is the photon survival probability for medium II. The formula (5.15) is valid if $n > 1$ (as shown in Fig. 5.6). If $n < 1$ then

$$\varphi_1(\boldsymbol{r}_j, \omega_j, \boldsymbol{\rho}^*) = \frac{q_{\text{II}} \exp[-(\sigma_{\text{I}}\rho_1 + \sigma_{\text{II}}\rho_2)]g(\mu)(1 - R)}{2\pi(\rho_1 + \rho_2)^2 p(\boldsymbol{\rho}^*)} \Delta_i(t^*)\Delta(\boldsymbol{l}^*).$$

To obtain the time of detection of the photon $t_j$, the time of photon collision at the point $\boldsymbol{r}_j$ must be added to $\rho_1/v_{\text{I}} + \rho_2/v_{\text{II}}$, where $v_{\text{I}}$ and $v_{\text{II}}$ are the velocities of light in media I and II, respectively. To estimate the contribution (5.15), it is necessary to calculate $R[\beta, \gamma(\beta)]$ at each collision point, because the coefficient $R$ in (5.15) is a function of the angles $\beta$ and $\gamma$ [see (5.14)] that depend on $\boldsymbol{r}_j$ and $\boldsymbol{\rho}^*$. It is convenient to tabulate the function $R[\beta, \gamma(\beta)]$ and to use linear interpolation.

Let $p_0(t)$ be the time-distribution density of emitted photons.

The most frequently applied approximations of the light pulse are

$$p_0^{(1)}(t) = \varkappa^2 t \exp(-\varkappa t), \quad p_0^{(2)}(t) = \frac{1}{2\pi\beta} \exp(-t^2/\beta^2), \quad t \geqslant 0.$$

The time $t_0$ of emission of the photon from the source is sampled as

$$t_0^{(1)} = -\frac{1}{\varkappa} \ln(\alpha_1\alpha_2), \quad t_0^{(2)} = \beta(-2\ln\alpha_1)^{1/2} \cos(2\pi\alpha_2).$$

The estimated time dependence $I(t)$ can be improved by use of convolution,

$$I(t) = \int_0^t p_0(t - t^*) I_\delta(t^*)\, dt^*, \tag{5.16}$$

where $I_\delta(t)$ is the time distribution of the light intensity for the initial density $p_0(t) = \delta(t)$. The corresponding algorithm is easily obtained by combining formulas (5.16) and (5.6).

Because

$$\begin{aligned} I(t) &= \int_{S_r} ds \int_{\Omega_r} \Phi[\boldsymbol{r}^*(s), \omega^*, t]\, d\omega^* \\ &= \int_{S_r} ds \int_{\Omega_r} \left[ \int_0^t p_0(t - t^*) \Phi_\delta[\boldsymbol{r}^*(s), \omega^*, t^*]\, dt^* \right] d\omega^*, \end{aligned}$$

and

$$\Phi_\delta(x^*) = \int_X \frac{k(x', x^*)}{\sigma} f(x')\,dx',$$

(see Sect. 3.8), we have

$$I(t) = \int_{S_r} ds \int_{\Omega_r} \left\{ \int_X \left[ \int_0^t \frac{p_0(t - t^*)k(x', x^*)}{\sigma} dt^* \right] f(x')\,dx' \right\} d\omega^*$$
$$= \int_X \varphi_1'(x', x^*) f(x')\,dx' = E \sum_{j=0}^{N} Q_j \varphi_1'(x_j, x_j^*),$$

where

$$\varphi_1'(x_j, x^*) = \frac{\exp[-\tau(r_j, r_j^*)]g\left[\left(\omega_j, \frac{r_j^* - r_j}{|r_j^* - r_j|}\right)\right]}{2\pi|r_j^* - r_j|^2 p(r_j^*)}$$
$$\times \left\{ \int_0^t p_0(t - t^*)\delta\left[t^* - \left(t_j + \frac{|r_j - r_j^*|}{v}\right)\right] dt^* \right\} \Delta(l^*)$$
$$= \frac{\exp[-\tau(r_j, r_j^*)]g\left[\left(\omega_j, \frac{r_j^* - r_j}{|r_j^* - r_j|}\right)\right]}{2\pi|r_j^* - r_j|^2 p(r_j^*)} p_0\left[t - \left(t_j + \frac{|r_j - r_j^*|}{v}\right)\right]\Delta(l^*).$$

Thus, simulating the photon trajectories, starting at $t = 0$ and averaging the random variable

$$\xi_1' = \sum_{j=1}^{N} Q_j \varphi_1'(x_j, x_j^*),$$

over all trajectories, we estimate the light intensity for any desired $t$. Furthermore, the estimate of the time dependence of $I(t)$ may be improved significantly by averaging analytically over one random variable, namely the time of emission of the photons from the source. From the expressions for the estimates $\varphi_1'(x', x^*)$ and $\varphi_1(x', x^*)$ we see that, in this case, we must calculate also $p_0[t - (t' + |r' - r^*|/v)]$ at each photon collision point in the medium. When the function $p_0(t)$ is complicated, a histogram $I_{\delta_i}$ of the time distribution of the light intensity is calculated for a source with a $\delta$ time distribution of emitted photons; then, the desired distribution is

$$I(t) = \sum_i I_{\delta_i} \int_{t_{i-1}}^{t_i} p_0(t - t')\,dt'. \tag{5.17}$$

When the estimates (5.6) and (5.13) are used, it is necessary to calculate the quantity $1 - \cos\gamma_r$ (in particular, to determine whether the scattered photon will hit the receiver field of view). Owing to round off-errors, the accuracy of the quantity $1 - \cos\gamma_r$ may be less than needed to estimate the desired functionals sufficiently precisely. Therefore, instead of calculating the quantity $1 - \cos\gamma_r$, it is expedient to expand it into a series, and use several terms, depending on the value of $\gamma_r$.

## 5.3 Estimation of the Intensity of Singly-Scattered Light

Approximate analytical formulas are available for evaluating the time distribution of illumination $E^{(1)}(t)$ and intensity $I^{(1)}(t)$ of singly scattered light of a pencil beam [see for example (5.21)]. These formulas are applicable only if $\gamma_r \gg \gamma_s$, and the size of a source is negligible.

Furthermore, these formulas were obtained under the assumption that the optical axes of the source and receiver lie in the same plane. However, in many real problems, this is not true. Therefore, there is need to improve Monte Carlo algorithms for estimating the intensity of singly scattered light.

Estimation of the intensity of singly scattered light consists of evaluation of

$$\begin{aligned} I_i^{(1)} &= \int_{t_{i-1}}^{t_i} dt^* \int_{S_r} ds \int_{\Omega_r} d\omega \int_X \frac{k(x, x^*)}{\sigma} f_1(x)\,dx \\ &= \int_{t_{i-1}}^{t_i} dt^* \int_{S_r} ds \int_{\Omega_r} d\omega^* \int_X f_1(r, \omega, t) \frac{q \exp\{-\tau[r, r^*(s)]\}}{2\pi|r - r^*(s)|^2} \\ &\times g\left[\left(\omega, \frac{r^*(s) - r}{|r^*(s) - r|}\right)\right] \cdot \delta\left[t^* - \left(t + \frac{|r - r^*(s)|}{v}\right)\right] \\ &\times \delta\left(\omega^* - \frac{r^*(s) - r}{|r^*(s) - r|}\right) dr\,d\omega\,dt, \quad i = 1, \ldots, n_t,\ t_0 = 0. \end{aligned}$$

Here $f_1(x)$ is the density of first collisions in the phase space $X$ of coordinates, directions and time. Because

$$\begin{aligned} f_1(r, \omega, t) &= \int_{S_s} ds \int_{\Omega_s} d\omega_0 \int_0^\infty k[r_0(s), \omega_0, t_0; r, \omega, t] f_0(r_0(s), \omega_0, t_0)\,dt_0 \\ &= \int_{S_s} ds \int_{\Omega_s} d\omega_0 \int_0^\infty \frac{\sigma \exp\{-\tau[r, r_0(s)]\}}{2\pi|r - r_0(s)|^2} \cdot g\left[\left(\omega_0, \frac{r - r_0(s)}{|r - r_0(s)|}\right)\right] \\ &\times \delta\left(\omega - \frac{r - r_0(s)}{|r - r_0(s)|}\right) \delta\left[t - \left(t_0 + \frac{|r - r_0(s)|}{v}\right)\right] \\ &\qquad \times f_0[r_0(s), \omega_0, t_0]\,dt_0, \end{aligned}$$

(see Sect. 3.7) then, using the polar coordinate system $(l, \mu, \varphi)$ we have from (5.13)

$$I_i^{(1)} = \int_{t_{i-1}}^{t_i} dt^* \int_{S_r} ds^* \int_{\Omega_r} d\omega^* \int_{S_s} ds \int_{\Omega_s} d\omega_0 \int_0^\infty f_0(\boldsymbol{r}_0(s), \omega_0, t_0)\, dt_0$$
$$\times \int_0^\infty \frac{\exp\{-\tau[\boldsymbol{r}(l), \boldsymbol{r}^*(s^*)]\}}{2\pi|\boldsymbol{r}(l) - \boldsymbol{r}^*(s^*)|^2} \cdot g\left[\left(\frac{\boldsymbol{r}(l) - \boldsymbol{r}_0(s)}{|\boldsymbol{r}(l) - \boldsymbol{r}_0(s)|}, \frac{\boldsymbol{r}^*(s^*) - \boldsymbol{r}(l)}{|\boldsymbol{r}^*(s^*) - \boldsymbol{r}(l)|}\right)\right]$$
$$\times \delta\left[t^* - \left(t_0 + \frac{|\boldsymbol{r}^*(s^*) - \boldsymbol{r}(l)| + |\boldsymbol{r}(l) - \boldsymbol{r}_0(s)|}{v}\right)\right] \sigma\, \mathrm{e}^{-\sigma l}$$
$$\times \delta\left(\omega^* - \frac{\boldsymbol{r}^*(s^*) - \boldsymbol{r}(l)}{|\boldsymbol{r}^*(s^*) - \boldsymbol{r}(l)|}\right) dl, \tag{5.18}$$

where $\boldsymbol{r}(l) = \boldsymbol{r}_0 + \omega_0 l$.

Naturally, the estimate of the intensity of singly scattered light is meaningless if the transmitted photon is not incident in the receiver field of view. Therefore, we assume that the density $f_1(\boldsymbol{r}, \omega, t)$ is chosen so that

$$\boldsymbol{n}_s \cdot (\boldsymbol{r} - \boldsymbol{r}^*)/|\boldsymbol{r} - \boldsymbol{r}^*| \geqslant \cos(\gamma_r), \tag{5.19}$$

i.e. $(\boldsymbol{r} - \boldsymbol{r}^*)/|\boldsymbol{r} - \boldsymbol{r}^*| \in \Omega_r$. Otherwise, the singly scattered light will not reach the receiver. The coordinate system is oriented so that its origin coincides with the detector and the axis $OZ$ is coincident with the optical axis of the detector (see Fig. 5.8). The receiver field of view is bounded by a surface

$$x^2 + y^2 = (z \tan \gamma_r + R_r)^2, \quad z \geqslant 0. \tag{5.20}$$

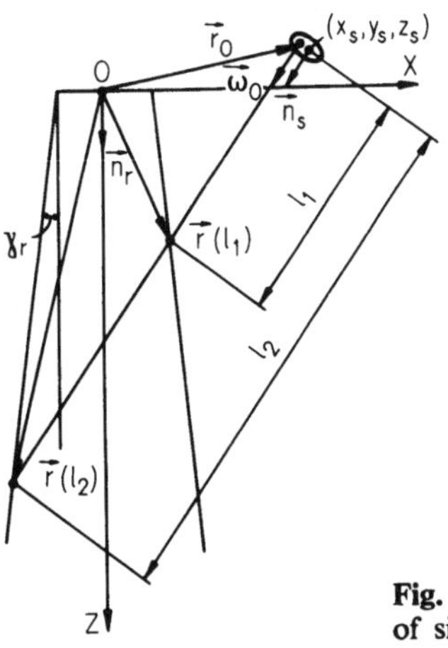


**Fig. 5.8.** Geometry of intensity estimation of singly-scattered light

Obviously, the inequality holds only if $|\mathbf{r}(l_1)| \leqslant |\mathbf{r}(l)| \leqslant |\mathbf{r}(l_2)|$, where $\mathbf{r}(l_1)$ and $\mathbf{r}(l_2)$ are points at which the ray $\mathbf{r}_0 + \boldsymbol{\omega}_0 l$ intersects the surface (5.20). The quantities $l_1$ and $l_2$ are solutions of the quadratic equation $Al^2 + Bl + C = 0$, where

$$A = a_0^2 + b_0^2 - c_0^2 \tan^2 \gamma_r,$$

$$B = 2(a_0 x_0 + b_0 y_0 - z_0 c_0 \tan^2 \gamma_r - c_0 \tan \gamma_r \cdot R_r),$$

$$C = (x_0^2 + y_0^2 + z_0^2 \tan^2 \gamma_r - 2z_0 R_r \tan \gamma_r - R_r^2),$$

that follows from the system

$$x^2 + y^2 = (z \tan \gamma_r + R_r)^2,$$

$$x = x_0 + a_0 l, \quad y = y_0 + b_0 l, \quad z = z_0 + c_0 l.$$

Here $x, y, z$ are the coordinates of the point $\mathbf{r}(l)$, $x_0, y_0, z_0$ are the coordinates of the point $\mathbf{r}_0$, and $a_0, b_0, c_0$ are the direction cosines of the vector $\boldsymbol{\omega}_0$.

If $l_1$ and $l_2$ are negative or complex, the ray $\mathbf{r}_0 + \boldsymbol{\omega}_0 l$ does not intersect the receiver field of view. Thus the random length $\lambda$ of the first run must be sampled from a distribution function defined on the interval $(l_1, l_2)$. The most natural choice is

$$p_\lambda(l) = \sigma \exp(-\sigma l)/[\exp(-\sigma l_1) - \exp(-\sigma l_2)].$$

If now the random point $\boldsymbol{\rho}^*$ on the detector surface is sampled from a density $|p(\mathbf{r}^*)|$ such that $[\boldsymbol{\rho}^* - \mathbf{r}(\lambda)]/|\boldsymbol{\rho}^* - \mathbf{r}(\lambda)| \in \Omega_r$ [for example, from the density (5.10)], then

$$I_i^{(1)} = M\varphi_1^{(1)},$$

$$\varphi_1^{(1)} = \frac{q \exp\{-\tau[\mathbf{r}(\lambda), \boldsymbol{\rho}^*]\}}{2\pi|\mathbf{r}(\lambda) - \boldsymbol{\rho}^*|^2} g\left[\left(\frac{\mathbf{r}(\lambda) - \boldsymbol{\rho}_0}{|\mathbf{r}(\lambda) - \boldsymbol{\rho}_0|}, \frac{\boldsymbol{\rho}^* - \mathbf{r}(\lambda)}{|\boldsymbol{\rho}^* - \mathbf{r}(\lambda)|}\right)\right] \times \frac{\sigma \exp(-\sigma\lambda)}{p_\lambda(\lambda) p(\boldsymbol{\rho}^*)} \cdot \Delta_i(t^*),$$

where $\boldsymbol{\rho}_0$ is a random point on the source surface, distributed according the density $p_{\rho}^{(0)}(\mathbf{r}_0)$. When $\gamma_r < \gamma_s$, a more effective algorithm could be obtained if $I_i^{(1)}$ were calculated by the use of the adjoint problem. In this case, we must substitute

$$p_{\omega}^{(0)}(\boldsymbol{\omega}_0) = \begin{cases} [2\pi(1 - \cos \gamma_r)]^{-1}, & \text{if} \quad \boldsymbol{\omega}_0 \in \Omega_r, \\ 0, & \text{if} \quad \boldsymbol{\omega}_0 \notin \Omega_r, \end{cases}$$

and

$$p_\rho^{(0)}(\boldsymbol{r}_0) = \begin{cases} (\pi R_r^2)^{-1}, & \text{if } \boldsymbol{r}_0 \in S_r, \\ 0, & \text{if } \boldsymbol{r}_0 \notin S_r, \end{cases}$$

in (5.18), where the integrations with respect to $\boldsymbol{r}^*$ and $\boldsymbol{\omega}^*$ must be performed over the surface $S_s$ and solid angle $\Omega_s$, respectively.

In conclusion, let us consider a numerical example in which the illumination $E^{(1)}(t)$ of a circle with $R_r = 0.05$ m by singly scattered light from a source that coincided with the circle was calculated.

The source with $R_s = 0.05$ m sends out an instantaneous light pulse that is transmitted within a circular solid angle $2\gamma_s = 3'$. In this case, the illumination $E^{(1)}(t)$ is described sufficiently precisely by (see [40])

$$E^{(1)}(t) = \frac{\sigma_s R_r^2 g(-1) v}{2} \cdot \frac{\exp(-\sigma v t)}{v^2 t^2} . \tag{5.21}$$

The illumination $E^{(1)}(t)$ of a unit surface, was calculated by means of the Monte Carlo algorithm to form a histogram $E_i^{(1)}$. For comparison, the integrals of (5.21)

$$E_i^{(1)} = \frac{1}{\pi R_r^2 (t_i - t_{i-1})} \int_{t_{i-1}}^{t_i} E^{(1)}(t) dt, \tag{5.22}$$

were evaluated by Simpson's rule to within 1%.

The indicatrix $g(\mu)$ for the sea water was used (see Fig. 5.9), and $\sigma = 0.1$ m$^{-1}$, $v = 0.225$ m ns$^{-1}$, $\sigma_s = 0.09$ m$^{-1}$. The time interval (10–700 ns) was divided

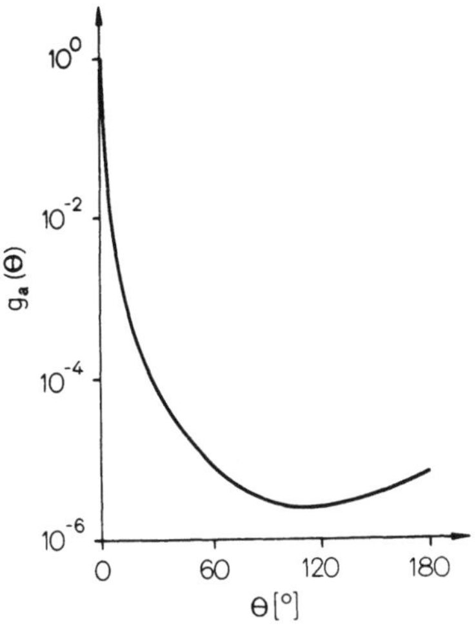


**Fig. 5.9.** Scattering indicatrix $g_a(\theta)$

**Table 5.1.** Computational results compared with formula (5.22)

| $n$ | $E_A$ | $E_{MC}$ | $\kappa$ |
|---|---|---|---|
| 1 | $0.30 \times 10^{-4}$ | $0.32 \times 10^{-4}$ | 3 |
| 2 | $0.80 \times 10^{-5}$ | $0.78 \times 10^{-5}$ | 3 |
| 3 | $0.32 \times 10^{-5}$ | $0.32 \times 10^{-5}$ | 4 |
| 4 | $0.15 \times 10^{-5}$ | $0.15 \times 10^{-5}$ | 4 |
| 5 | $0.81 \times 10^{-6}$ | $0.84 \times 10^{-6}$ | 4 |
| 6 | $0.46 \times 10^{-6}$ | $0.44 \times 10^{-6}$ | 4 |
| 7 | $0.28 \times 10^{-6}$ | $0.26 \times 10^{-6}$ | 5 |
| 8 | $0.17 \times 10^{-6}$ | $0.17 \times 10^{-6}$ | 5 |
| 9 | $0.11 \times 10^{-6}$ | $0.11 \times 10^{-6}$ | 5 |
| 10 | $0.22 \times 10^{-7}$ | $0.23 \times 10^{-7}$ | 2 |
| 11 | $0.70 \times 10^{-9}$ | $0.71 \times 10^{-9}$ | 4 |
| 12 | $0.35 \times 10^{-10}$ | $0.39 \times 10^{-10}$ | 7 |
| 13 | $0.22 \times 10^{-11}$ | $0.27 \times 10^{-11}$ | 11 |
| 14 | $0.15 \times 10^{-12}$ | $0.13 \times 10^{-12}$ | 23 |
| 15 | $0.11 \times 10^{-13}$ | $0.13 \times 10^{-13}$ | 36 |

into subintervals: 10–100 ns, in steps of 10 ns; 100–700 ns, in steps of 100 ns. The computational results are shown in Table 5.1; the results obtained from (5.22) are in column $E_A$ and the Monte Carlo results are in column $E_{Mc}$; $\kappa$ is the standard deviation of the Monte Carlo estimate, and $n$ was the number of intervals in the histogram. The results agree well.

## 5.4 Approximate Asymptotic Solutions of the Spherical Milne Problem with Anisotropic Scattering

To estimate the time distribution of photon flux in the neighborhood of a point $(x^*, y^*, z^*)$ at a large distance from a light source, it seems plausible to use the asymptotic solution of the adjoint spherical Milne problem (see, for example [14]) that is given by

$$W(r, \mu_r) = \exp(-r/L)a(\mu_r)/r, \tag{5.23}$$

where $r = |\boldsymbol{r}|$, $\boldsymbol{r} = (x - x^*, y - y^*, z - z^*)$, $\mu_r = \boldsymbol{\omega} \cdot \boldsymbol{r}/r$ and $\boldsymbol{\omega}$ is the new direction of motion of the photon after the collision. There is a connection between the diffusion length $L$ and the function $a(\mu_r)$

$$\left(1 + \frac{\mu_r}{\sigma L}\right) a(\mu_r) = \frac{q}{2\pi} \int_{\Omega} g(\omega, \omega') a(\mu_r') \, d\omega'. \tag{5.24}$$

Here $\Omega$ is the direction space. It is difficult to use $W(r, \mu_r)$ because of the divisor $r$ in (5.23). Therefore, in our modified simulation, an approximate function

$$W_0(r, \mu_r) = \exp\left(-\frac{r}{L}\right) a(\mu_r), \tag{5.25}$$

will be used.

Making use of the explicit form of the integral transfer equation, we obtain the following scheme for simulating the Markov chain that corresponds to the function (5.25). The free-path length $\lambda$ is sampled from a density function proportional to

$$f_\lambda(l) = \exp(-\sigma l) \exp[-r(l)/L], \tag{5.26}$$

where

$$r(l) = [r^2(0) + l^2 + 2l(\boldsymbol{r}(0) \cdot \omega]^{1/2}.$$

Here $\boldsymbol{r}(l) = (x - x^*, y - y^*, z - z^*)$, $\boldsymbol{r}(l) = \boldsymbol{r}(0) + l\omega$, $|\boldsymbol{r}(l)| = r(l)$, and $(x, y, z)$ is a new collision point. Because

$$\mu_r(l) = \omega \cdot \boldsymbol{r}(l)/r(l) = \frac{\partial r(l)}{\partial l}, \quad r(l) = \int_0^l \mu_r(t)\, dt + r(0). \tag{5.27}$$

Therefore,

$$f_\lambda(l) = \exp\left\{\int_0^l [\sigma + \mu_r(t)/L]\, dt\right\}, \tag{5.28}$$

within a constant factor.

This implies that the Markov chain describes the photon transfer process in a fictitious medium in which the total extinction coefficient is $\sigma + \mu_r/L$. It follows from (5.27) and (5.28) that the free-path length $\lambda$ can be obtained from

$$\lambda + \frac{r(\lambda) - r(0)}{\sigma L} = -\frac{\ln \alpha}{\sigma}. \tag{5.29}$$

This equation can be easily solved by use of (5.26). Bias is eliminated by introducing the weight factor,

$$Q(\lambda) = \exp\left\{\frac{r(\lambda) - r(0)}{L}\right\}\left(1 + \frac{\mu_r(l)}{\sigma L}\right)^{-1}.$$

A similar algorithm for free-path-length simulation is obtained when a spherical exponential transformation is used (see Sect. 2.4). In our case, however, we use the parameter $L^{-1}$ that is defined by (5.24). The scattering must be simulated according to a density proportional to

$$f(\omega) = g(\mu)a(\mu_r). \tag{5.30}$$

Furthermore the weight must be multiplied by

$$Q(\mu) = I(\mu_r')/a(\mu_r),$$

where

$$I(\mu_r') = \frac{1}{2\pi}\int_\Omega g(\omega, \omega')a(\mu_r)\,d\omega.$$

To simplify the algorithm, we find the asymptotic solution in a transport approximation. The scattering is then isotropic with the effective mean free-path length

$$\bar{l} = \{\sigma[q(1-\bar{\mu}) + 1 - q]\}^{-1}.$$

and the effective survival probability

$$\bar{q} = q(1-\bar{\mu})(1-q\bar{\mu})^{-1},$$

where $\bar{\mu} = \int_{-1}^{+1} \mu g(\mu) d\mu$. Then $L$ is the solution of a transcendental equation

$$\frac{2\bar{l}}{L\bar{q}} = \ln\left(\frac{\bar{\sigma}L+1}{\bar{\sigma}L-1}\right), \quad a(\mu_r) = (1+\mu_r/\bar{\sigma}L)^{-1}, \quad \bar{\sigma} = \bar{l}^{-1}. \tag{5.31}$$

In this case,

$$\begin{aligned} I(\mu_r') &= \frac{1}{2\pi}\int_\Omega \frac{g(\omega\cdot\omega')\,d\omega}{1+\mu_r'\bar{\sigma}L} \\ &= \int_{-1}^{+1} \frac{g(\mu)\,d\mu}{\left[\left(\frac{\mu}{\bar{\sigma}L}+\mu_r'\right)^2 + ((\bar{\sigma}L)^2 - 1)(1-\mu_r'^2)(\bar{\sigma}L)^{-2}\right]^{1/2}}. \end{aligned}$$

If the scattering indicatrix is tabulated and linear interpolation

$$g(\mu) = g_n + (g_{n+1} - g_n)(\mu - \mu_n)/(\mu_{n+1} - \mu_n),$$

where $\mu_1 = 1$, $\mu_N = -1$, $g_n = g(\mu_n)$, $\mu_{n+1} \leqslant \mu \leqslant \mu_n$, $\mu_1 > \mu_2 > \cdots > \mu_n > \cdots > \mu_N$ ($N$ is the number of interpolation points) is used, then the integral

$I(\mu_r')$ is

$$I(\mu_r') = (\bar{\sigma}L - 1)\sum_{n=1}^{N-1}\Bigg[(g_n - g_{n-1})(b_2 - b_1)\bar{\sigma}L + g_n\mu_{n-1} - g_{n+1}\mu_n + \mu_r'(g_n - g_{n-1})\bar{\sigma}L\cdot\ln\left(\frac{\mu_{n+1}/\bar{\sigma}L + b_1 + \mu_r'}{\mu_n/\bar{\sigma}L + b_2 + \mu_r'}\right)\Bigg](\mu_n - \mu_{n-1})^{-1}.$$

where $\mu_r' \neq -1$, $\mu_r' \neq 1$,

$$b_1 = \frac{1}{\bar{\sigma}L}[(\mu_{n+1} + \mu_r')(\bar{\sigma}^2L^2 - 1)(1 - \mu_r'^2)]^{1/2},$$

$$b_2 = \frac{1}{\bar{\sigma}L}[(\mu_n + \mu_r')(\bar{\sigma}^2L^2 - 1)(1 - \mu_r'^2)]^{1/2},$$

$$I(-1) = (\bar{\sigma}L - 1)\sum_{n=1}^{N-1}\Bigg[(g_n - g_{n-1})(\mu_{n+1} - \mu_n)\bar{\sigma}L + (g_{n+1}\mu_n - g_n\mu_{n+1} + g_n - g_{n+1})\ln\left(\frac{\bar{\sigma}L\cdot\mu_{n+1} - 1}{\bar{\sigma}L\cdot\mu_n - 1}\right)\Bigg]\Bigg/(\mu_n - \mu_{n+1}),$$

$$I(1) = (\bar{\sigma}L - 1)\sum_{n=1}^{N-1}\Bigg[(g_n - g_{n+1})(\mu_{n+1} - \mu_n)\bar{\sigma}L + (g_{n+1}\mu_n - g_n\mu_{n+1} + g_n - g_{n+1})\ln\left(\frac{\bar{\sigma}L\cdot\mu_{n+1} + 1}{\bar{\sigma}L\cdot\mu_n + 1}\right)\Bigg]\Bigg/(\mu_n - \mu_{n+1}).$$

To sample $\mu$ from the density (5.30), the rejection method can be applied. To this end, we choose $\mu$ from the density $g(\mu)$. If the indicatrix is tabulated, then $\mu$ is simulated as

$$\mu = \mu_n - \frac{g_n\Delta\mu_n + [g_n^2(\Delta\mu_n)^2 - 2\Delta\mu_n(g_n - g_{n-1})\kappa]^{1/2}}{g_n - g_{n-1}},$$

if $\kappa \leqslant 0$,

where

$$\kappa = \alpha + \sum_{i=1}^{n}\frac{g_{i-1} + g_i}{2}\Delta\mu_i;\ \Delta\mu_i = \mu_i - \mu_{i-1} < 0.$$

Rejection is performed according to

$$G(\mu_r) = \frac{\bar{\sigma}L - 1}{1 + \mu_r/\bar{\sigma}L}.$$

Furthermore, the weight factor is given by

$$Q = I(\mu_r')/G(\mu_r),$$

where

$$I(\mu) = I_{i-1} - \left[\frac{m(1 + \mu_r')}{2} - i + 1\right](I_{i-1} - I_i);$$

$$i = E\left(\frac{\mu_r' \cdot m}{2}\right) + \frac{m}{2} + 1;$$

$m + 1$ is the number of interpolation points, $I_i = I(\mu_r')$ for $\mu_r' = 2(i - 1)/(m - 1)$; $E$ is the sign of the integer part.

The rejection probability $p(\mu_r)$ is given by

$$p(\mu_r') = 1 - I(\mu_r')/\bar{\sigma}L.$$

For values $\mu_r' \approx -1$ (i.e., when the rejection probability is large) the quantity $\omega$ may be sampled from approximate density

$$f_1(\omega) = pg(\mu) + (1 - p)\frac{\bar{q}}{2(1 + \mu_r/\bar{\sigma}L)},$$
$$p = \frac{\bar{q}}{\bar{q} + 2g(\mu_r')(1 + \mu_r/\bar{\sigma}L)}. \qquad (5.32)$$

In this case, $\mu$ is sampled from $g(\mu)$ with probability $p$, while with probability $1 - p$, the quantity $\mu_r$ is calculated

$$\mu_r = \bar{\sigma}L\left[\left(1 - \frac{1}{\bar{\sigma}L}\right)\exp\left\{\frac{2\alpha}{\bar{q}\bar{\sigma}L}\right\} - 1\right].$$

Here the resulting bias is eliminated by introducing the weight factor $g(\omega', \omega)/f_1(\omega)$.

Consider now the Monte Carlo algorithm for estimating the diffusion length $L$. It is known (see, for example, [14]) that, for constant $\sigma$ and $q$, the total photon flux within a medium can be represented in the form

$$\Phi(\tau) = \Phi_{as}(\tau) + \Phi_{tr}(\tau), \qquad (5.33)$$

where $\Phi_{as}(\tau) \sim \exp(-\tau/\sigma L)$, $\Phi_{tr}(\tau) = O(e^{-\tau})$. Consequently, at a large optical depth $\tau$, the quantity $\Phi_{tr}(\tau)$ in (5.33) can be neglected. Therefore,

$$\frac{\Phi(\tau)}{\Phi(\tau_0)} \approx \exp\left[\frac{-\tau + \tau_0}{\sigma L}\right].$$

Thus

$$L \approx -\frac{1}{\sigma}\frac{-\tau+\tau_0}{\ln \Phi(\tau) - \ln \Phi(\tau_0)},$$

and

$$L \approx -1\left[\sigma \frac{\partial \ln \Phi(\tau)}{\partial \tau}\bigg|_{\tau=\tau_0}\right]^{-1}, \tag{5.34}$$

as $\tau \to \tau_0$. Introduce a parameter $\lambda = \tau/\tau_0$. The expression (5.34) can be rewritten as

$$L \approx -\left[\sigma \frac{\partial \ln \Phi(\tau)}{\partial \lambda}\bigg|_{\lambda=1} \cdot \frac{\partial \lambda}{\partial \tau}\right]^{-1} = \frac{-\tau_0}{\sigma} \frac{\Phi(\tau_0)}{\left.\dfrac{\partial \Phi(\tau)}{\partial \lambda}\right|_{\lambda=1}}. \tag{5.35}$$

Because $L$ is completely defined by the optical characteristics $\sigma$, $q$, and $g(\mu)$ [see (5.24)], and does not depend on the geometric parameters of the problem, to calculate $\Phi(\tau)$ and $\partial\Phi/\partial\lambda|_{\lambda=1}$ in (5.35), the solution of the appropriate model problem can be used. It follows from the theorem of optical mutuality [14] that, to estimate $L$ more precisely, a source with an angular distribution that corresponds to a $(-\mu)$ must be used. In order to examine the accuracy of (5.35), $\Phi(\tau_0)$ and $\partial\Phi/\partial\lambda|_{\lambda=1}$ were calculated for the following problem by use of the Monte Carlo method.

Consider a half-space filled with an optical homogeneous medium. It is assumed that the scattering is isotropic, and that the survival probability $q$ is 0.8 (for this case the exact value of the quantity $\sigma L$ is known: $\sigma L = 1.4077$). A stationary point source is positioned inside the medium at a distance of $\tau_0/\sigma$ from the plane $z = 0$. The total flux $\Phi(\tau_0)$ of photons flying into the half-space $z < 0$ in unit time was evaluated. The following estimate was used (see Sect. 3.11):

$$\left.\frac{\partial \Phi(\tau)}{\partial \lambda}\right|_{\lambda=1} \approx \sum_{i=1}^{M} \frac{q^{n_i}(n_i - \tau_i)}{M}, \tag{5.36}$$

where $n_i$ is the number of the collision of the photon in the $i$th trajectory until the photon escapes from the half-space $z > 0$, and $\tau_i$ is the total optical length of the trajectory. Computations with $M = 10000$ show that the error of the estimate of $\sigma L$ is less than 0.2% already for $\tau_0 = 2.5$.

To investigate the efficiency of the algorithm described in this section, the following problem was solved. Photons were assumed to be emitted uniformly from a circle $x^2 + y^2 \leqslant 25 \times 10^{-8}$, $z = 0$. The emission was assumed to be isotropic within a circular conic solid angle. The angle between the source element and the axis (parallel to the $z$-axis) was taken to be 30″. The time distribution of the emitted photons was taken to be $\kappa^2 t \exp(-\kappa t)$, where $\kappa = 9$. The photon velocity in the medium was taken to be $v = 0.023$, the extinction coefficient $\sigma = 1$, and $q = 0.9$ (in dimensionless units). The time distribution $E(t)$ of the average number of photons from a circle of radius 0.1 situated at a point (0, 0,20) perpendicular to the $z$-axis was calculated. The indicatrix for sea water was tabulated (average cosine of the scattering angle $\mu \approx 0.9$, see Fig. 5.9).

To check the applicability of the transport approximation, the residual of (5.24) was calculated. The values of the left- and right-hand sides of (5.24) are shown in Table 5.2 for three different points. The results show good agreement (the difference is less than 7%).

In addition, the quantity $\sigma L$ was calculated by the Monte Carlo method based on (5.35) for various values of $\tau$. For $\tau$ from 1 to 20 in unit steps, 100000 samplings were used. For $\tau = 1$, $\sigma L \approx 3.4$. Then $\sigma L$ increases as $\tau$ increases, and becomes constant (4.8) when $\tau \geqslant 5$. For this case, (5.31) gives $\sigma L = 5.4$.

Consequently, the solution of (5.31) is somewhat greater than the true value of $\sigma L$ for sea water. However the difference is less than 12%. The time distribution of irradiance $E(t)$ evaluated by the described algorithm, as well as the corresponding results obtained by the direct-simulation method are shown in Table 5.3. In it $\{t_k\}$ are the histogram points (in dimensionless units); $E_k$ is the estimate of irradiance; and $\kappa_{ak}$, $\kappa_{nk}$ are the relative standard errors of the estimate $E_k$ [%] for the method described in this section and the direct-simulation method,

**Table 5.2.** Residual of Equation (5.24)

| $\mu_r$ | $a(\mu_r)(1 + \mu_r/\sigma L)$ | $(q/2\pi) \int_\Omega [g(\omega, \omega')a(\mu_r')]d\omega'$ |
|---|---|---|
| −1 | 24.8 | 23.0 |
| 0 | 1.0 | 1.003 |
| +1 | 0.602 | 0.578 |

**Table 5.3.** Comparison with direct simulation

| $t_k$ | $E_k \cdot 10^7$ | $t_a \cdot \kappa_{ak}^2$ | $t_n \cdot \kappa_{nk}^2 \cdot 10^{-5}$ |
|---|---|---|---|
| 20.70 | 0.296 | 120 | 2.82 |
| 21.15 | 0.947 | 115 | 0.88 |
| 21.60 | 1.160 | 109 | 0.72 |
| 22.05 | 0.984 | 119 | 0.85 |
| 22.50 | 0.831 | 183 | 1.00 |
| 22.95 | 0.612 | 201 | 1.36 |
| 23.40 | 0.440 | 227 | 1.89 |
| 23.85 | 0.242 | 289 | 3.69 |
| 24.30 | 0.222 | 306 | 3.71 |
| 24.75 | 0.210 | 365 | 4.00 |
| 25.20 | 0.189 | 446 | 4.41 |
| 25.65 | 0.108 | 693 | 7.71 |

respectively. The calculations of $E_k$ required 300 seconds of BESM-6 computer time with the direct-simulation method (with 36,000 trajectories) and 1100 seconds with the method described in this section. To obtain the results to within a given error, as Table 5.3 shows, the direct-simulation method requires $10^3$ times as much computer time as the method using the asymptotic solution.

## 5.5 Estimation of the Intensity of Light Reflected by a Medium

Consider an important practical problem, namely estimating the time distribution of intensity $I(t)$ of a light signal reflected by the atmosphere or sea water. There are difficulties in the Monte Carlo calculations of this problem because the atmosphere and sea water have strongly anisotropic scattering, i.e., $g(1)/g(-1) \approx 10^2$–$10^6$.

The local-estimate method requires a large number of samplings because of wide fluctuations of the factor $g(\mu)$ in (5.6). It is extremely difficult to estimate $I(t)$ for large values of $t$. In order to reduce errors in $I(t)$ calculations, *Golubitzki* et al. [41] proposed a modified local estimate based on analysis of trajectories according to certain indications. The following example illustrates such a technique. Assume that the system is symmetric so that the source and the receiver are interchangeable but not coincident (e.g., $\gamma_s = \gamma_r$, $R_s = R_r$, and the distance between the source and receiver is $l_\delta > 0$). Consider an arbitrary trajectory $\boldsymbol{x}_0, \ldots, \boldsymbol{x}_{n-1}, \boldsymbol{x}_n$, starting at a point on the source surface and terminating in the receiver. Define a function $P_n$ of the trajectory as

$$P_n = P_n(\omega_{n-1} \cdot \omega_n) = \begin{cases} 1, & \text{if } (\omega_{n-1} \cdot \omega_n) \leqslant \mu_0, \\ 0, & \text{otherwise,} \end{cases}$$

where $\mu_0$ is a given number, $-1 \leqslant \mu_0 \leqslant +1$. From the source–receiver symmetry, one half of the contribution to the estimation of the intensity of multiply scattered light (of $n$th order) is due to trajectories for which $P_n = 1$.

Therefore (see Sect. 5.3),

$$I_i = 2E\xi, \quad \xi = \sum_{j=1}^{N} Q_j \varphi(\boldsymbol{x}_j, \boldsymbol{\rho}^*) \cdot P_j\left(\omega_{j-1} \cdot \frac{\boldsymbol{\rho}^* - \boldsymbol{r}_j}{|\boldsymbol{\rho}^* - \boldsymbol{r}_j|}\right), \tag{5.37}$$

$$\varphi(\boldsymbol{x}_j, \boldsymbol{\rho}^*) = \frac{\exp\left[-\tau(\boldsymbol{r}_j, \boldsymbol{\rho}^*)\right] g\left[\left(\omega_{j-1} \cdot \dfrac{\boldsymbol{\rho}^* - \boldsymbol{r}_j}{|\boldsymbol{\rho}^* - \boldsymbol{r}_j|}\right)\right]}{2\pi |\boldsymbol{r}_j - \boldsymbol{\rho}^*|^2 p(\boldsymbol{\rho}^*)} \Delta(\boldsymbol{l}^*) \cdot \Delta_i(t^*), \tag{5.38}$$

where $\boldsymbol{l}^* = (\boldsymbol{r}_j - \boldsymbol{\rho}^*)/|\boldsymbol{r}_j - \boldsymbol{\rho}^*|$, $t^* = t_j + |\boldsymbol{\rho}^* - \boldsymbol{r}_j|/v$ and the random point $\boldsymbol{\rho}^*$ on the receiver surface is chosen according to the density $p(\boldsymbol{r}^*)$, $\boldsymbol{r}^* \in S_r$. Therefore, fluctuations of $g(\mu)$ is (5.38) can be significantly reduced by appropriate choice of $\mu_0$. This technique may be applied to an arbitrary system that also simulates the trajectories of the adjoint transfer equation. In that case, instead of doubling in (5.37), it is necessary to add the results for the direct and adjoint trajectories. The disadvantage of this method is that estimation of angular and spatial distributions of $I(\boldsymbol{r}, \boldsymbol{\omega}, t)$ is not possible, because, to simulate the adjoint trajectories, it is necessary to determine the receiver parameters (i.e., $R_r$, $\gamma_r$, $l_\delta$).

The technique described improves the local estimate only if $l_\delta > 0$. A pecularity of the estimation of $I(t)$ for atmosphere and sea water was first described by *Golubitzki* et al. [41]. For example, consider the problem of estimating the intensity of doubly scattered light in the symmetric system. Because of the strong anisotropy of the scattering indicatrix, trajectories in which the photon's new direction after the first collision is within the forward half-sphere are most frequently sampled. Trajectories in which the scattering angle after the first collision is such that $\pi/2 \leqslant \theta \leqslant \pi$ are seldom sampled. However for the symmetric system, both types of trajectories make equal contributions to the estimate of the intensity of doubly scattered light.

Therefore, the results may be essentially underestimated when only a small number of samples is used. Furthermore, this error increases as the order of multiplicity of scattering increases. In order to improve the calculations, an attempt could be made to apply the above-described technique.

For that purpose, for each scattering, an indication of the usefulness of each trajectory must be defined, and contributions must be taken only from certain trajectories, according to specific combinations of their usefulness; the contribution is then multiplied by the corresponding weight. When the intensity of doubly scattered light is estimated, the usefulness of this algorithm is obvious: the contribution must be calculated from trajectories for which the scattering angle afte the first collision is such that $0 \leqslant \theta = \pi/2$; that contribution is doubled. When the intensity of multiply scattered light is estimated, difficulties arise in defining the usefulness. It is simpler to use a method that uses the approximate importance based on a modification of the scattering-angle simulation. The cosine of the scattering angle is sampled from a density that is proportional to the function $g(\mu)a(\mu_r)$, where $a(\mu_r)$ is the function defined in Sect. 5.4, $\mu_r = \boldsymbol{\omega}\cdot\boldsymbol{r}/|\boldsymbol{r}|$, where $\boldsymbol{\omega}$ is the direction of motion of the photon after scattering, $\boldsymbol{r}$ is the radius vector (with origin at the detector) of the collision point. Furthermore, the photon weight must be multiplied by

$$\int_{-1}^{+1} \frac{g(\mu')a[\mu_r(\mu')]\,d\mu'}{a(\mu_r)}.$$

The weight $[a(\mu_r)]^{-1}$ can be compensated by choosing the free-path length from the density function (5.28). If this technique is applied for example, to sea water (with parameters $q = 0.9$, $\sigma L = 4.8$ and the scattering indicatrix shown in Fig. 5.1), the probability that the scattering angle $\theta$ is in the interval $(-\pi/2, \pi)$ is 30 times as large as the corresponding probability for the case of direct simulation. This technique also allows the intensity $I(t)$ to be estimated more precisely when the source and receiver coincide ($l_\delta = 0$).

There have been many successful attempts in connection with approximate and asymptotic estimates of the time distribution of intensity $I(t)$ of light (pulsed, from a spatially bounded source) reflected by a turbid medium for various time intervals. For instance, *Beljantcev* et al. [40] obtained the time distribution of the intensity for a large reception angle, immediately following the light pulse.

The theory of asymptotic solutions for long durations was discussed in detail in [42]. Of the approximate, asymptotic, and numerical solutions, for engineering purposes, the first two kinds are the most useful. These solutions are defined, however, only within a constant factor and describe only the qualitative behavior of the time distribution of $I(t)$.

Furthermore, there are also difficulties in studying the limits of applicability of approximate and asymptotic solutions and in estimating the accuracy of these solutions. Often, the Monte Carlo technique is the only method of improving them and extending their range of validity. Because detailed discussion of approximate and asymptotic solutions is beyond of the scope of this book, we consider here only solutions that have been used to solve some practical problems of laser sounding of sea water. Assume that a detector, situated near the source, records only singly scattered light. Then the light intensity recorded at the time $t$ is given by

$$I(t) = A \int_0^t \varphi(t')p_t^{(0)}(t-t')\,\mathrm{e}^{-v\sigma t'}\,dt', \tag{5.39}$$

where $p_t^{(0)}$ is the time-distribution density of the emitted photons, and $A$ is a parameter that depends on the characteristics of the medium, the power of the source, and the receiver's sensitivity. The function $\varphi$ takes account of the geometry of the problem. Different regions of the medium (where the light beam is propagated) make different contributions to $I(t)$. For instance, in (5.21),

$$A = \frac{\sigma_s R_s^2 \cdot g(-1)}{2v}, \quad \varphi(t) = t^{-2}.$$

Investigation of the limits of applicability of (5.39) is important, because this formula can be used to infer information about the optical characteristics of the medium, on the basis of $I(t)$ measurements [42]. Obviously, (5.39) is applicable only to small values of $t$, and especially when a large angular reception aperture is considered. Also, the accuracy of determination of $\varphi(t)$ affects the applicability of (5.39). The crude approximation of the geometry of the problem may result in noticeable error in $I(t)$ calculations for small values of $t$, for which the multiple-scattering contribution is still not large. Notice that $\varphi(t)$ for definite parameters of a problem can be easily estimated on the basis of preliminary Monte Carlo calculations of $I(t)$.

We now consider an example in which, in order to investigate the limits of applicability (5.39) to various geometries and optical characteristics of the medium, the total intensity of light reaching the receiver was calculated by the Monte Carlo method. Calculations were done for the case of coincident circular receiver and source regions of radii 0.005 m, source aperture angle $2\gamma_s = 1'$, and $p_t^{(0)}(t) = \kappa^2 \cdot t \cdot \exp(-\kappa t)$, $\kappa = 0.2\ \mathrm{ns}^{-1}$. The scattered light was determined at angles $2\gamma_r = 1'$ and $2\gamma_r = \pi$ (in the latter case the illumination was determined). The parameters were taken to be: $\sigma = 0.1\ \mathrm{m}^{-1}$, $q = 0.9$, 0.7, and the scattering indicatrix was assumed to have a relatively large forward peak, which is the case for sea water (the mean cosine of the scattering angle $\mu \approx 0.9$, see Fig. 5.9). The relative standard deviations of the calculations were less than 7–10% for small $t$ ($< 300$ ns), whereas for large $t$ ($> 700$ ns) they were about 30%. The calculated time distributions of intensity $I(t)$ (averaged over an angle of 1′) and illumination $E(t)$ of the back-scattered (into the receiver) light are shown in Figs. 5.10, 11, in relative units.

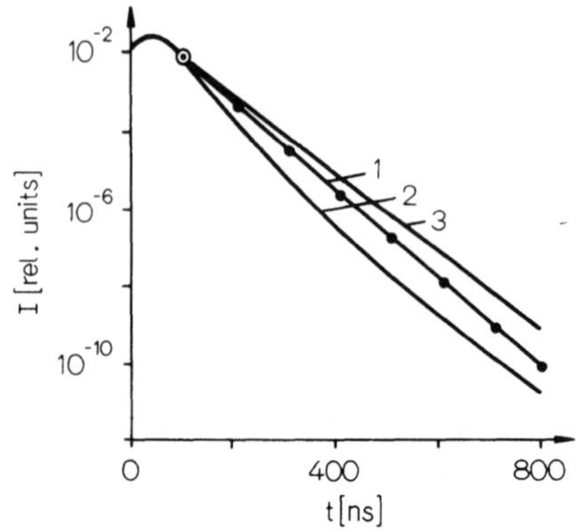


**Fig. 5.10.** Intensity of back-scattered light as a function of time (*1*) Monte Carlo results [points were obtained according to (5.39)], (*2, 3*) Formula (5.39) with $\varphi(t) = t^{-2}$ and $\varphi(t) = const$, respectively

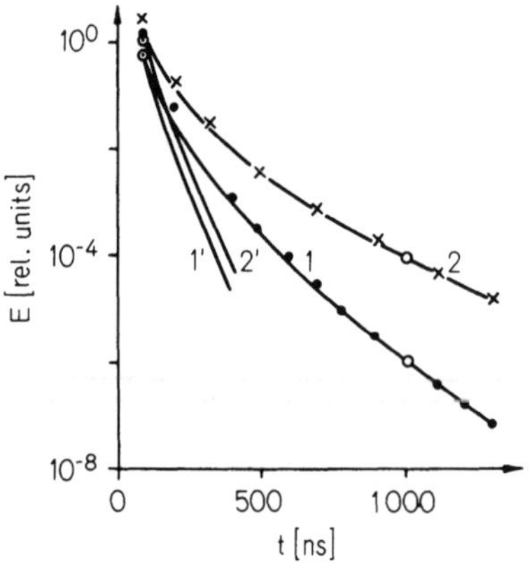


**Fig. 5.11.** Illumination as function of time (*1, 2*) Monte Carlo results, (*1′, 2′*) Formula (5.39), (*1. 1′*) $q = 0.7$; (*2. 2′*) $q = 0.9$; Points and crosses correspond to (5.40)

The parameter $A$ can be obtained by comparison of these results with the values of intensity calculated by the Monte Carlo method for $t = 100$ ns, for the function $\varphi$ taken from [43]. As Fig. 5.9 shows, the single-scattering approximation is valid for all time intervals under consideration. It should be noted that the value $t = 800$ ns corresponds, in this case, to a sufficiently large dimensionless variable $\sigma v t = 18$. Evidently, the increase of the angle $\gamma_r$ leads to an increase of the contribution of multiply scattered light.

Illumination $E(t)$ due to singly scattered light, as a function of time, is shown in Fig. 5.11 (compared with curves 1 and 2 for $t = 100$ ns). These results shown that (5.39) is not valid for the earliest times.

The influence of the error in approximating the function $\varphi(t)$, on the calculation of $I(t)$ from (5.39) is shown in Fig. 5.10, where curve 2 corresponds to $\varphi(t) = t^{-2}$, and curve 3 to $\varphi(t) = \text{const}$. An asymptotic solution that describes the intensity of the back-scattered light for $t \gg \rho/\sigma v$,

$$I(t) = At^{-5/2} \exp[-(\sigma - \sigma_s)vt], \tag{5.40}$$

was obtained by *Sege* et al. [42] by use of a diffusion approximation for the point source. Here $\rho$ is the optical distance from the collision point to the source, $A$ is a constant that depends on the characteristics of the receiver, the source and the medium. Here arises a question of the time of occurrence of the asymptotic condition. The larger the angular reception aperture, the sooner the asymptotic condition occurs. This can be clearly seen by a comparison of the results of Figs. 5.10, 11. Singly scattered light makes the dominant contribution to the intensity of light recorded at a small backward reception angle until $\sigma v t \approx 18$ (dimensionless units). The multiple-scattering contribution to the illumination is already large for early times. The results plotted as crosses and points in Fig. 5.10 were obtained under the assumption that the decrease of the illumination $E(t)$ is described by (5.40).

The quantity $A$ was obtained by comparison of results calculated by the Monte Carlo method with those obtained from (5.40) at $t = 1000$ ns. The asymptotic condition occurs, as can be easily seen, if $\sigma v t > 10$. This estimate is valid, of course, only to within the accuracy of the Monte Carlo calculations. The change of intensity of reflected light during a time interval shorter than that for which the asymptotic condition (5.40) is fulfilled, and while (5.39) is also not applicable, can be estimated in fact only by use of the Monte Carlo method.

## 5.6 Calculation of the Signal-to-Noise Ratio

Consider a typical laser sounding problem, namely estimating the signal-to-noise ratio, which is one of the most important characteristics of any lidar system. Assume that a scattering and absorbing layer $h \leqslant z \leqslant H$ is illuminated by a pencil beam. A circular plane body (object) is situated in the plane

$$a_T(x - x_T) + b_T(y - y_T) + C_T(z - z_T) = 0.$$

Here $(x_T, y_T, z_T)$ is the center of the object; $R_T$ is the radius of the object, and $\tau$ is the optical distance from $(x_T, y_T, z_T)$ to the source.

Spatial orientation of the body is determined by a unit vector $\boldsymbol{n}_T = (a_T, b_T, c_T)$ that is perpendicular to the body's surface. The light reflection from the body is characterized by an angle function $p_t(\omega)$ and albedo $A_T$ (reflection coefficient). Assume for a moment that there is no object in the medium. Then, denote the intensity of the backscattered light reaching the receiver at the time $t$ by $I_0(t)$ [if $2\gamma_r = \pi$, the illumination $E_0(t)$ is recorded]. Denote the function that describes the time distribution of the intensity of the recorded light in the presence of an object by $I_1(t)$ [respectively $E_1(t)$, if the illumination is recorded]. If there is no another noise except the noise due to backscattering, then the signal-to-noise ratio is characterized by $\xi(t) = \Delta I(t)/I_0(t)[\eta(t) = \Delta E(t)/E_0(t)$, respectively if $2\gamma_r = \pi]$, where $\Delta I(t) = I_1(t) - I_0(t)$ and $\Delta E(t) = E_1(t) - E_0(t)$.

In $\xi(t)$ calculating, it is convenient to estimate $I_0(t)$ and the difference $\Delta I(t)$ separately. To estimate $\Delta I(t)$, the algorithm described in Sect. 5.5 will be used. The free-path length $\lambda$ after the photon's collision at a point $(x, y, z)$ in direction $\omega'$ is simulated according to

$$\lambda = \frac{-B + \sqrt{B^2 - AC}}{A},$$

$$A = 1 - \sigma^2 L^2, \quad B = r_0(\mu_r' + \sigma L) + \beta\sigma L^2, \quad C = -2Lr_0\beta - (\beta L)^2,$$

where

$$\beta = -\ln(\alpha), \quad \mu_r' = \omega' \cdot \boldsymbol{r}_0/r_0, \quad r_0 = |\boldsymbol{r}_0|, \quad \boldsymbol{r}_0 = (x - x_T, y - y_T, z - z_T).$$

It is, in fact, the solution of the (5.29) where (5.26) is taken into account. The scattering is simulated according to the density (5.30). The photon weight after the $\lambda$ simulation must be multiplied by

$$Q(\lambda, \omega) = I(\mu_r') \exp\left[\frac{r(\lambda) - r_0}{L}\right],$$

where $r(\lambda) = |\boldsymbol{r}(\lambda)|$, $\boldsymbol{r}(\lambda) = \boldsymbol{r}_0 + \omega\lambda$. Assume that the trajectory so constructed intersects the surface of the object at a point $(x_1, y_1, z_1)$ in the direction $\omega_1$, and that the photon weight is $Q_1$. In this case, the trajectory splits into two trajectories: the trajectory in the direction $\omega_1$, and the trajectory reflected in the direction $\omega_2$. Further, these trajectories are simulated according to the above-described scheme with $\mu_r' = \omega \cdot \boldsymbol{r}_0'/r_0$, where $r_0' = |\boldsymbol{r}_0'|$ is the distance from the receiver's position to the collision point. The new direction $\omega_2$ after the reflection may be sampled (see Sect. 5.4) from a density proportional to

$$\frac{p_T(\omega)}{1 + \mu_r/\sigma L},$$

where $\mu_r = \omega \cdot \boldsymbol{r}_T/r_T$, $r_T = |\boldsymbol{r}_T|$ is the distance from the point $(x_T, y_T, z_T)$ to the receiver center. The photon weight after the reflection is given by

$$Q_2 = Q_1 \cdot A_T \exp\left[\frac{r_1 - r_0}{L}\right]\left(1 + \frac{\mu_r}{\sigma L}\right) \int_0^{2\pi} \int_0^1 \frac{p[\omega(\mu, \varphi)]}{1 + \mu_r(\mu, \varphi)/\sigma L} d\mu \, d\varphi,$$

where $r_1 = |\boldsymbol{r}_1|, r_0 = |\boldsymbol{r}_0|$ are the distances from the object center to the points $(x_1, y_1, z_1)$ and $(x, y, z)$, respectively. Here $(x, y, z)$ is the last collision point before intersection with the surface of the object. In particular, if the Lambert reflection law is assumed, i.e., $p_T(\omega) = \mu/\pi$; $\mu = \omega \cdot \boldsymbol{n}_T$, and the body is oriented perpendicular to the optical axis of the detector, which passes through the center $(x_T, y_T, z_T)$ of the object, then the simulating formula for choice of $\mu$ is

$$\mu = \sigma L\{1 - \exp[\alpha \ln(1 - 1/\sigma L)]\}.$$

To increase the efficiency of this method, assume emission from the intersection point $(x_1, y_1, z_1)$ in the direction $\omega_1$ of $n_1$ independent photons, each having a weight $Q_1' = Q/n_1$ and $n_2$ photons with weights $Q_2' = Q_2/n_2$ in the direction $\omega_2$ (see Sect. 3.13). We shall shortly consider a method of finding appropriate values of $n_1$ and $n_2$.

In order to calculate the contributions from the trajectories, it is convenient to use, in view of the small size of the receiver, the local estimate (5.6). In this case, we have

$$\Delta I_i = I_1^{(i)} - I_0^{(i)} = M(\xi_1^{(i)} - \xi_0^{(i)}),$$

$$\xi_1^{(i)} = Q_2' \sum_{k=n}^{N_1} Q_k \frac{\exp[-\tau(\boldsymbol{r}_k, \boldsymbol{\rho}_1^*)]}{2\pi|\boldsymbol{r}_k - \boldsymbol{\rho}_1^*|2p(\boldsymbol{\rho}_1^*)} g\left[\left(\omega_k \cdot \frac{\boldsymbol{\rho}_1^* - \boldsymbol{r}_k}{|\boldsymbol{\rho}_1^* - \boldsymbol{r}_k|}\right)\right] \Delta(\boldsymbol{l}_1^*)\Delta^{(i)}(t_1^*),$$

$$\xi_0^{(i)} = Q_1' \sum_{j=n}^{N_0} \frac{\exp[-\tau(\boldsymbol{r}_j, \boldsymbol{\rho}_0^*)]}{2\pi|\boldsymbol{r}_j - \boldsymbol{\rho}_0^*|^2 p(\boldsymbol{\rho}_0^*)} g\left[\left(\omega_j \cdot \frac{\boldsymbol{\rho}_0^* - \boldsymbol{r}_j}{|\boldsymbol{\rho}_0^* - \boldsymbol{r}_j|}\right)\right] \Delta(\boldsymbol{l}_0^*)\Delta^{(i)}(t_0^*),$$

where $\boldsymbol{r}_m$ is the radius vector of the $m$th collision point, $\boldsymbol{\rho}_1^*$ and $\boldsymbol{\rho}_0^*$ are two random points on the surface of thr receiver sampled from the density $p(\boldsymbol{r}^*)$,

$$\boldsymbol{l}_1^* = (\boldsymbol{r}_k - \boldsymbol{\rho}_1^*)/|\boldsymbol{r}_k - \boldsymbol{\rho}_1^*|, \quad \boldsymbol{l}_0^* = (\boldsymbol{r}_j - \boldsymbol{\rho}_0^*)/|\boldsymbol{r}_j - \boldsymbol{\rho}_0^*|,$$

$$t_1^* = t_k + |\boldsymbol{r}_k - \boldsymbol{\rho}_1^*|/v, \quad t_0^* = t_j + |\boldsymbol{r}_j - \boldsymbol{\rho}_0^*|/v;$$

$n$ is the number of the collision that follows the intersection of the surface of the object by the trajectory. Algorithms for estimating $I_0(t)$ were described in Sect. 5.5.

The function $\xi(t)$ can be estimated by use of the Monte Carlo method for a finite time interval $t_0 \leqslant t \leqslant T$, where $t_0 = (\rho_1 + \rho_2)/v$ and $\rho_1$, $\rho_2$ are the minimal distances from the object to the source and the detector, respectively. Therefore the photon paths may be terminated. This is done by calculating $t = (|\boldsymbol{r}_j - \boldsymbol{r}_\mathrm{T}| + \rho_2)/v + t_j$ at each collision point until the trajectory intersects the body's surface; $t = |\boldsymbol{r}_j|/v + t_j$ after this intersection. Here $\boldsymbol{r}_j$ and $\boldsymbol{r}_\mathrm{T}$ are the radius vectors (with origin at the detector center) of the collision point and the body center, respectively; $t_j$ is the time at which the collision occured. If $t > T$, the trajectory is terminated, for it can not make a contribution to the desired result. Termination of the trajectories implies that the fictitious absorption is increased, which leads in turn to a decrease of the quantity $\sigma L$. For example, for sea water ($q = 0.9$), estimation according to (5.35) gives $\sigma L \approx 4.8$. However analysis of the $\Delta I$ results shows that the optimal value of $\sigma L$ is 3 if the photon trajectories are terminated for $T - t_0 = 1.35/\sigma v$.

An analogous result gives the estimate of $\sigma L$ according to (5.35) if in calculating $\Phi(\tau_0)$ and $\partial\Phi/\partial\lambda|_{\lambda=1}$ only trajectories with total optical thickness less than $\tau_0 + 1.35$ are taken into account.

Notice that, as arguments of Sect. 3.13 show, the optimal values of $n_1$ and $n_2$ are close to $1/P_1$ where $P_1$ is the probability that the photon reaches the object. The quantity $P_1$ can be approximately evaluated in preliminary calculations. Calculations show that this technique provides a satisfactory estimate of values $v_1$ and $v_2$. Use of splitting has significantly improved the estimation of $\Delta I(t)$.

## 5.7 Monte Carlo Solution of Some Practical Problems. Comparison with Experimental Data

In this section, we present results for some atmospheric and hydro-optic problems calculated by use of algorithms described in the present chapter.

The time distribution of intensity $I(t)$ of the light reflected by a semi-infinite, optically homogeneous medium was calculated for coincident source and receiver, each having a radius of 0.005 m. The aperture angle of the source was taken to be $2\gamma_r = 1'$, and $2\gamma_r = 30'$, $1°$, $5°$, $10°$.

The time distribution of the power of the source was described by $\kappa^2 t \cdot \exp(-\kappa t)$, $\kappa = 0.2\ \mathrm{ns}^{-1}$. The calculations were for media with $\sigma = 0.1\ \mathrm{m}^{-1}$, $q = 0.9$, 0.7. The scattering indicatrix was as shown in Fig. 5.9 (it corresponds to an average model of light scattering by sea water). The results (Table 5.4) were obtained for the time interval (20, 700 ns) in the form of a histogram, i.e., the integrals

$$I_i = [2\pi(1 - \cos\gamma_r)(t_i - t_{i-1})]^{-1} \int_{t_{i-1}}^{t_i} I(t)\,dt,$$

were evaluated.

**Table 5.4.** Function $I_0(t)$ for various values of $\gamma_r$. $\beta_i$ is the relative standard deviation of the estimate

| $i$ | $t_i$ ns | $\gamma_r = 30'$ $I_i$ | $\beta_i$ | $\gamma_r = 15'$ $I_i$ | $\beta_i$ | $\gamma_r = 2°30'$ $I_i$ | $\beta_i$ | $\gamma_r = 5°$ $I_i$ | $\beta_i$ |
|---|---|---|---|---|---|---|---|---|---|
| | | | | $q = 0.7$ | | | | | |
| 1 | 30 | $0.831 \times 10^{-7}$ | 0.11 | $0.831 \times 10^{-7}$ | 0.11 | $0.831 \times 10^{-7}$ | 0.11 | $0.831 \times 10^{-7}$ | 0.11 |
| 2 | 40 | $0.167 \times 10^{-5}$ | 0.02 | $0.166 \times 10^{-5}$ | 0.02 | $0.168 \times 10^{-5}$ | 0.02 | $0.168 \times 10^{-5}$ | 0.02 |
| 3 | 50 | $0.196 \times 10^{-5}$ | 0.02 | $0.195 \times 10^{-5}$ | 0.02 | $0.204 \times 10^{-5}$ | 0.02 | $0.207 \times 10^{-5}$ | 0.02 |
| 4 | 60 | $0.126 \times 10^{-5}$ | 0.02 | $0.124 \times 10^{-5}$ | 0.02 | $0.132 \times 10^{-5}$ | 0.02 | $0.137 \times 10^{-5}$ | 0.02 |
| 5 | 70 | $0.757 \times 10^{-6}$ | 0.02 | $0.740 \times 10^{-6}$ | 0.02 | $0.818 \times 10^{-6}$ | 0.02 | $0.858 \times 10^{-6}$ | 0.03 |
| 6 | 80 | $0.443 \times 10^{-6}$ | 0.02 | 0.43[illegible] $\times 10^{-6}$ | 0.02 | $0.491 \times 10^{-6}$ | 0.02 | $0.525 \times 10^{-6}$ | 0.03 |
| 7 | 90 | $0.253 \times 10^{-6}$ | 0.02 | $0.242 \times 10^{-6}$ | 0.02 | $0.297 \times 10^{-6}$ | 0.02 | $0.320 \times 10^{-6}$ | 0.03 |
| 8 | 100 | $0.162 \times 10^{-6}$ | 0.02 | $0.154 \times 10^{-6}$ | 0.02 | $0.199 \times 10^{-6}$ | 0.03 | $0.224 \times 10^{-6}$ | 0.04 |
| 9 | 200 | $0.330 \times 10^{-7}$ | 0.01 | $0.304 \times 10^{-7}$ | 0.01 | $0.453 \times 10^{-7}$ | 0.02 | $0.547 \times 10^{-7}$ | 0.04 |
| 10 | 300 | $0.130 \times 10^{-8}$ | 0.01 | $0.109 \times 10^{-8}$ | 0.01 | $0.242 \times 10^{-8}$ | 0.03 | $0.352 \times 10^{-8}$ | 0.06 |
| 11 | 400 | $0.838 \times 10^{-10}$ | 0.03 | $0.632 \times 10^{-10}$ | 0.02 | $0.180 \times 10^{-9}$ | 0.06 | $0.267 \times 10^{-9}$ | 0.11 |
| 12 | 500 | $0.654 \times 10^{-11}$ | 0.05 | $0.500 \times 10^{-11}$ | 0.05 | $0.201 \times 10^{-10}$ | 0.15 | $0.298 \times 10^{-10}$ | 0.15 |
| 13 | 600 | $0.743 \times 10^{-12}$ | 0.13 | $0.472 \times 10^{-12}$ | 0.10 | $0.247 \times 10^{-11}$ | 0.13 | $0.382 \times 10^{-11}$ | 0.09 |
| 14 | 700 | $0.627 \times 10^{-13}$ | 0.11 | $0.385 \times 10^{-13}$ | 0.12 | $0.208 \times 10^{-12}$ | 0.09 | $0.375 \times 10^{-12}$ | 0.11 |
| | | | | $q = 0.9$ | | | | | |
| 1 | 30 | $0.107 \times 10^{-6}$ | 0.11 | $0.107 \times 10^{-6}$ | 0.11 | $0.107 \times 10^{-6}$ | 0.11 | $0.107 \times 10^{-6}$ | 0.11 |
| 2 | 40 | $0.215 \times 10^{-5}$ | 0.02 | $0.214 \times 10^{-5}$ | 0.02 | $0.216 \times 10^{-5}$ | 0.02 | $0.218 \times 10^{-5}$ | 0.02 |
| 3 | 50 | $0.254 \times 10^{-5}$ | 0.02 | $0.251 \times 10^{-5}$ | 0.02 | $0.266 \times 10^{-5}$ | 0.02 | $0.273 \times 10^{-5}$ | 0.02 |
| 4 | 60 | $0.163 \times 10^{-5}$ | 0.02 | $0.160 \times 10^{-5}$ | 0.02 | $0.173 \times 10^{-5}$ | 0.02 | $0.181 \times 10^{-5}$ | 0.03 |
| 5 | 70 | $0.989 \times 10^{-6}$ | 0.02 | $0.961 \times 10^{-6}$ | 0.02 | $0.109 \times 10^{-5}$ | 0.02 | $0.116 \times 10^{-5}$ | 0.03 |
| 6 | 80 | $0.581 \times 10^{-6}$ | 0.02 | $0.560 \times 10^{-6}$ | 0.02 | $0.664 \times 10^{-6}$ | 0.02 | $0.721 \times 10^{-6}$ | 0.03 |
| 7 | 90 | $0.336 \times 10^{-6}$ | 0.02 | $0.318 \times 10^{-6}$ | 0.02 | $0.410 \times 10^{-6}$ | 0.02 | $0.450 \times 10^{-6}$ | 0.03 |
| 8 | 100 | $0.217 \times 10^{-6}$ | 0.02 | $0.202 \times 10^{-6}$ | 0.02 | $0.279 \times 10^{-6}$ | 0.03 | $0.324 \times 10^{-6}$ | 0.04 |
| 9 | 200 | $0.450 \times 10^{-7}$ | 0.01 | $0.406 \times 10^{-7}$ | 0.01 | $0.667 \times 10^{-7}$ | 0.02 | $0.837 \times 10^{-7}$ | 0.05 |
| 10 | 300 | $0.200 \times 10^{-8}$ | 0.01 | $0.153 \times 10^{-8}$ | 0.01 | $0.407 \times 10^{-8}$ | 0.04 | $0.641 \times 10^{-8}$ | 0.06 |
| 11 | 400 | $0.133 \times 10^{-9}$ | 0.03 | $0.936 \times 10^{-10}$ | 0.02 | $0.336 \times 10^{-9}$ | 0.06 | $0.536 \times 10^{-9}$ | 0.09 |
| 12 | 500 | $0.109 \times 10^{-10}$ | 0.05 | $0.780 \times 10^{-11}$ | 0.05 | $0.424 \times 10^{-10}$ | 0.13 | $0.678 \times 10^{-10}$ | 0.12 |
| 13 | 600 | $0.134 \times 10^{-11}$ | 0.12 | $0.785 \times 10^{-12}$ | 0.11 | $0.634 \times 10^{-11}$ | 0.16 | $0.107 \times 10^{-10}$ | 0.08 |
| 14 | 700 | $0.143 \times 10^{-12}$ | 0.12 | $0.693 \times 10^{-13}$ | 0.12 | $0.763 \times 10^{-12}$ | 0.14 | $0.158 \times 10^{-11}$ | 0.11 |

**Table 5.5.** Functions $\Delta E(t)$ and $\Delta I(t)$

| $i$ | $\tau = 20$ | | $\tau = 25$ | | $\tau = 30$ | |
|---|---|---|---|---|---|---|
| | $\Delta E_i$ | $\Delta I_i$ | $\Delta E_i$ | $\Delta I_i$ | $\Delta E_i$ | $\Delta I$ |
| 1 | $0.13 \times 10^{-16}$ | $0.44 \times 10^{-23}$ | $0.12 \times 10^{-20}$ | $0.51 \times 10^{-27}$ | $0.48 \times 10^{-28}$ | — |
| 2 | $0.11 \times 10^{-15}$ | $0.87 \times 10^{-23}$ | $0.14 \times 10^{-18}$ | $0.11 \times 10^{-26}$ | $0.12 \times 10^{-24}$ | — |
| 3 | $0.12 \times 10^{-14}$ | $0.92 \times 10^{-23}$ | $0.94 \times 10^{-18}$ | $0.48 \times 10^{-25}$ | $0.77 \times 10^{-20}$ | $0.75 \times 10^{-31}$ |
| 4 | $0.33 \times 10^{-14}$ | $0.98 \times 10^{-23}$ | $0.15 \times 10^{-16}$ | $0.83 \times 10^{-24}$ | $0.98 \times 10^{-18}$ | $0.12 \times 10^{-29}$ |
| 5 | $0.94 \times 10^{-14}$ | $0.16 \times 10^{-22}$ | $0.60 \times 10^{-16}$ | $0.44 \times 10^{-23}$ | $0.96 \times 10^{-17}$ | $0.60 \times 10^{-29}$ |
| 6 | $0.24 \times 10^{-13}$ | $0.22 \times 10^{-22}$ | $0.20 \times 10^{-15}$ | $0.10 \times 10^{-22}$ | $0.39 \times 10^{-16}$ | $0.85 \times 10^{-29}$ |
| 7 | $0.39 \times 10^{-13}$ | $0.23 \times 10^{-22}$ | $0.42 \times 10^{-15}$ | $0.13 \times 10^{-22}$ | $0.84 \times 10^{-16}$ | $0.12 \times 10^{-28}$ |
| 8 | $0.81 \times 10^{-13}$ | $0.21 \times 10^{-22}$ | $0.76 \times 10^{-15}$ | $0.13 \times 10^{-22}$ | $0.11 \times 10^{-15}$ | $0.68 \times 10^{-28}$ |
| 9 | $0.14 \times 10^{-12}$ | $0.18 \times 10^{-22}$ | $0.16 \times 10^{-14}$ | $0.12 \times 10^{-22}$ | $0.23 \times 10^{-15}$ | $0.41 \times 10^{-27}$ |
| 10 | $0.17 \times 10^{-12}$ | $0.16 \times 10^{-22}$ | $0.32 \times 10^{-14}$ | $0.10 \times 10^{-22}$ | $0.32 \times 10^{-15}$ | $0.59 \times 10^{-26}$ |
| 11 | $0.18 \times 10^{-12}$ | $0.13 \times 10^{-22}$ | $0.61 \times 10^{-14}$ | $0.70 \times 10^{-23}$ | $0.39 \times 10^{-15}$ | $0.12 \times 10^{-25}$ |
| 12 | $0.19 \times 10^{-12}$ | $0.11 \times 10^{-22}$ | $0.90 \times 10^{-14}$ | $0.61 \times 10^{-23}$ | $0.47 \times 10^{-15}$ | $0.13 \times 10^{-25}$ |
| 13 | $0.20 \times 10^{-12}$ | $0.82 \times 10^{-23}$ | $0.11 \times 10^{-13}$ | $0.46 \times 10^{-23}$ | $0.54 \times 10^{-15}$ | $0.12 \times 10^{-25}$ |
| 14 | $0.21 \times 10^{-12}$ | $0.63 \times 10^{-23}$ | $0.12 \times 10^{-13}$ | $0.35 \times 10^{-23}$ | $0.59 \times 10^{-15}$ | $0.11 \times 10^{-25}$ |
| 15 | $0.22 \times 10^{-12}$ | $0.47 \times 10^{-23}$ | $0.13 \times 10^{-13}$ | $0.26 \times 10^{-23}$ | $0.65 \times 10^{-15}$ | $0.87 \times 10^{-26}$ |
| 16 | $0.24 \times 10^{-12}$ | $0.34 \times 10^{-23}$ | $0.15 \times 10^{-13}$ | $0.19 \times 10^{-23}$ | $0.77 \times 10^{-15}$ | $0.69 \times 10^{-26}$ |
| 17 | $0.26 \times 10^{-12}$ | $0.25 \times 10^{-23}$ | $0.16 \times 10^{-13}$ | $0.14 \times 10^{-23}$ | $0.17 \times 10^{-14}$ | $0.53 \times 10^{-26}$ |
| 18 | $0.27 \times 10^{-12}$ | $0.18 \times 10^{-23}$ | $0.17 \times 10^{-13}$ | $0.10 \times 10^{-23}$ | $0.26 \times 10^{-14}$ | $0.41 \times 10^{-26}$ |
| 19 | $0.34 \times 10^{-12}$ | $0.13 \times 10^{-23}$ | $0.18 \times 10^{-13}$ | $0.81 \times 10^{-24}$ | $0.29 \times 10^{-14}$ | $0.30 \times 10^{-26}$ |
| 20 | $0.40 \times 10^{-12}$ | $0.93 \times 10^{-24}$ | $0.19 \times 10^{-13}$ | $0.81 \times 10^{-24}$ | $0.29 \times 10^{-14}$ | $0.22 \times 10^{-26}$ |

| | | | | | | |
|---|---|---|---|---|---|---|
| 21 | $0.54 \times 10^{-12}$ | $0.66 \times 10^{-24}$ | $0.20 \times 10^{-13}$ | $0.52 \times 10^{-24}$ | $0.30 \times 10^{-14}$ | $0.16 \times 10^{-26}$ |
| 22 | $0.59 \times 10^{-12}$ | $0.47 \times 10^{-24}$ | $0.21 \times 10^{-13}$ | $0.40 \times 10^{-24}$ | $0.31 \times 10^{-14}$ | $0.12 \times 10^{-26}$ |
| 23 | $0.58 \times 10^{-12}$ | $0.33 \times 10^{-24}$ | $0.22 \times 10^{-13}$ | $0.31 \times 10^{-24}$ | $0.34 \times 10^{-14}$ | $0.87 \times 10^{-27}$ |
| 24 | $0.54 \times 10^{-12}$ | $0.23 \times 10^{-24}$ | $0.23 \times 10^{-13}$ | $0.23 \times 10^{-24}$ | $0.37 \times 10^{-14}$ | $0.62 \times 10^{-27}$ |
| 25 | $0.47 \times 10^{-12}$ | $0.16 \times 10^{-24}$ | $0.25 \times 10^{-13}$ | $0.17 \times 10^{-24}$ | $0.36 \times 10^{-14}$ | $0.45 \times 10^{-27}$ |
| 26 | $0.42 \times 10^{-12}$ | $0.11 \times 10^{-24}$ | $0.26 \times 10^{-13}$ | $0.13 \times 10^{-24}$ | $0.34 \times 10^{-14}$ | $0.32 \times 10^{-27}$ |
| 27 | $0.39 \times 10^{-12}$ | $0.80 \times 10^{-25}$ | $0.27 \times 10^{-13}$ | $0.91 \times 10^{-25}$ | $0.32 \times 10^{-14}$ | $0.23 \times 10^{-27}$ |
| 28 | $0.37 \times 10^{-12}$ | $0.56 \times 10^{-25}$ | $0.28 \times 10^{-13}$ | $0.65 \times 10^{-25}$ | $0.30 \times 10^{-14}$ | $0.16 \times 10^{-27}$ |
| 29 | $0.34 \times 10^{-12}$ | $0.39 \times 10^{-25}$ | $0.30 \times 10^{-13}$ | $0.48 \times 10^{-25}$ | $0.28 \times 10^{-14}$ | $0.11 \times 10^{-27}$ |
| 30 | $0.29 \times 10^{-12}$ | $0.27 \times 10^{-25}$ | $0.33 \times 10^{-13}$ | $0.34 \times 10^{-25}$ | $0.27 \times 10^{-14}$ | $0.80 \times 10^{-28}$ |
| 31 | $0.25 \times 10^{-12}$ | $0.19 \times 10^{-25}$ | $0.37 \times 10^{-13}$ | $0.24 \times 10^{-25}$ | $0.26 \times 10^{-14}$ | $0.56 \times 10^{-28}$ |
| 32 | $0.22 \times 10^{-12}$ | $0.13 \times 10^{-25}$ | $0.34 \times 10^{-13}$ | $0.17 \times 10^{-25}$ | $0.26 \times 10^{-14}$ | $0.39 \times 10^{-28}$ |
| 33 | $0.20 \times 10^{-12}$ | $0.96 \times 10^{-26}$ | $0.29 \times 10^{-13}$ | $0.12 \times 10^{-25}$ | $0.24 \times 10^{-14}$ | $0.28 \times 10^{-28}$ |
| 34 | $0.19 \times 10^{-12}$ | $0.65 \times 10^{-26}$ | $0.27 \times 10^{-13}$ | $0.87 \times 10^{-25}$ | $0.23 \times 10^{-14}$ | $0.19 \times 10^{-28}$ |
| 35 | $0.18 \times 10^{-12}$ | $0.46 \times 10^{-26}$ | $0.26 \times 10^{-13}$ | $0.61 \times 10^{-25}$ | $0.21 \times 10^{-14}$ | $0.13 \times 10^{-28}$ |
| 36 | $0.16 \times 10^{-12}$ | $0.32 \times 10^{-26}$ | $0.25 \times 10^{-13}$ | $0.43 \times 10^{-25}$ | $0.20 \times 10^{-14}$ | $0.98 \times 10^{-29}$ |
| 37 | $0.12 \times 10^{-12}$ | $0.22 \times 10^{-26}$ | $0.24 \times 10^{-13}$ | $0.30 \times 10^{-25}$ | $0.17 \times 10^{-14}$ | $0.87 \times 10^{-29}$ |
| 38 | $0.10 \times 10^{-12}$ | $0.16 \times 10^{-26}$ | $0.22 \times 10^{-13}$ | $0.21 \times 10^{-25}$ | $0.15 \times 10^{-14}$ | $0.80 \times 10^{-29}$ |
| 39 | $0.97 \times 10^{-13}$ | $0.11 \times 10^{-06}$ | $0.21 \times 10^{-13}$ | $0.14 \times 10^{-25}$ | $0.14 \times 10^{-14}$ | $0.75 \times 10^{-29}$ |
| 40 | $0.94 \times 10^{-13}$ | $0.76 \times 10^{-27}$ | $0.20 \times 10^{-13}$ | $0.10 \times 10^{-25}$ | $0.12 \times 10^{-14}$ | $0.71 \times 10^{-29}$ |

In calculations of $\Delta E(t)$ and $\Delta I(t)$ (see Sect. 5.6) for the system described in the previous section the object's radius $R_T$ was taken to be 0.5 m and the albedo $A_T = 0.8$. The object was assumed to be a Lambert reflector. To estimate $\Delta I(t)$ and $\Delta E(t)$, the following integrals were evaluated

$$\Delta E_i = \frac{1}{t_i - t_{i-1}} \int_{t_{i-1}}^{t_i} \Delta E(t)\,dt, \quad \Delta I_i = \frac{1}{t_i - t_{i-1}} \int_{t_{i-1}}^{t_i} \Delta I(t)\,dt,$$

where $t_i = t_0 + i\Delta t$, $t_0 = 2\tau/\sigma v$, $\Delta t = 2$ ns, $i = 1, 2, \ldots, 40$.

The results (Table 5.5) were obtained for three values of $\tau$: 20, 25, 30. The relative standard errors of estimates of $\Delta E_i$ and $\Delta I_i$ are less than 30%. The cited values of $\Delta I(t)$ were obtained for $2\gamma_r = 1'$. A quantity $\eta = \Delta E_m / E_{0,m}$ may be used to define the efficiency of the lidar system. Here $\Delta E_m$ and $E_{0,m}$ are the values of $\Delta E(t)$ and $E_0(t)$ at the instant that corresponds to the maximum value of $E_1(t)$ if $t \geqslant t_0$. Estimates of $\eta$ are shown in Table 5.6 for three values of $A_T$: 0.8, 0.3, 0.03, and three values of $\tau$: 20, 25, 30.

The function $E_0(t)$ was calculated according to the asymptotic formula (5.40), for which the coefficient $A$ was estimated by the Monte Carlo method in the time interval 300–700 ns. The result was $A = (4.3 \pm 0.4) \times 10^{-2}$; therefore, $E_0(t) = 4.3 \times 10^{-2} \cdot t^{-5/2} \cdot \exp[-vt\sigma(1-q)]$. To estimate $\eta$ for large values of $\tau$, use the approximate formula

$$\eta \approx C \cdot \tau^{1/2} \cdot \exp\left[\left(-\frac{1}{\sigma L} + q - 1\right)\tau\right]. \tag{5.41}$$

The dependence of $\eta$ on the values of $\tau$ is shown in Fig. 5.12. The solid line corresponds to the Monte Carlo results. The points represent the results of

**Table 5.6.** $\eta = \Delta E_m / E_{0,m}$ as function of $\tau$ and $A_T$, [%]

| $\tau$ | $A_T = 0.8$ | $A_T = 0.3$ | $A_T = 0.03$ |
|---|---|---|---|
| 20 | 11.7 ± 4.8 | 4.4 ± 1.8 | 0.44 ± 0.18 |
| 25 | 3.4 ± 0.8 | 1.3 ± 0.3 | 0.13 ± 0.03 |
| 30 | 1.5 ± 0.6 | 0.5 ± 0.2 | 0.05 ± 0.02 |

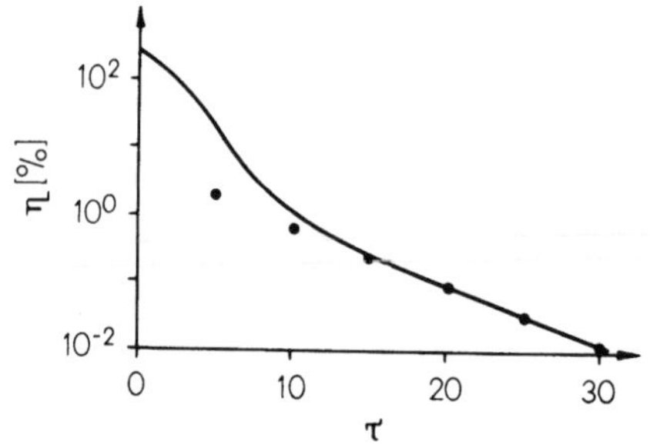


**Fig. 5.12.** $\eta$ as function of $\tau$

calculations by use of (5.41), where $q = 0.9$ and $\sigma L = 4.8$. It is seen that (5.41) may be applied, in this case, to optical thicknesses $\tau > 10$.

In order to estimate the intensity $I(t)$ of the light reflected by the upper layers of the plane-parallel inhomogeneous atmosphere, the atmosphere was divided into 90 layers each of thickness 1 km, having constant coefficients of aerosol scattering $\sigma_a$ and molecular scattering $\sigma_M$.

The altitude profiles of the coefficients $\sigma_a$ and $\sigma_M$ (at $\lambda = 0.65\ \mu m$) are shown in Fig. 5.13 [45]. The absorption of light at this wavelength is negligible and was therefore not taken into account. The scattering indicatrix used was

$$g(\mu) = \frac{\sigma_M g_M(\mu) + \sigma_a g_a(\mu)}{\sigma_M + \sigma_a},$$

where $g_M(\mu) = (3/8)(1 + \mu^2)$, and $g_a(\mu)$, the indicatrix of aerosol scattering, is shown in Fig. 5.14. Curve 1 was used for altitudes below 7 km, and curve 2 was used above 7 km. The source and receiver were assumed to be placed on the Earth's surface; $l_\delta$ is the distance between their positions. Their optical axes are

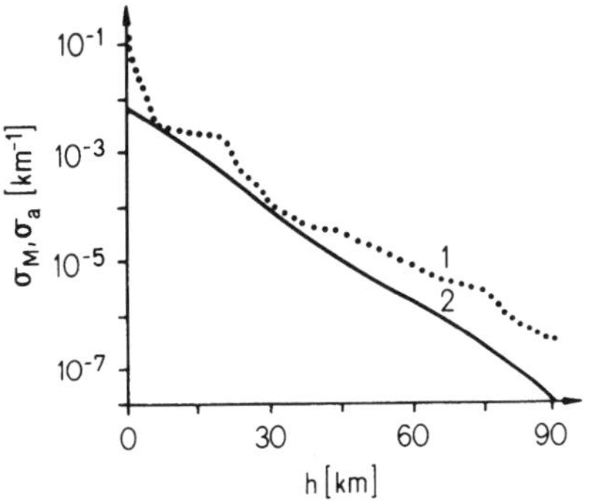


**Fig. 5.13.** Altitude profiles of coefficients $\sigma_a(1)$ and $\sigma_M(2)$

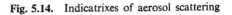

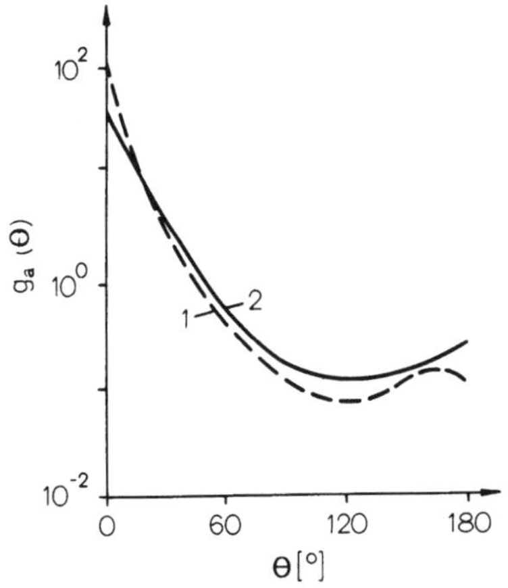


**Fig. 5.14.** Indicatrixes of aerosol scattering

parallel. The angle between the optical axis and the normal to the Earth's surface is $\theta$. The calculations were carried out for a system with $R_s = 0.5$ cm, $R_r = 10$ cm, $\gamma_s = 10'$, $\gamma_r = 7'$. The computational results for $\theta = 0$ and $85°$ are shown in Fig. 5.15.

We now compare the experimental data and the Monte Carlo calculations of the following problem. The source and receiver are situated in the ground layer of the atmosphere; $l_\delta$ is the distance between them. The optical axes of the source and receiver lie in a plane that is parallel to the Earth's surface ($z_0$ is the distance from this plane to the Earth's surface); $\gamma$ is the angle between these axes. The Earth's surface is assumed to be a Lambert reflector with albedo $P_a$. Two different indicatrices were used in the calculations (see Fig. 5.16). The intensity $I(t)$ of the light reflected by the ground layer of the atmosphere was calculated for various values of $t$ (the time at which the back-scattered signal is recorded in $\mu$s), as in Fig. 5.17 where $P_a = 0.8$, $z_0 = 1$ m. The solid lines are experimental, the points are the Monte Carlo results of calculation of $I(t)/I_m(t)$. Here $I_m$ is the maximum of $I(t)$.

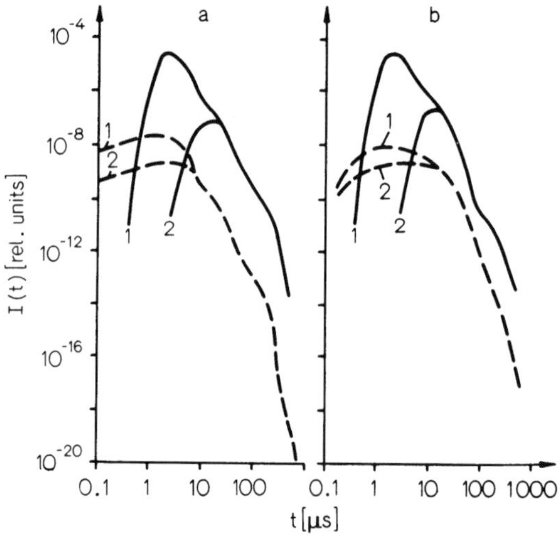


**Fig. 5.15.** $I(t)$ for $l_\delta = 0.4$ m (*1*) and $l_\delta = 3$ m (*2*); $\gamma_r = 2'20''$; $\theta = 0°$ (**a**) and $\nu = 85°$ (**b**). Solid lines show single scattering; dotted line for the multiple-scattering case

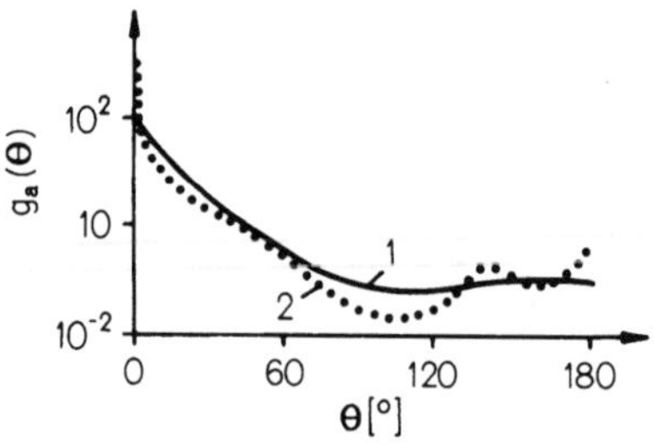


**Fig. 5.16.** Scattering indicatrix (*1*) clear sky, (*2*) fog

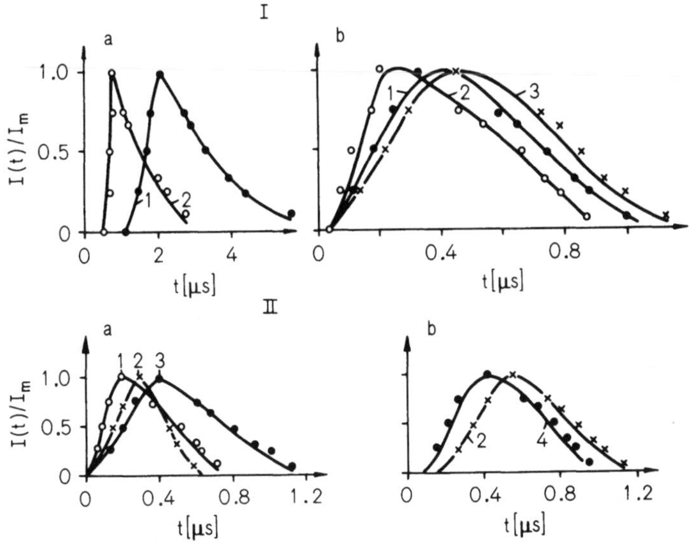


**Fig. 5.17I, II.** Ratio $I(t)/I_m$
**(I)** dependence on angle of view $\gamma$ for $\gamma_s = 10'$, **(a)** clear sky, $\sigma = 0.1\ \mathrm{km}^{-1}$, $l_\delta = 1.4$ m, $\gamma_r = 15'$, **(b)** fog, $\sigma = 45\ \mathrm{km}^{-1}$, $l_\delta = 0.4$ m, $\gamma_r = 5'$; *(1)* $\gamma = 0$, *(2, 3)* $\gamma = 15'$
**(II)** dependence on reception angle $\gamma_r$, thick fog ($\sigma = 50\ \mathrm{km}^{-1}$) for $l_\delta = 0.4$ m, $\gamma = 0$, **(a)** $\gamma_r = 10'$; **(b)** $\gamma_r = 0.5'$; *(1)* $\gamma_r = 45''$, *(2)* $\gamma_r = 2'30''$, *(3)* $\gamma_r = 5'$, *(4)* $\gamma_r = 15'$

Calculations of $I(t)$ for various values of $P_a$ and $z_0$ show that the change of $P_a$ and $z_0$ does not essentially affect the quantity $I(t)$ except for very small values of $z_0$. Errors of the experimental measurements are less than 30%, and the errors of the calculation results are less than 20%. The results agree satisfactorily. The best agreement is observed in the foward part of the reflected signal.

# 6. Monte Carlo Algorithms for Estimating the Correlation Function of Strong Light Fluctuations in a Turbulent Medium

Light scattering in the clear atmosphere is caused by random fluctuations of the refraction index of the component gases. These fluctuations are due to air motion caused by turbulence in the presence of temperature gradients. The quantitative theory of these phenomena is based on the wave equation with random variations of refractive index. It is impossible to obtain the exact solution of this problem. Therefore, various approximate methods must be used. Perhaps the most frequently applied is the *Rytov* method [46]. When the statistical characteristics of the refractive index are investigated, the results of turbulence theory may be applied. Small fluctuations of the refractive index may lead to strong fluctuations of light intensity. The qualitative picture of the strong fluctuations of the light intensity is clear enough.

There are, however, difficulties in the quantitative description of these effects, because the perturbation theory due to *Rytov* is not applicable here. A new approach, based on the interpretation of light propagation in a turbulent medium as a Markov random process has been developed by *Tatarsky* [47]. The differential equations for the statistical moments proposed in this paper are also applicable to the case of strong light fluctuations. Explicit expressions for the first- and second-order moments can be obtained from these equations. To calculate the higher-order moments, it is necessary to apply numerical methods. In this chapter, we construct effective Monte Carlo algorithms for estimating the fourth-order moment of the field of light wave in turbulent medium and, as a special case, the correlation function of the strong light fluctuations.

It should be noted that the method presented in this chapter depends on the idea of sampling the velocities of the photons, not from the sphere of directions, but from a larger space, in which the magnitudes of their velocities are allowed to vary [see (6.23)]. To emphasize this, throughout this chapter we speak of "(fictitious) particles" rather than "photons." The justification of this approach is in the results we obtain. (A paper by *Tatarsky* et al. dealing with this method is shortly to appear in *Optica Acta*.)

## 6.1 Statement of the Problem; Reformulation of the Integro-Differential Equation into Integral Form

Assume that a plane wave is incident on a half-space filled with a medium having an inhomogeneous refractive index. Introduce the cartesian system of spatial coordinates $\boldsymbol{r}(x, y, z)$ so that the $x$-axis is coincident with the direction of initial wave propagation. Consider the fourth-order moment of the field of light wave,

$$\tilde{\Gamma} = \langle u(\boldsymbol{r}_1)u(\boldsymbol{r}_2)u^*(\boldsymbol{r}_1 + \boldsymbol{r}_2)u^*(\boldsymbol{r}_1 - \boldsymbol{r}_2)\rangle,$$

where $u$ is the complex amplitude of the light field, $\boldsymbol{r} = \boldsymbol{r}(x, y, z)$. The second-order differential equation for $\tilde{\Gamma}$ can be reformulated, using the Fourier transform with respect to one of the coordinates, as [47]:

$$\frac{\partial\varphi(x, \boldsymbol{r}, \boldsymbol{\kappa})}{\partial x} + \boldsymbol{\kappa}\cdot\nabla_{\boldsymbol{r}}\varphi(x, \boldsymbol{r}, \boldsymbol{\kappa}) + \mu_1|\boldsymbol{r}|^{5/3}\varphi(\boldsymbol{k}, \boldsymbol{r}, \boldsymbol{\kappa})$$
$$= \mu_2 \iint_{-\infty}^{\infty} |\boldsymbol{\kappa}|^{-11/3}(1 - \cos(\boldsymbol{\kappa}'\cdot\boldsymbol{r}))\varphi(x, \boldsymbol{r}, \boldsymbol{\kappa} - \boldsymbol{\kappa}')\,d\boldsymbol{\kappa}', \tag{6.1}$$

with initial condition

$$\varphi(x, \boldsymbol{r}, \boldsymbol{\kappa})|_{x=0} = \delta(\boldsymbol{\kappa}). \tag{6.2}$$

It is known [47] that

$$\lim_{|\boldsymbol{r}|\to\infty} \varphi(x, \boldsymbol{r}, \boldsymbol{\kappa}) = \frac{1}{2\pi}\int_0^{\infty} I_0(\rho\cdot|\boldsymbol{\kappa}|)\,\mathrm{e}^{-\mu_1 x\rho^{5/3}}\cdot\rho\,d\rho. \tag{6.3}$$

Here

$$\mu_1 = \frac{\pi a}{2\alpha}, \quad \mu_2 = \frac{\pi a}{\alpha}, \quad A = 0.033, \quad a = 0.47,$$
$$\alpha = 3\pi^2 A[11\Gamma(11/6)\cos(\pi/12)]^{-1}, \quad \boldsymbol{r} = \boldsymbol{r}(y, z), \quad x \geqslant 0,$$

and $\boldsymbol{\kappa} = (\kappa_1, \kappa_2)$ is a special parameter.

Let $B(x, \boldsymbol{r})$ be a correlation function of the light fluctuations. Then

$$B(xL, \boldsymbol{r}l) = \iint_{-\infty}^{\infty} \varphi(x, \boldsymbol{r}, \boldsymbol{\kappa})\,d\boldsymbol{\kappa} - 1. \tag{6.4}$$

The quantity $L$ is obtained from

$$\alpha C_\varepsilon^2 k^{7/6} L^{11/6} = 1,$$

where $k$ is the wave number, $C_\varepsilon$ is a structural characteristic of the fluctuations of the refractive index, and $l = (L/k)^{1/2}$. The value $B(x, 0)$ is the variance of the intensity fluctuations of the field of the light wave. For convenience, we put

$$t = x, \quad \boldsymbol{v} = \boldsymbol{\kappa}, \quad \boldsymbol{v}' = \boldsymbol{\kappa}', \quad r = |\boldsymbol{r}|, \quad v = |\boldsymbol{v}|, \quad \boldsymbol{\Omega} = \boldsymbol{v}/v.$$

Equation (6.1) and the initial condition (6.2) are then written as

$$\frac{\partial\varphi(t, \boldsymbol{r}, \boldsymbol{v})}{\partial t} + v\boldsymbol{\Omega}\cdot\boldsymbol{\nabla}_r\varphi(t, \boldsymbol{r}, \boldsymbol{v}) + \mu_1 r^{5/3}\cdot\varphi(t, \boldsymbol{r}, \boldsymbol{v})$$
$$= \iint_{-\infty}^{\infty} \mu_2 v'^{-11/3}[1 - \cos(\boldsymbol{v}'\cdot\boldsymbol{r})]\varphi(t, \boldsymbol{r}, \boldsymbol{v} - \boldsymbol{v}')\,d\boldsymbol{v}', \tag{6.5}$$

$$\varphi(t, \boldsymbol{r}, \boldsymbol{v})|_{t=0} = \delta(\boldsymbol{v}).$$

By making the change of variable $\boldsymbol{v}'' = \boldsymbol{v} - \boldsymbol{v}'$ in the right-hand side of (6.5) we get, for $\Phi = v\cdot\varphi$,

$$\frac{1}{v}\frac{\partial\Phi}{\partial t} + \boldsymbol{\Omega}\cdot\boldsymbol{\nabla}_r\Phi + \frac{\mu_1 r^{5/3}}{v}\Phi = \iint_{-\infty}^{\infty} \frac{\mu_2}{v''}|\boldsymbol{v} - \boldsymbol{v}''|^{-11/3}$$
$$\times \{1 - \cos[(\boldsymbol{v} - \boldsymbol{v}'')\cdot\boldsymbol{r}]\}\Phi(t, \boldsymbol{r}, \boldsymbol{v}'')\,d\boldsymbol{v}'', \tag{6.6}$$

with the initial condition

$$\Phi(t, \boldsymbol{r}, \boldsymbol{v})|_{t=0} = v\cdot\delta(\boldsymbol{v}).$$

Because

$$\iint_{-\infty}^{\infty} |\boldsymbol{v}|^{-11/3}[1 - \cos(\boldsymbol{v}\cdot\boldsymbol{r})]\,d\boldsymbol{v} = \frac{a\cdot r^{5/3}}{2A},$$

(6.6) can be transformed into

$$\frac{1}{v}\frac{\partial\Phi}{\partial t} + \boldsymbol{\Omega}\cdot\boldsymbol{\nabla}_r\Phi + \sigma\Phi = \iint_{-\infty}^{\infty} \sigma_s g(\boldsymbol{r}, \boldsymbol{v}' \to \boldsymbol{v})\,\Phi(t, \boldsymbol{r}, \boldsymbol{v}')\,d\boldsymbol{v}', \tag{6.7}$$

where

$$\sigma = \sigma(r, v) = \frac{\mu r^{5/3}}{v}, \quad \sigma_s = \sigma_s(r, v) = \frac{\mu r^{5/3}}{v}, \quad \mu = 2.3724,$$

$$\iint_{-\infty}^{\infty} g(\boldsymbol{r}, \boldsymbol{v}' \to \boldsymbol{v})\,d\boldsymbol{v}' = \iint_{-\infty}^{\infty} \frac{2A}{ar^{5/3}}|\boldsymbol{v} - \boldsymbol{v}'|^{-11/3}\{1 - \cos[(\boldsymbol{v} - \boldsymbol{v}')\cdot\boldsymbol{r}]\}\,d\boldsymbol{v}' = 1.$$

Note that the equation (6.7) coincides with the equation of photon transfer (see Sect. 1.2) with the following exception: in (6.7), $v$ is a variable, and the vectors $\boldsymbol{r}$, $\boldsymbol{v}$, $\boldsymbol{\omega}$ are of dimension 2. Consequently, it seems convenient to interpret the equation (6.7) as an equation of transfer of fictitious particles. This interpretation makes it possible to use the Monte Carlo method for solving the equation (6.7). Thus, (6.5) may be regarded as a nonstationary equation of particle transfer in a plane medium with the total cross section $\sigma(r, v) = \mu r^{3/3}/v$, the scattering cross section $\sigma_s(r, v) = \mu r^{3/3}/v$ (i.e., pure scattering) and the scattering indicatrix $g(\boldsymbol{r}, \boldsymbol{v}' \to \boldsymbol{v})$. The variable $t$ is time, $\boldsymbol{r}$ = spatial coordinates, and $\boldsymbol{v}$ = velocity. The functions $\Phi(t, \boldsymbol{r}, \boldsymbol{v})$ can be interpreted as the flux and the density of particles, respectively [48].

We recall the general scheme for simulating trajectories in Monte Carlo solution of problems of transfer theory:

a) initial state of the trajectory is chosen from the source distribution density;
b) free-path length is sampled;
c) escape from the medium is examined or, if desired, another type of termination of the trajectory is tested (e.g., lifetime outflow);
d) coordinates of a new collision point are calculated;
e) type of collision is chosen (absorption or scattering; in our case, we assume no absorption in the medium);
f) new velocity and direction of particle motion are sampled;
g) return to b).

The Monte Carlo technique enables us to estimate various linear functionals of the solution of transfer equation (e.g., the integral of the flux or the flux at a given point of the phase space [48]). The correlation function may be regarded as a linear functional of the solution of (6.7). Therefore, the correlation function may be estimated by the Monte Carlo method.

## 6.2 Calculation of the Flux of Particles that have not Undergone Collision; Density of First Collisions

A special case of (6.5),

$$\frac{\partial \varphi(t, \boldsymbol{r}, \boldsymbol{v})}{\partial t} + \boldsymbol{v} \cdot \nabla_{\boldsymbol{r}} \varphi(t, \boldsymbol{r}, \boldsymbol{v}) + \mu r^{5/3} \varphi(t, \boldsymbol{r}, \boldsymbol{v}) = 0,$$

with the initial condition

$$\varphi(t, \boldsymbol{r}, \boldsymbol{v})|_{t=0} = \delta(\boldsymbol{v}),$$

defines the density of particles that have not undergone a collision,

$$\varphi(t, \boldsymbol{r}, \boldsymbol{v}) = \exp(-\mu r^{5/3} t) \delta(\boldsymbol{v}). \tag{6.8}$$

Let $f(t, \boldsymbol{r}, \boldsymbol{v})$ be the collision density at a point $(t, \boldsymbol{r}, \boldsymbol{v})$. By (6.8) and from the known relation $f(t, \boldsymbol{r}, \boldsymbol{v}) = \sigma(\boldsymbol{r}, \boldsymbol{v}) \varphi(t, \boldsymbol{r}, \boldsymbol{v})$ we obtain an expression for the density of first collisions

$$f_0(t, \boldsymbol{r}, \boldsymbol{v}) = \frac{\mu r^{5/3}}{v} v \exp(-\mu r^{5/3} t)\, \delta(\boldsymbol{v}). \tag{6.9}$$

On the average, there is one collision per unit area, because

$$\int_0^\infty \mu r^{5/3} \exp(-\mu r^{5/3} t)\, dt = 1.$$

If, instead of the physical density $f_0(t, \boldsymbol{r}, \boldsymbol{v})$ a probability density $p_0(t, \boldsymbol{r}, \boldsymbol{v})$ is used to simulate the first collisions then the initial weight of each particle must be taken as $Q_0 = f_0(t, \boldsymbol{r}, \boldsymbol{v})/p_0(t, \boldsymbol{r}, \boldsymbol{v})$. The contribution from a trajectory to the calculated functional is then multiplied by this weight.

## 6.3 Simulation Algorithms

*Simulation of the Initial Parameters of the Trajectory*

The transfer process is considered in a circle $D$ whose radius $R$ is sufficiently large. The particle position may be defined by $|\boldsymbol{r}| = r$, and the direction of its motion may be specified by $\boldsymbol{v}\cdot\boldsymbol{r}/(|\boldsymbol{v}|\,|\boldsymbol{r}|)$ because the total cross section $\sigma$ depends only on $|\boldsymbol{r}| = r$, and the scattering indicatrix $g(\boldsymbol{r}, \boldsymbol{v}' \to \boldsymbol{v})$ depends on $r$ and on the cosine between $\boldsymbol{r}$ and $\boldsymbol{v}$.

Divide the circle $D$ into $m$ concentric circles $K_i$; $r_i$ is the radius of $i$th circle ($i = 0, \ldots, m$; $r_0 = 0$, $r_m = R$). Denote the index of a circular layer $D_i = K_{i-1} \backslash K_i$ by $i$. An essential contribution to the variance $B(t, 0)$ makes the particles originated near the point $r = 0$. Therefore it is convenient to sample the initial point from the density

$$f_r(x) = p\cdot f_1(x) + (1 - p)\cdot f_2(x).$$

Here $p < 1$, and

$$f_1(x) = (1 - \beta)x^{-\beta}, \quad 0 < x \leqslant 1, \quad f_2(x) = 2x(R^2 - 1), \quad x > 1.$$

In simulating the direct trajectories, we have used $p = 1/2$, $\beta = 1/2$, which corresponds to the simulating algorithm,

$$r = (\alpha_2)^{2/3}, \quad Q_0^{(r)} = \frac{8}{3\pi r^{1/2}} \quad \text{if} \quad \alpha_1 < p, \quad \text{and}$$

$$r = [\alpha_3\cdot(R^2 - 1) + 1]^{1/2}, \quad Q_0^{(r)} = 2\pi(R^2 - 1), \quad \text{otherwise}.$$

Here, as usual, $\alpha$ with, or without indexes are random variables uniformly distributed between 0 and 1.

The time distribution of the first collisions for fixed $r$ is (see Sect. 6.3):

$$f_0(t) = \mu r^{5/3} \exp(-\mu r^{5/3}\cdot t), \quad 0 \leqslant t < \infty.$$

Because the functional $B(t, r)$ is calculated for $t_k$, $k = 1, \ldots, n_t$, the photon trajectory may be constructed only for $t \leqslant T = \max_k \{t_k\}$.

Therefore, it is here convenient to apply the following probability density $p_1(t)$: $p_1(t) = \mu r^{5/3} \cdot [\exp(-\mu r^{5/3} t)](1 - \mu r^{5/3} T)^{-1}, 0 \leqslant t \leqslant T$. The simulating formula takes the form,

$$t = -\ln\{1 - \alpha \cdot [1 - \exp(-\mu r^{5/3} T)]\}/\mu r^{5/3}, \quad \text{and}$$

$$Q_0 = 1 - \exp(-\mu r^{5/3} T).$$

The distribution of the velocity $\boldsymbol{v}$ before the first collision has the density $f_0(\boldsymbol{v}) = \delta(\boldsymbol{v})$. Consequently, for the first collision, we put $\boldsymbol{v}_0 = 0$, and take $\psi_0 = 0$, where $\psi_0$ is the angle between the vectors $\boldsymbol{v}$ and $\boldsymbol{r}_0$. The total initial weight of the photon is

$$Q_0 = Q_0^{(r)} \cdot Q_0^{(t)}.$$

*Free-Path Length Simulation*

It is known [48] that the distribution density of the free-path length $l$ is

$$f_l(s) = \sigma[r(s), v] \exp\left\{-\int_0^s \sigma(r(u), v]\, du\right\}. \tag{6.10}$$

Integrating (6.10) yields

$$F_l(s) = 1 - \exp[-\tau(s)], \quad \text{where} \quad \tau(s) = \int_0^s \sigma[r(u), v]\, du,$$

is the optical length of the interval $(0, s)$. The quantity $l$ may be obtained from $\tau(l) = -\ln \alpha$, which can be solved easily in the case of piecewise-constant cross section,

$$\sigma_i(r, v) = \frac{1}{r_{i+1} - r_i} \int_{r_i}^{r_{i+1}} \frac{\mu s^{5/3}}{v} ds = \frac{3\mu(r_{i+1}^{8/3} - r_i^{8/3})}{8v\,(r_{i+1} - r_n)}, \quad r_i \leqslant r \leqslant r_{i+1}.$$

The free-path length is simulated:

a) Let $l_{i_1}, \ldots, l_{i_k}$ be the lengths of the straight path of the particle in the direction $\boldsymbol{v}$ inside the layers $D_{i_1}, \ldots, D_{i_k}$, respectively; $l_{ij} = 0, \sigma_{i_1}, \ldots, \sigma_{i_k}$ are the corresponding cross sections.

b) By successive subtraction, index $i_j$ is found from

$$M = \sum_{m=0}^{j-1} \sigma_{i_m} \cdot l_{i_m} < -\ln \alpha < \sum_{m=0}^{j} \sigma_{i_m} \cdot l_{i_m}; \tag{6.11}$$

c) If no number $i_j$ satisfies the inequalities (6.11), then the particle escapes from the medium. Otherwise

$$l = \sum_{m=0}^{j-1} l_{i_m} + \frac{-\ln \alpha - M}{\sigma_{i_j}};$$

$t = l/v$ is the time between two successive collisions.

*Simulation of the New Velocity $\boldsymbol{v}$*

Assume that a photon moving in a direction $\boldsymbol{v}'$ has undergone a collision at the point $\boldsymbol{r}$. Let $\psi'$ be the angle between $\boldsymbol{v}'$ and $\boldsymbol{r}$. The distribution of the new velocity $\boldsymbol{v}$ is defined by the scattering indicatrix,

$$g(\boldsymbol{r}, \boldsymbol{v}' \to \boldsymbol{v}) = \frac{2A}{ar^{5/3}} \cdot |\boldsymbol{v} - \boldsymbol{v}'|^{-11/3}\{1 - \cos [(\boldsymbol{v} - \boldsymbol{v}')\cdot \boldsymbol{r}]\}.$$

Introducing now a polar coordinate system $(v_p, \psi_p)$ with origin at the point $\boldsymbol{r}$ and axis coinciding with $\boldsymbol{r}$, we get

$$g(\boldsymbol{r}, v_p, \psi_r) = \frac{2A}{a} \cdot v_p^{-8/3} \cdot [1 - \cos (v_p \cdot \cos \psi_r)], \tag{6.12}$$

where $v_p = |\boldsymbol{v} - \boldsymbol{v}'| \cdot r$, $\psi_r$ is the angle between $\boldsymbol{v}_p = \boldsymbol{v} - \boldsymbol{v}'$ and $\boldsymbol{r}$.

The distribution density of $\xi$ and $\eta$ is defined by (6.12). Integrating (6.12) over $0 \leqslant \psi_r \leqslant 2\pi$ yields the marginal distribution density of $v_p$,

$$g_v(x) = C \cdot x^{-8/3} \cdot [1 - I_0(x)].$$

Here, $I_0(x)$ is the Bessel function, and $C = 4A\pi/a$. Let

$$F_v(x) = \int_0^x g_v(y)\, dy.$$

Samples from the distribution $F_v(x)$ may be obtained by solving $F_v(v_p) = \alpha$, i.e., $v_p = F_v^{-1}(\alpha)$. The function $F_v^{-1}(a)$ was tabulated for $0 \leqslant a \leqslant 1$, $\Delta a = 0.001$ by successive solution of

$$C \int_0^x y^{-8/3}(1 - I_0(y))\, dy = i\Delta a, \quad i = 0, 1, \ldots, 999, \tag{6.13}$$

by Newton's method. The integral in (6.13) was evaluated by Simpson's rule with relative error $\varepsilon = 10^{-3}$. For $\alpha \leqslant 0.1$, $I_0$ was expanded in the power series $I_0(x) \approx 1 - (x/2)^2$. Linear interpolation was used between the tabulated values of $v_p$.

We turn now to the simulation of angle $\psi_r$. Substituting in (6.12) a sample of $v_p$ yields the conditional distribution density of $\psi_r$.

The following modification of the rejection method was used (see [48]). Let $f_\xi(x)$ be the density function of the random variable $\xi$, and let

$$C_1 f_\xi(x) \leqslant C_2 g_\eta(x), \tag{6.14}$$

where $g_\eta(x)$ is a density function of an auxiliary random variable $\eta$. Then, $\xi$ may be sampled as follows:

a) $\eta_0$ is sampled from the density $g_\eta(x)$, then $\beta$ is calculated: $\beta = \alpha_1 \cdot C_2 \cdot g(\eta_0)$

b) if $\beta < C_1 \cdot f(\eta_0)$ then $\xi = \eta_0$, otherwise the procedure is repeated.

In our case, this modification is based on the inequalities:

$$1 - \cos(v_p \cdot \cos\psi_r) \leqslant \frac{v_p^2 \cdot \cos^2\psi_r}{2}, \quad \text{if} \quad v_p^2 \leqslant 8,$$

$$1 - \cos(v_p \cdot \cos\psi_r) \leqslant 2, \qquad \text{if} \quad v_p^2 > 8.$$

If $v_p^2 \leqslant 8$, $\psi_r$ is sampled from $g_\eta(\psi_r) \sim \cos^2(\psi)$.
If $v_p^2 > 8$, $\psi_r$ is sampled from $g_\eta(\psi_r) = 1/2\pi$, $\psi_r = 2\pi \cdot \alpha$.
The average probability that rejection occurs is given by:

$$p = 1 - \frac{C_1}{C_2} = \begin{cases} 1 - 4[1 - I_0(v_p)]/v_p^2, & \text{if} \quad v_p^2 \leqslant 8, \\ 1 - [1 - I_0(v_p)]/2\pi, & \text{if} \quad v_p^2 > 8; \end{cases}$$

i.e., the efficiency of this algorithm is reasonably high.

It is easy to see, by geometric agruments, that, when $v_p$ and $\psi_r$ are sampled, the parameters of the particle are

$$v = \left[\left(\frac{v_p}{r}\right)^2 + v'^2 + 2v' \cdot \frac{v_p}{r} \cdot \cos(\psi' - \psi_r)\right]^{1/2}, \quad \psi = \arccos\gamma,$$

where $\gamma = (v_p \cos\psi_r + v' \cos\psi')/v$. We recall that $\psi$ is the angle between $\boldsymbol{v}$ and $\boldsymbol{r}$.

*Estimation of the number of Particles for Given Instants*

The appropriate representation is

$$B(t, r) = \int_V \varphi(t, \boldsymbol{r}, \boldsymbol{v})\, d\boldsymbol{v} + \exp(-\mu r^{5/3} \cdot t) - 1, \tag{6.15}$$

where $\varphi(t, \boldsymbol{r}, \boldsymbol{v})$ is the density of scattered photon, and $\exp\{-\mu r^{5/3} \cdot t\}$ is the density of nonscattered photons. It is not difficult to calculate the instant at which a photon intersects the boundaries of the layers $D_i$.

Therefore, it is also easy to calculate, using a simple logical algorithm, the number of photons in a given layer at given instants $t_1, \ldots, t_n$. The function $B(t, \mathbf{r})$ for $j$th layer is estimated by

$$B_j(t_i) = \frac{Q_0 \cdot \chi_{ij}}{\pi(r_{j+1}^2 - r_j^2)} + I_{ij}(t_i) - 1,$$

where

$$\chi_{ij} = \begin{cases} 1 & \text{if the photon is in the } j\text{th layer at the instant } t_j, \\ 0 & \text{otherwise.} \end{cases}$$

$Q_0$ is the corresponding weight.

The contribution due to the nonscattered radiation is

$$I_{r_j}(t_i) = \frac{1}{\pi(r_{j+1}^2 - r_j^2)} \int_0^{2\pi} d\psi \int_{r_j}^{r_{j+1}} r \exp(-\mu r^{5/3} t)\, dr. \tag{6.16}$$

The integral in (6.16) was evaluated by Simpson's rule. The algorithm enables us to calculate the function $B(t, \mathbf{r})$ simultaneously for several values of $t$. The computer time depends to a great extent on $R$, the radius of the circle. Therefore, we have taken $R = R_0$, where $R_0$ was chosen from the condition that, for given $T$, the function $B(t, \mathbf{r})$ must be sufficiently close to the asymptotic solution if $r > R_0$. The values $R_0(T)$ were proposed by *Tatarsky* [47].

The contribution that corresponds to the asymptotic solution could be taken into account by simulating an additional photon source on the boundary of the circle. It is, however, too difficult to construct such an algorithm, because the asymptotic solution is complicated.

Therefore, the asymptotic solution on the boundary of the circle was approximately taken into account. The photon was reflected backward from the boundary $R_0(t)$ in the same direction and with the same velocity. Comparisons of the computational results with the experimental data show that such an algorithm for calculating the contribution from the region $r > R_0$ does not estimate the result sufficiently precisely.

In order to estimate $B(t, \mathbf{r})$ at the point $\mathbf{r} = 0$, a special local estimate was constructed. This estimate seems to be ineffective, because its value rapidly increases as $\mathbf{r} \to 0$. Thus direct calculation of the number of photons described in this section does not guarantee accurate evaluation of the function $B(t, \mathbf{r})$. Therefore, a Monte Carlo algorithm based on simulating the adjoint trajectories and analytical averaging was constructed that provides effective calculation of the asymptotic results. This algorithm is discussed in the next section.

## 6.4 Use of the Asymptotic Solution based on the Theorem of Optical Mutuality and the Method of Expected Values

*Use of the Theorem of Optical Mutuality*

Suppose that it is desired to estimate a functional $I_p = (\Phi, p)$ where $p$ is a non-negative function, and $\Phi(t, \boldsymbol{r}, \boldsymbol{v})$ is the solution of the transfer equation

$$\frac{1}{v}\frac{\partial\Phi(t,\boldsymbol{r},\boldsymbol{v})}{\partial t} + \frac{1}{v}\boldsymbol{v}\cdot\nabla_{\boldsymbol{r}}\Phi(t,\boldsymbol{r},\boldsymbol{v}) + \sigma(\boldsymbol{r},v)\Phi(t,\boldsymbol{r},\boldsymbol{v}) = \int_V \sigma_s(\boldsymbol{r},\boldsymbol{v}')g(\boldsymbol{r},\boldsymbol{v}'\to\boldsymbol{v})\Phi(t,\boldsymbol{r},\boldsymbol{v}')\,d\boldsymbol{v}' + q(t,\boldsymbol{r},\boldsymbol{v}), \tag{6.17}$$

with the boundary condition $\Phi(t, \boldsymbol{r}, \boldsymbol{v}) = 0$ if $\boldsymbol{v}\cdot\boldsymbol{n} > 0$. Here, $\boldsymbol{n}_r$ is the normal to the boundary. It is known that $I_p = (\Phi, p) = (\Phi^*, q)$, where $\Phi^*(t, \boldsymbol{r}, \boldsymbol{v})$ is the solution of the adjoint transfer equation

$$-\frac{1}{v}\frac{\partial\Phi^*(t,\boldsymbol{r},\boldsymbol{v})}{\partial t} - \frac{1}{v}\boldsymbol{v}\cdot\nabla_{\boldsymbol{r}}\Phi^*(t,\boldsymbol{r},\boldsymbol{v}) + \sigma(\boldsymbol{r},v)\Phi^*(t,\boldsymbol{r},\boldsymbol{v}) = \int_V \sigma_s(\boldsymbol{r},v)g(\boldsymbol{r},\boldsymbol{v}\to\boldsymbol{v}')\Phi^*(t,\boldsymbol{r},\boldsymbol{v})\,d\boldsymbol{v}' + p(t,\boldsymbol{r},\boldsymbol{v}),$$

with the boundary condition $\Phi^*(t, \boldsymbol{r}, \boldsymbol{v}) = 0$ if $\boldsymbol{v}\cdot\boldsymbol{n} < 0$. Changing the signs of $\boldsymbol{v}$ and $\boldsymbol{v}'$, we obtain, for $\Phi_1(t, \boldsymbol{r}, \boldsymbol{v}) = \Phi^*(t, \boldsymbol{r}, -\boldsymbol{v})$,

$$\frac{1}{|\boldsymbol{v}|}\frac{\partial\Phi_1}{\partial t}(t,\boldsymbol{r},\boldsymbol{v}) + \frac{1}{v}\boldsymbol{v}\cdot\nabla_{\boldsymbol{r}}\Phi_1(t,\boldsymbol{r},\boldsymbol{v}) + \sigma(\boldsymbol{r},v)\cdot\Phi_1(t,\boldsymbol{r},\boldsymbol{a}) = \int_V \frac{\sigma_s(\boldsymbol{r},\boldsymbol{v})}{\sigma_s(\boldsymbol{r},\boldsymbol{v}')}\cdot\sigma_s(\boldsymbol{r},\boldsymbol{v}')g(\boldsymbol{r},\boldsymbol{v}\to\boldsymbol{v}')\Phi_1(t,\boldsymbol{r},\boldsymbol{v}')\,d\boldsymbol{v}' + p(t,\boldsymbol{r},-\boldsymbol{v}).$$

To simulate the transfer process according to this equation, it is necessary to multiply the photon weight after scattering, as comparison with (6.17) shows, by

$$\sigma_s(\boldsymbol{r},\boldsymbol{v})/\sigma_s(\boldsymbol{r},\boldsymbol{v}').$$

Now,

$$I_p = (\Phi^*, q) = \int_0^\infty\int_R\int_V \Phi^*(t,\boldsymbol{r},\boldsymbol{v})q(t,\boldsymbol{r},\boldsymbol{v})\,d\boldsymbol{v}\,d\boldsymbol{r}\,dt = \int_0^\infty\int_R\int_V \Phi_1(t,\boldsymbol{r},\boldsymbol{v})\cdot q(t,\boldsymbol{r},-\boldsymbol{v})\,d\boldsymbol{v}\,d\boldsymbol{r}\,dt. \tag{6.18}$$

Thus, in order to estimate the functional $I_p$, the process of transfer with the source density $p_1(t, \boldsymbol{r}, \boldsymbol{v}) = p(t, \boldsymbol{r}, -\boldsymbol{v})$ can be simulated and the quantity (6.18) that defines weighted receiver's readings [with the weight function $q(t, \mathbf{r}, -\mathbf{v})$] can be calculated. This is, in fact, the statement of the theorem of optical mutuality [49]. To utilize this theorem, it remains only to represent (6.18) as a functional of the collision density $f_1^*(t, \mathbf{r}, \mathbf{v}) = f^*(t, \mathbf{r}, -\mathbf{v})$,

$$I_p = \int_0^\infty \int_R \int_V f_1^*(t, \mathbf{r}, \mathbf{v}) \frac{q(t, \mathbf{r}, -\mathbf{v})}{\sigma(\mathbf{r}, \mathbf{v})} d\mathbf{v}\, d\mathbf{r}\, dt;$$

i.e., it is necessary to calculate the quantity $q(t, \mathbf{r}, -\mathbf{v})/\sigma(\mathbf{r}, \mathbf{v})$ at each collision point.

*Method of Expected Values in Radiative Transfer Theory*

Consider again a problem of estimating, by use of the Monte Carlo method, the linear functional

$$I_p = (f, p) = \int_0^\infty \int_R \int_V f(t, \mathbf{r}, \mathbf{v}) \cdot p(t, \mathbf{r}, \mathbf{v})\, d\mathbf{v}\, d\mathbf{r}\, dt = E\xi_p.$$

Here $f(t, \mathbf{r}, \mathbf{v})$ is the collision density function, and $p(t, \mathbf{r}, \mathbf{v})$ is a nonnegative function. Assume that the importance function $f^*(t, \mathbf{r}, \mathbf{v})$ (i.e., the solution of the adjoint transfer equation) is known precisely enough in a domain $\mathscr{D}$ of the phase space. The importance function $f^*(t_0, \mathbf{r}_0, \mathbf{v}_0)$ is the conditional expected value of the contribution to the calculated functional relative to the hypothesis that the trajectory originates at the point $(t_0, \mathbf{r}_0, \mathbf{v}_0)$ [48]. Let $\zeta$ be a part of the trajectory from the source position to the first intersection with $\mathscr{D}$, and let $\eta$ be the remainder of the trajectory after this intersection (if the trajectory does not interesect the domain $\mathscr{D}$, the contribution to the result does not depend on $\eta$). It is known that

$$I_p = E_\zeta E_\eta(\xi_p|\zeta). \tag{6.19}$$

Here

$$E_\zeta[\varphi] = \int_{-\infty}^{\infty} \varphi(x)\, dF(x),$$

where $F(x)$ is the distribution of $\zeta$, and $\xi_p$ is the random estimate of the functional $I_p$. Thus, if the quantity $E_\eta(\xi_p|\zeta)$ could be calculated for each value of $\zeta$, the multiplicity of the integral in (6.19) would be decreased. Furthermore, the variance of the estimate of the functional $I_p$ is then decreased [48]. In our case, the trajectory, after its first intersection with the domain $\mathscr{D}$, must be terminated, and the quantity $f^*(t_0, \mathbf{r}_0, \mathbf{v}_0)$ must be stored. There are many methods for estimating the solution of the direct transfer equation in certain domains of phase space. The theorem of optical mutuality may be applied to utilize those partial solutions. Then, the adjoint equation is regarded as the direct one, and *vice versa*, that is, the above conditional expected value is equal to the solution of the direct equation.

*Simulation of Adjoint Trajectories by Taking into Account the Asymptotic Solution; Estimation of the Function $B(t, \mathbf{r})$*

We shall first describe an algorithm for estimating the asymptotic solution [see (6.3)],

$$\Psi(x, \boldsymbol{\kappa}) = \frac{1}{2\pi}\int_0^\infty I_0(\kappa\rho)\exp(-\mu\rho^{5/3}\cdot x)\rho\, d\rho. \tag{6.20}$$

By making the change of variables $t = \rho\cdot x^{3/5}$, we find that

$$\Psi(x, \boldsymbol{\kappa}) = \frac{1}{2\pi x^{6/5}}\cdot\int_0^\infty I_0\left(\frac{\kappa}{x^{3/5}}\cdot t\right)\mathrm{e}^{-\mu t^{5/3}}\cdot t\, dt.$$

Let $c = \kappa x^{-3/5}$.

The function $\Psi(x, \boldsymbol{\kappa})$ may be written in the form:

$$\begin{aligned}\Psi(x, \boldsymbol{\kappa}) &= \frac{1}{2\pi x^{6/5}}\sum_{i=0}^{\infty}\int_{t_i}^{t_{i+1}} I_0(c\cdot t)\cdot\mathrm{e}^{-\mu t^{5/3}} t\, dt\\ &\approx \frac{1}{2\pi x^{6/5}}\sum_{i=0}^{\infty}\exp\left[-\mu\left(\frac{t_{i+1}+t_i}{2}\right)^{5/3}\right]\cdot\int_{t_i}^{t_{i+1}} I_0(c\cdot t)\cdot t\, dt\end{aligned}$$

Let $z = c\cdot t$, then

$$\begin{aligned}\Psi(x, \boldsymbol{\kappa}) &\approx \frac{1}{2\pi x^{6/5}}\sum_{i=0}^{N}\exp\left[-\mu\left(\frac{t_{i+1}+t_i}{2}\right)^{5/3}\right]\int_{c\cdot t_i}^{c\cdot t_{i+1}} czI_0(z)\, dz\\ &= \frac{c}{2\pi\kappa^2}\sum_{i=0}^{N}\exp\left[-\mu\left(\frac{t_{i+1}+t_i}{2}\right)^{5/3}\right][t_{i+1}I_1(c\cdot t_{i+1}) - t_i I_1(c\cdot t_i)]\\ &= \frac{c}{2\pi\kappa^2}\cdot\tilde{\Psi}(c).\end{aligned} \tag{6.21}$$

The function $\tilde{\Psi}(c)$ was tabulated in the interval $1 \leqslant c \leqslant 100$ with step $\Delta c = 1$, and in the interval $100 \leqslant c \leqslant 1000$ with step $\Delta c = 10$. In the interval $0 \leqslant c \leqslant 1$, the function $\tilde{\Psi}(c)/c$ was tabulated, because $\tilde{\Psi}(c) \sim c$ as $c \to 0$. The integration with respect to $t$ was performed over the interval (0, 5), where $\Delta t = t_i - t_{i-1} = 0.5 \times 10^{-3}$, i.e., $N = 10{,}000$. Some values of

$$\tilde{\tilde{\Psi}}(c) = \begin{cases} c^{-1}\cdot\tilde{\Psi}(c) & \text{if } c < 1,\\ \tilde{\Psi}(c) & \text{if } c \geqslant 1,\end{cases}$$

are given in Table 6.1.

**Table 6.1.** Values of $\Psi(c)$

| $c$ | 0 | 0.5 | 1 | 2 | 5 | 10 | 100 | 1000 |
|---|---|---|---|---|---|---|---|---|
| $\tilde{\tilde{\Psi}}(c)$ | 0.196 | 0.189 | 0.173 | 0.244 | 0.081 | $0.69 \times 10^{-2}$ | $0.99 \times 10^{-5}$ | $0.21 \times 10^{-7}$ |

We shall now discuss the algorithm for estimating the function $B(t, \boldsymbol{r})$, based on simulation of adjoint trajectories. Using $\Phi(t, \boldsymbol{r}, \boldsymbol{v}) = v\varphi(t, \boldsymbol{r}, \boldsymbol{v})$, we rewrite (6.15) in the form

$$B(t^*, \boldsymbol{r}^*) = \int_V \frac{\Phi(t^*, \boldsymbol{r}^*, \boldsymbol{v})}{v} d\boldsymbol{v} + e^{-\mu r^{*5/3} \cdot t^*} - 1.$$

where $\Phi(t^*, \boldsymbol{r}^*, \boldsymbol{v}^*)$ is the flux of scattered photons at $(t^*, \boldsymbol{r}^*, \boldsymbol{v})$. Write the integral as

$$I(t^*, \boldsymbol{r}^*) = \int_V \frac{\Phi(t^*, \boldsymbol{r}^*, \boldsymbol{v})}{v} d\boldsymbol{v} = (\Phi, \delta_{\boldsymbol{r}^*, t^*}/v),$$

where $\delta_{\boldsymbol{r}^*, t^*}(r, t) \equiv \delta(\boldsymbol{r} - \boldsymbol{r}^*)\delta(t - t^*)$.

We have, by the theorem of optical mutuality,

$$(\Phi, \delta_{\boldsymbol{r}^*, t^*}/v) = (\Phi^*_{\boldsymbol{r}, t}, s),$$

where $s = s(t, \boldsymbol{r}, \boldsymbol{v})$ is the distribution density of particles just after their initial scattering, and $\Phi^*(t, \boldsymbol{r}, \boldsymbol{v})$ is the solution of the adjoint equation for the source density $\delta_{\boldsymbol{r}_*, t_*}/v$. Keeping in mind that $s(t, \boldsymbol{r}, \boldsymbol{v}) = f_0(t, \boldsymbol{r}, \boldsymbol{v}) \cdot g(\boldsymbol{r}, 0 \to \boldsymbol{v})$, where $f_0(t, \boldsymbol{r}, \boldsymbol{v})$ is the density of first collisions, we have (see Sect. 6.3)

$$\begin{aligned} s(t, \boldsymbol{r}, \boldsymbol{v}) &= \mu r^{5/3} e^{-\mu r^{5/3} t} \cdot g(\boldsymbol{r}, 0 \to \boldsymbol{v}) \\ &= \frac{2\mu A}{a} \cdot e^{-\mu r^{5/3} t} \cdot v^{-11/3} \cdot [1 - \cos(\boldsymbol{v} \cdot \boldsymbol{r})]. \end{aligned} \tag{6.22}$$

Because $I(t^*, \boldsymbol{r}^*) = (\Phi^*_{\boldsymbol{r}, t}, s) = (\Phi^*_{\boldsymbol{r}, t}\sigma, s/\sigma)$, it is necessary to calculate the contribution $s/\sigma$ to the functional $I(t^*, \boldsymbol{r}^*)$ in each state of the adjoint trajectory. As examination of the expressions for the density $q$ and the asymptotic function $\Psi(t, \boldsymbol{v})$ shows particles with small velocities make an essential contribution to the result. Therefore the initial velocity was sampled from the density

$$f_0(v) = \frac{(1 - \rho) \cdot v^{-\rho}}{V^{1-\rho}}, \quad 0 \leqslant v \leqslant V = 100. \tag{6.23}$$

As calculations show, $\rho = 2/3$ is the most appropriate choice. The initial particle weight is given by

$$Q_0 = 2\pi \times V^{1-\rho} \cdot v^{\rho}/(1 - \rho).$$

The adjoint trajectories originate at point $\boldsymbol{r}$, at the time $t_0 = t^*$. Next, the run times are successively subtracted, as follows from the adjoint nonstationary transfer equation, from the initial time $t$. It is also possible to originate the trajectories at the time $t_0 = 0$ but the run times are then added to $t_0$.

The contribution $s/\sigma$ is then calculated for $t^* - t$, where $t^*$ is a given time, and $t$ is the collision time. Such a technique gives an estimate of the function $B(t^*, \boldsymbol{r})$, simultaneously for various values of $t^*$. As examination of the transfer equation (6.17) shows, the particle weight after each scattering must be multiplied by

$$\frac{\sigma_s(v)}{\sigma_s(v')} = \frac{\mu r^{5/3} \cdot v^{-1}}{\mu r^{5/3} \cdot v'^{-1}} = \frac{v'}{v}.$$

Here $v'$, $v$ are the particle velocities before and after scattering. We now show that the method of expected values permits calculation of the contribution that corresponds to the asymptotic solution. Let us suppose that a particle has undergone a scattering at a point $(t, \boldsymbol{r}, \boldsymbol{v})$ of the phase space. In order to test the condition $r > R_0 \cdot (t^* - t)$, the quantity $R_0(t^* - t)$ was calculated by linear interpolation between the values proposed by *Tatarsky* [47] in Table 6.2.

If the inequality $r > R_0(t^* - t)$ holds, then the trajectory is terminated and the expected value of the contribution is stored. The theorem of optical mutuality yields the expected value,

$$\Phi(t^* - t, \boldsymbol{r}, \boldsymbol{v}) = v \cdot \varphi(t^* - t, \boldsymbol{r}, \boldsymbol{v}). \tag{6.24}$$

If $r \geqslant R_0(t^* - t)$ then, instead of $\varphi(t^* - t, \boldsymbol{r}, \boldsymbol{v})$ in (6.24), the asymptotic solution may be used. Values linearly interpolated between the tabulated values of $\hat{\Psi}(c)$ were used to calculate the asymptotic function. For $r < R_0(t^* - t)$, the contribution $s(t^* - t, \boldsymbol{r}, \boldsymbol{v})/\sigma(\boldsymbol{r}, \boldsymbol{v})$ was calculated.

Thus the trajectories in the Monte Carlo calculations of the function $B(t^*, \boldsymbol{r}^*)$ were simulated according to scheme:

a) A particle is "emitted" from the point $\boldsymbol{r}^*$ at the time $t_0$. Its initial velocity $\boldsymbol{v}_0$ has the distribution density (6.23). The initial particle weight is given by

**Table 6.2.** Values of $R_0(t)$

| $t$ | 0.5 | 1 | 2 | 3 | 4 | 5 | 6 | 7 | 8 | 9 | 10 |
|---|---|---|---|---|---|---|---|---|---|---|---|
| $R_0(t)$ | 2 | 10 | 20 | 26 | 30 | 33 | 35 | 36.5 | 38 | 39 | 40 |

$Q_0 = 2\pi \cdot V^{1-\rho} \cdot v^{\rho}/(1-\rho)$. The free-path length $l$ (see Sect. 6.4) and the new collision point $\boldsymbol{r}$ are calculated. The collision time is given by $t = l/v$.
b) The new velocity is sampled from the indicatrix

$$g(\boldsymbol{r}, \boldsymbol{v} \to \boldsymbol{v}'); \quad Q' = Q \cdot v'/v.$$

c) If $t^* - t < 0$, then the trajectory is terminated; go to a). Otherwise the inequality $r \geqslant R_0(t^* - t)$ is tested.
d) If $r \geqslant R_0(t^* - t)$, the trajectory is terminated, and the asymptotic value $v\,\Psi(t^* - t, \boldsymbol{v})Q$ is stored.
Then go to a)
e) If $r < R_0(t^* - t)$, the contribution

$$\begin{aligned}\tilde{I} &= Q's(t^* - t, \boldsymbol{r}, \boldsymbol{v}) \cdot \sigma(v) \\ &= Q' \cdot \frac{2A}{a \cdot r^{5/3}} \exp\left[-\mu r^{5/3}(t^* - t)\right](1 - \cos(\boldsymbol{v} \cdot \boldsymbol{r}))v^{-8/3},\end{aligned}$$

is stored. Then go to b).

The method described was used to calculate $B(t^*, 0)$. In this case, the contribution due to unscattered radiation is 1.

## 6.5 Computational Results

The function $B(t, 0)$ was calculated by the method of Sect. 6.4 for $t = 0.5$; 1; 2; 3; 4; $R_0$ was taken to be 40. The following subdivision was used: 0–0.1, in steps of 0.01; 0.1–1, in steps of 0.1; 1–40, in steps of 1. Calculations show that this subdivision is sufficiently accurate. The standard error of the present calculations is less than 1%. All calculations were performed in 60 minutes of BESM-6 computer time. The number of trajectories $N = 100\,000$. The Monte Carlo results agree well with the experimental data (see Fig. 6.1) reported by *Grachova* et al. [50].

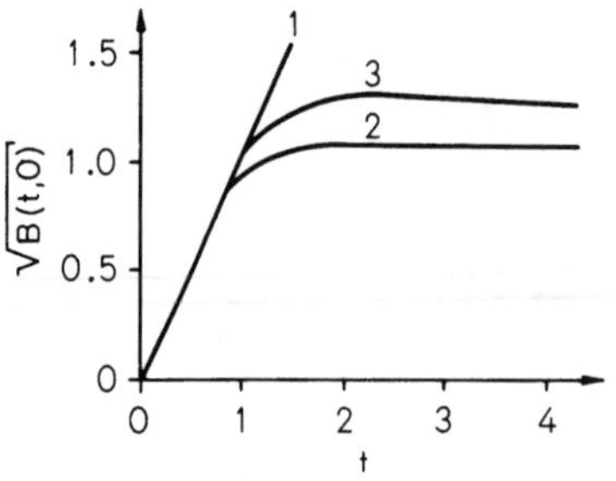


**Fig. 6.1.** The function $B(t, 0)$: (*1*) Perturbation method, (*2*) Monte Carlo method, (*3*) Experimental data

Of the numerical methods used for estimating $B(t, 0)$, only the method of Sect. 6.4 is satisfactorily accurate. Increase of the asymptotic radii $R_0(t)$ does not affect the estimate of the function $B(t, 0)$. Consequently, the choice of these asymptotic radii is valid.

# References


1. G. V. Rosenberg: Usp. Fiz. Nauk *56*, 77–109 (1955) [Russian]
2. S. Chandrasekhar: *Radiative Transfer* (Oxford University Press, Oxford 1955)
3. H. C. van de Hulst: *Light Scattering by Small Particles* (Wiley, New York 1957) p. 520
4. V. N. Ogibin, E. S. Kuropatenko: Zh. Vychisl. Mat. Mat. Fiz. *8*, 212–216 (1968) [Russian]
5. S. M. Ermakov: *The Monte Carlo Method and Related Problems* (Nauka, Moscow 1971) p. 327 [Russian]
6. W. A. Coleman: Nucl. Sci. Eng. *32*, 76–81 (1968)
7. V. S. Vladimirov: *Mathematical Problems of the Single-Velocity Theory of Particle Transfer.* Tr. Mat. inst. Akad. Nauk. SSSR *61*, 157 (1961) [Russian]
8. G. I. Marchuk: *Design of Nuclear Reactors* (Atomizdat, Moscow 1961) p. 666 [Russian]
9. S. M. Ermakov, V. G. Zolotulehin: "Monte Carlo Applications to the Nuclear Shielding Problem", in *Voprosy Fiz. Zazc. React.* ed. by D. L. Broder (Atomizdat, Moscow 1963) pp. 171–182 [Russian]
10. G. A. Mikhailov: Theor. Verovatn. Le Primen. *15*, 142–144 (1970) [Russian]
11. A. I. Hisamutdinov: Zh. Vychisl. Mat. Mat. Fiz. *10*, 1269–1280 (1970) [Russian]
12. T. H. Holton: SIAM (Soc. Ind. Appl. Math.) Rev. *12*, 1–63 (1970)
13. A. I. Hisamutdinov: Zh. Vychisl. Mat. Mat. Fiz. *10*, 999–1005 (1970) [Russian]
14. B. Davison: *Neutron Transport Theory* (Clarendon Press, Oxford 1958) p. 520
15. M. N. Kalos: Nucl. Sci. Eng. *16*, 111–117 (1963)
16. W. Feller: *An Introduction to Probability Theory and Its Applications,* Vol. *2* (Wiley, New York 1966)
17. I. F. Podlivaev, J. I. Rusu: Zh. Vychisl. Mat. Mat. Fiz. *12*, 252–256 (1972) [Russian]
18. M. N. Kalos, H. A. Steinberg: Nucl. Sci. Eng. *44*, 406–412 (1971)
19. B. M. Golubitzki, M. V. Tantashev: Zh. Vychisl. Mat. Mat. Fiz. *12*, 249–251 (1972) [Russian]
20. V. N. Ogibin: "On the Splitting Method and Roulette Technique in the Monte Carlo Calculations of the Particle Transfer Problem", in *Method Monte Carlo v Probleme Perenosa Izluchenij,* ed. by G. I. Marchuk (Atomizdat, Moscow 1967) pp. 72–82 [Russian]
21. L. V. Kantorovich, G. P. Akilov: *Functional Analysis in Normed Spaces* (Fizmatgiz, Moscow 1959) p. 684 [Russian]
22. G. A. Mikhailov: Zh. Vychisl. Mat. Mat. Fiz. *9*, 1145–1152 (1969) [Russian]
23. D. G. Collins, W. G. Blattner, M. B. Wells, H. G. Horak: Appl. Opt. *11*, 2684–2696 (1972)
24. G. I. Marchuk: *Equations for the Value of Information From the Meteorological Satellites and Formulations of Inverse Problems.* Kosm. Issled. *2*, (3), 462–477 (1964) [Russian]
25. A. N. Tichonov: Dokl. Akad. Nauk SSSR *151*, 501–504 (1963) [Russian]
26. V. F. Turchin, V. Z. Nosik: Izv. Akad. Nauk SSSR. Fiz. Atm. Okeana *5*, 29–38 (1969) [Russian]
27. V. F. Turchin: Zh. Vychisl. Mat. Mat. Fiz. *7*, 1270–1284 (1967) [Russian]
28. T. A. Germogenova: Izv. Akad. Nauk SSSR. Ser. Geofiz, *6*, 854–856 (1962) [Russian]
29. G. V. Rosenberg: *Twilight* (Fizmatgiz, Moscow 1963) p. 380 [Russian]
30. G. V. Rosenberg: Usp. Fiz. Nauk *95*, 159–208 (1968) [Russian]
31. L. Elterman: *UV, Visible, and IR Altenuation for Alitudes to 50 km* (Office of Aerospace Research, United States Air Force, 1968) p. 50

32. L. S. Ivlev, S. I. Popova: Izv. Vyssh. Uchebn. Zaved. Fiz. *5,* 91–98 (1972) [Russian]
33. L. S. Ivlev, S. I. Popova: Izv. Akad. Nauk SSSR, Fiz. Atmos. Okeana *10,* 1034–1043 (1973) [Russian]
34. K. J. Kontratjev (ed.): *Radiation Characteristics of the Atmosphere and Ocean* (Gydrometeoizdat, Leningrad 1969) p. 564 [Russian]
35. T. A. Germogenova, T. A. Sushkevich: "Solution of the Transfer Equation by Means of the Mean Flux Method", in *Voprosy Fiz. Zasc. Reakt. Vyp. 3,* ed. by D. L. Broder (Atomizdat, Moscow 1979) pp. 24–36, [Russian]
36. A. B. Sandomirski, G. V. Rosenberg, N. P. Altovskaja, T. A. Sushkevich: Izv. Akad. Nauk SSSR. Fiz. Atmos. Okeana *7,* 737–749 (1971) [Russian]
37. K. L. Coulson, I. V. Dave, Z. Sekera: *Tables Related to Radiation Emerging from a Planetary Atmosphere with Rayleigh Scattering* (Univ. of California Press, Berkeley, Los Angeles 1960) p. 548
38. A. I. Ivanov, G. Sh. Livshitz, V. E. Pavlov, B. T. Tasherov, J. A. Teutel: *Light Scattering in the Atmosphere,* Vol. *2* (Nauka, Alma-Ata 1968) p. 116 [Russian]
39. A. P. Ivanov: *Optics of the Scattering Mediums* (Nauka i Technika, Minsk 1969), p. 592 [Russian]
40. A. M. Beljantsev, L. S. Dolin, V. A. Saveljev: Izv. Vyssh. Ushebn. Zaved. Radiofiz. *10,* 489–498 (1967) [Russian]
41. B. M. Golubitzki, T. M. Jadco, M. V. Tantashev: Izv. Akad. Nauk SSSR. Fiz. Atmos. Okeana *8,* 1226–1229 (1972) [Russian]
42. E. P. Sege, I. L. Katsev: *Time Asymptotical Solutions of the Radiative Transfer Equation and Their Applications.* Reprint (I. F. Akad. Nauk SSSR, Minsk 1973) p. 62 [Russian]
43. G. K. Iljich, I. L. Katzev, V. D. Kozlov: Izv. Akad. Nauk SSSR. Ser. Fiz. Mat. Nauk, *5,* 96–100 (1969) [Russian]
44. E. P. Sege, A. P. Ivanov, I. L. Katzev, B. A. Kargin, S. V. Kuznetsov, G. A. Mikhailov: "Some Questions Connected with the Optical Pulse Location in Natural Scattering Mediums", in *Tezisy 10 th Vsesojuz. Konferen. po Rasprostr. Radiovoln* (Nauka, Moscow 1972) pp. 337–341 [Russian]
45. L. I. Koprova: Izv. Akad. Nauk SSSR. Fiz. Atmos. Okeana *7,* 622–632 (1971) [Russian]
46. S. M. Rytov: Izv. Akad. Nauk SSSR. Ser. Fiz. *2,* 223–259 (1937) [Russian]
47. V. I. Tatarsky: Zh. Eksp. Teor. Fiz. *56,* 2106–2117 (1969) [Russian]
48. G. A. Mikhailov: *Some Questions of the Monte Carlo Techniques* (Nauka, Novosibirsk 1974) p. 142 [Russian]
49. G. I. Marchuk, V. V. Orlov: "On the Conjugate Functions Theory", in *Neitronnaya Fizika* (Gosatomizdat, Moscow 1961) pp. 30–45 [Russian]
50. M. E. Grachŏva, A. S. Gurvich, S. S. Kashkarov, V. V. Pokasov: Zh. Eksp. Teor. Fiz., *67,* 2035–2046 (1974) [Russian]; also in: *Laser Beam Propagation in the Atmosphere,* Topics in Applied Physics, Vol. *25,* ed. by J. W. Strohbehn (Springer, Berlin, Heidelberg, New York 1978) pp. 8–107
51. G. I. Gorchakov, G. V. Rosenberg: Izv. Akad. Nauk SSSR, Ser. Fiz. Atmos. Okeana *3,* 611–620 (1967) [Russian]

# Subject Index

## Laser Beam Propagation in the Atmosphere

Editor: J. W. Strohbehn

1978. 78 figures, 1 table. XII, 325 pages
(Topics in Applied Physics, Volume 25)
ISBN 3-540-08812-1

Contents:
*J. W. Strohbehn:* Introduction. Laser Beam Propagation in the Atmosphere. – *S. F. Clifford:* The Classical Theory of Wave Propagation in a Turbulent Medium. – *J. W. Strohbehn:* Modern Theories in the Propagation of Optical Waves in a Turbulent Medium. – *M. E. Gracheva, A. S. Gurvich, S. S. Kashkarov, V. V. Pokasov:* Similarity Relations and Their Experimental Verification for Strong Intensity Fluctuations of Laser Radiation. – *A. Ishimaru:* The Beam Wave Case and Remote Sensing. – *J. H. Shapiro:* Imaging and Optical Communication Through Atmospheric Turbulence. – *J. L. Walsh, P. B. Ulrich:* Thermal Blooming in the Atmosphere. – Subject Index.

## Laser Monitoring of the Atmosphere

Editor: E. D. Hinkley

1976. 84 figures. XV, 380 pages
(Topics in Applied Physics, Volume 14)
ISBN 3-540-07743-X

Contents:
*E. D. Hinkley:* Introduction. – *S. H. Melfi:* Remote Sensing for Air Quality Management. – *V. E. Zuev:* Laser -Light Transmission Through the Atmosphere. – *R. T. H. Collis, P. B. Russell:* Lidar Measurement of Particles and Gases by Elastic Backscattering and Differential Absorption. – *H. Inaba:* Detection of Atoms and Molecules by Raman Scattering and Resonance Fluorescence. – *E. D. Hinkley, R. T. Ku, P. I. Kelley:* Techniques for Detection of Molecular Pollutants by Absorption of Laser Radiation. – *R. T. Menzies:* Laser Hererodyne Detection Techniques.

## Optical and Infrared Detectors

Editor: R. J. Keyes

1977. 115 figures, 13 tables. XI, 305 pages
(Topics in Applied Physics, Volume 19)
ISBN 3-540-08209-3

Contents:
*R. J. Keyes:* Introduction. – *P. W. Kruse:* The Photon Detection Process. – *E. H. Putley:* Thermal Detectors. – *D. Long:* Photovoltaic and Photoconductive Infrared Detectors. – *H. R. Zwicker:* Photoemissive Detectors. – *A. F. Milton:* Charge Transfer Devices for Infrared Imaging. – *M. C. Teich:* Nonlinear Heterodyne Detection.